Integrative Neuroscience and Personalized Medicine

Integrative Neuroscience and Personalized Medicine

Edited by

Evian Gordon, PhD, MBBCH
The Brain Resource Company
Sydney, Australia and San Franscisco, CA, USA
The University of Sydney, Australia, and
BRAINnet Foundation, USA

Stephen H. Koslow, PhD
The Brain Resource Company
Sydney, Australia and San Francisco, CA, USA
BRAINnet Foundation, USA
Biomedical Synergy
West Palm Beach, FL, and
American Foundation for Suicide Prevention
New York, NY, USA

2011

OXFORD
UNIVERSITY PRESS

Oxford University Press, Inc., publishes works that further
Oxford University's objective of excellence
in research, scholarship, and education.

Oxford New York
Auckland Cape Town Dar es Salaam Hong Kong Karachi
Kuala Lumpur Madrid Melbourne Mexico City Nairobi
New Delhi Shanghai Taipei Toronto

With offices in
Argentina Austria Brazil Chile Czech Republic France Greece
Guatemala Hungary Italy Japan Poland Portugal Singapore
South Korea Switzerland Thailand Turkey Ukraine Vietnam

Published by Oxford University Press, Inc.
198 Madison Avenue, New York, New York 10016

www.oup.com

Library of Congress Cataloging-in-Publication Data

Integrative neuroscience and personalized medicine/edited by Evian Gordon, Stephen H. Koslow.
p.; cm.
Includes bibliographical references and index.
ISBN 978-0-19-539380-4
1. Neurogenetics. 2. Genetic polymorphisms. 3. Brain—Diseases—Genetic aspects. I. Gordon, Evian. II. Koslow, Stephen H.
[DNLM: 1. Mental Disorders—therapy. 2. Genetic Markers. 3. Individualized Medicine—methods. 4. Nervous System Diseases—therapy. 5. Pharmacogenetics. WM 140 I59 2011]
QP356.22.I58 2011
612.8—dc22 2010017894

9 8 7 6 5 4 3 2
Printed in the United States of America
on acid-free paper

Foreword

Neurological or brain disorders represent a huge health burden worldwide. Although there is a long history of knowledge about and study of brain and psychiatric disorders, until approximately forty years ago, we viewed these as illnesses of interest for which therapeutics were almost universally unavailable. Until recently, there were no therapeutic treatment options available for patients who suffered from depression, suicidal thoughts, anxiety, post-traumatic stress disorder (PTSD), schizophrenia, brain tumors, Alzheimer's disease, Parkinson's disease, Huntington's disease, or diseases of movement, cognition, or sensation. Even now, therapeutics that can completely cure the patient of the illness are nonexistent, or are effective only in some patients.

There is increasing interest in personalized approaches in all fields of medicine, including brain-related diseases. The initial goal was to discover genetic signatures in order to individualize disease prediction, diagnosis, treatment efficacy, treatment course, and prognosis. Research has been based on the premise that clinicians could match markers to specific disease outcomes. However, much about the genetic bases of psychiatric and brain disorders remains unknown, and significant research is necessary to identify the multiple genetic and protein markers for these diseases. Furthermore, the combined use of ever more sophisticated brain imaging technologies, along with the ability to identify genomic, proteomic, or metabolomic markers, may vastly aid in elucidating the diagnosis, treatment prediction, and course of brain disorders.

We know more about some neurological conditions than others. For example, it is clear that stroke is usually secondary to either thrombotic or hemorrhagic insults. Yet, there are few clear prognostic indicators, and treatments are still being developed. We can, for example, stave off some damage from ischemic stroke by infusion of tissue plasminogen activator (tPA) within the first few hours after the event. But, at present, this is the only FDA-approved drug to treat stroke, and it is not without side effects.

Huntington's disease is always caused by a CAG (cytosine-adenine-guanine), repeat in a gene on chromosome 4, resulting in an abnormal polyglutamine-containing sequence in the protein huntington. This insoluble mutant protein accumulates within neurons and results in cell death. Thus, Huntington's is a disorder that is more therapeutically approachable than other neurological disorders because of the relatively unified understanding of some of the major causes of its pathology. Although the precise genetic deficit was discovered in 1993, thus far, this has led to no significant advances in therapy. This clearly indicates that we need to expand our understanding of gene regulation and control, and also suggests that we need to include other types of biological and epigenetic information in our studies.

Huntington's disease is exceptional in that it originates in the mutation of one gene, and so much is known about its genetic etiology. Increasing amounts of genetic information are being uncovered about Alzheimer's disease, yet numerous genes on different chromosomes have been implicated in familial disease, and the available treatments are minimal. Despite the elucidation of some mechanisms of some psychiatric and neurological disorders, in fact, the overall group of brain-related neurological and psychiatric disorders remains in need of substantial clarification.

Brain and psychiatric illnesses are increasing in prevalence. It is difficult to render a specific diagnosis and categorization of a specific treatment plan of these illnesses using current, often subjective methods. Furthermore, treatment modalities often are based on trial and error. For one patient with a specific disorder, a given treatment may be therapeutic, while the same treatment may have no effect on another individual with the same diagnosis. This lack of clarity challenges and complicates our capacity to do research. Therefore, discovery of specific biomarkers to allow better characterization of a particular disorder is now a critical, although challenging, goal of the brain and psychiatric disorders research community.

Successful research, however, may require much larger patient sample sizes than are usually used in studying brain disorders, as larger sample sizes have been more successful in studies of other medical disorders. For example, databases of samples from thousands of patients have been used to elucidate the role of the BRCA1 and 2 gene mutations in the etiology of hereditary breast cancer. Similar numbers should be used to study brain disorders. The goal of identifying a biological genotype and/or phenotype, or biomarker, is a high priority. The identification of a biological genotype or phenotype may, indeed, supersede current diagnostic schemes typically used in determining treatments. Such identification fosters appropriate diagnosis, prediction, and therapeutic treatment and dramatically changes the landscape of our approach to psychiatric illnesses.

There are encouraging findings in genomics and molecular genetics, giving clinicians the ability to hone treatments to fit specific characteristics of tumors or other diseases. These tools should be available to those of us studying the function of the brain. To help us gain access to genomics data and bring it into the brain disorders clinic, we should:

- Advocate for a lower cost for sequencing a complete human genome, bringing the price down to $100 or less.
- Collect and integrate potentially complementary data, including brain-imaging data, across scales and disciplines.
- Make all of these data accessible and transparent to investigators.
- Advocate for standardization of some of the investigational tools, so that data may be combined, integrated, and compared, thus maximizing the statistical power and comparability both within and across studies.

In this way, scientists can gain a deeper understanding of the pathogenesis and possible mechanisms of brain disorders. This increased understanding should, in turn, lead to discoveries of new, targeted, therapeutic and biological genotypes and phenotypes.

Because of the complexity of neurological or brain disorders, investigators from multiple disciplines should be encouraged to work together to discover genes, elucidate genetic mechanisms of disease, and discover treatments and cures. Our goal, however, is not just to accumulate data. It is to provide a new understanding of disease and its impact on the individual patient, something we refer to as "integrative brain health." This is a concept with potentially far-reaching effects, and it will allow us to translate useful knowledge that will enable clinicians to care for the patient and his or her distinct brain disorder in an integrated manner, utilizing new diagnostic tools and new pharmaceutical (and other) treatments which constitute the best management plan for that patient and that illness.

We will be able to take knowledge we already have, such as information on the maturation of the cerebral cortex, and determine its relationship to the etiology of brain disorders that become manifest in adolescence. For example, as the cortex matures, its size increases, and it experiences a concurrent increase in myelination. We are learning, too, that the gray matter gradually declines as pruning takes place. We need to distill newly gained information from multiple modalities—e.g., molecular genetics, neuroimaging, blood and CSF-related biomarkers, behavioral assessment, neuropsychological approaches, and brain electrical activity—into applicable clinical use. Furthermore, we need to learn how to use our emerging understanding of the brain and its evolution, and how factors from multiple domains interact, to advance our ability to predict, diagnose systematically, and effectively treat neurological and brain disorders.

Given the current scale and morbidity of brain disorders, "personalized medicine" should be more than a lofty goal to be attained in the distant future. Imagine having the ability to say to a patient at increased risk for a brain disorder, and even possibly to one suffering from a given illness, that we, as clinicians, can take specific productive steps to prevent the onset of the disorder, or to effect a cure! To be able to say, with a high probability, that X is the disorder and that the diagnosis immediately signifies specific symptoms, function, and dysfunction, and that there is a specific personalized effective treatment, would radically change the way brain disorders and psychiatric diseases are treated—and even cured. For decades, researchers, clinicians, patients, families and all others concerned with the mental health and brain disorder field have yearned to have at their fingertips a plethora of treatments that are likely to match the specific causes of disease. We are getting ever closer to achieving that goal. This book serves to highlight the factors that seem likely to help expedite personalized medicine, ushering it into mainstream clinical practice.

We should applaud those who are working valiantly to clarify the workings of the brain. We should encourage them to pool knowledge, to break down their silos, to be as imaginative as possible, but also to build systematic and pragmatic platforms that aggregate data and information to enhance the productive capacity of scientists worldwide. These are all noble efforts.

Personalized medicine is a goal we want to achieve. Personalized medicine offers one of the greatest opportunities to improve health care, treat in a more tailored and effective way, reduce the use of treatments of no value, and minimize the unfortunate accompanying side effects of treatments. It holds out the possibility of contributing to the reduction of health care costs, while simultaneously increasing quality of health care and quality of life.

This book assembles many leaders in the field, who bring the reader up-to-date information on the status of our knowledge. They will point to likely directions in research and also articulate not just hopes for the future, but also what is more likely to be clinically available for testing and iterative evidence-based refining *now*, and what requires more long-term research.

Herbert Pardes, MD
President and CEO
New York-Presbyterian Hospital
New York, NY

Contents

Contributors

Edward Abrahams, PhD
Personalized Medicine Coalition
Washington, DC

Nili Avidan, PhD
Multiple Sclerosis & Brain Research Center
Carmel Medical Center; and
Pharmacogenetics and Translational Genetics Center
Rappaport Faculty of Medicine & Research Institute
Technion-Israel Institute of Technology
Haifa, Israel

Eugene Baker, PhD
Employee Assistance Programs
OptumHealth Behavioral Solutions
Golden Valley, MN

Jacob S. Ballon, MD
Department of Psychiatry
College of Physicians and Surgeons
Columbia University; and
New York Psychiatric Institute
New York, NY

Elisabeth B. Binder, MD, PhD
Department of Psychiatry and Behavioral Sciences
Department of Human Genetics
Emory University School of Medicine
Atlanta, GA; and
Max Planck Institute of Psychiatry
Munich, Germany

Edward T. Bullmore, MD
Brain Mapping Unit & Clinical Neurosciences Institute
Department of Psychiatry
University of Cambridge; and
GSK Clinical Unit Cambridge
Addenbrooke's Hospital
Cambridge, UK

Opher Caspi, MD, PhD
Rabin Medical Center
Department of Integrative Medicine
Davidoff Cancer Center
Peta-Tikva, Israel

Simon D. Clarke, MD
Center for Research in Adolescents' Health
Department of Adolescent Medicine, Westmead Hospital and
Children's Hospital at Westmead
Brain Dynamics Centre
Westmead Millennium Institute and Sydney Medical School
New South Wales, Australia

William R. Crum, DPhil
Centre for Neuroimaging Sciences
Institute of Psychiatry
King's College London
London, UK

Ulrich Ettinger, PhD
University of Munich
Munich, Germany

Alex Fornito, PhD
Brain Mapping Unit & Clinical Neurosciences Institute
Department of Psychiatry
University of Cambridge
Cambridge, UK; and
Melbourne Neuropsychiatry Center
Department of Psychiatry
University of Melbourne
Melbourne, Australia

Richard Gevirtz, PhD, BCIAC
California School of Professional Psychology
Alliant International University
San Diego, CA

Charles F. Gillespie, MD, PhD
Department of Psychiatry and Behavioral Sciences
Emory University School of Medicine
Atlanta, GA

Ragy R. Girgis, MD
Department of Psychiatry
College of Physicians and Surgeons
Columbia University; and
New York Psychiatric Institute
New York, NY

Evian Gordon, PhD, MBBCH
The Brain Resource Company
Sydney, Australia and San Franscisco, CA, USA
The University of Sydney, Australia, and
BRAINnet Foundation, USA

Ronald Grunstein, MD
Sydney Medical School
University of Sydney; and
NHMRC Centre for Integrating Research and Understanding of Sleep (CIRUS)
Woolcock Institute of Medical Research
Department of Respiratory and Sleep Medicine
Royal Prince Alfred Hospital
Sydney, Australia

Paul E. Holtzheimer, MD
Department of Psychiatry and Behavioral Sciences
Emory University School of Medicine
Atlanta, GA

Matthew J. Kempton, PhD
Centre for Neuroimaging Sciences
Institute of Psychiatry
King's College London
London, UK

Michael R. Kohn, MD
Center for Research in Adolescents' Health
Department of Adolescent Medicine, Westmead Hospital and
Children's Hospital at Westmead
Brain Dynamics Centre
Westmead Millennium Institute and Sydney Medical School
New South Wales, Australia

Stephen H. Koslow, PhD
The Brain Resource Company
Sydney, Australia and San Francisco, CA, USA
BRAINnet Foundation, USA
Biomedical Synergy
West Palm Beach, FL, and
American Foundation for Suicide Prevention
New York, NY, USA

Izabella Lejbkowicz, PhD
Multiple Sclerosis & Brain Research Center
Carmel Medical Center; and
Pharmacogenetics and Translational Genetics Center
Rappaport Faculty of Medicine & Research Institute
Technion-Israel Institute of Technology
Haifa, Israel

Jeffrey A. Lieberman, MD
Department of Psychiatry
College of Physicians and Surgeons
Columbia University; and
New York Psychiatric Institute
New York, NY

Ellen M. Migo, PhD
Centre for Neuroimaging Sciences
Institute of Psychiatry
King's College London
London, UK

Ariel Miller, MD, PhD
Multiple Sclerosis & Brain Research Center
Carmel Medical Center; and
Pharmacogenetics and Translational Genetics Center
Rappaport Faculty of Medicine & Research Institute
Technion-Israel Institute of Technology
Haifa, Israel

Charles B. Nemeroff, MD, PhD
Department of Psychiatry and Behavioral Sciences
University of Miami Miller School of Medicine
Miami, FL

Tamar Paperna, PhD
Multiple Sclerosis & Brain Research Center
Carmel Medical Center; and
Pharmacogenetics and Translational Genetics Center
Rappaport Faculty of Medicine & Research Institute
Technion-Israel Institute of Technology
Haifa, Israel

Herbert Pardes, MD
President and CEO
New York-Presbyterian Hospital
New York, NY

Alan F. Schatzberg, MD
Kenneth T. Norris Jr. Professor and Chairman
Department of Psychiatry and Behavioral Sciences
Stanford University School of Medicine
Stanford, CA

Dan Segal, MSc
The Brain Resource Company
Sydney, Australia and San Francisco, CA, USA

Mike Silver, PhD
Synaptix Communications
Lexington, MA

Elsebeth Staun-Ram, PhD
Multiple Sclerosis & Brain Research Center
Carmel Medical Center; and
Pharmacogenetics and Translational Genetics Center
Rappaport Faculty of Medicine & Research Institute
Technion-Israel Institute of Technology
Haifa, Israel

David Whitehouse, MD
CMO Catalyst Health Solutions Inc.
Rockville, MD

David E. Williams, MBA
MedPharma Partners
Boston, MA

Leanne M. Williams, PhD
Brain Dynamics Centre
Westmead Millennium Institute and
Sydney Medical School at University of Sydney
Australia; and
BRAINnet Foundation
San Francisco, CA, USA

Steve C.R. Williams, PhD
Centre for Neuroimaging Sciences
Institute of Psychiatry
King's College London
London, UK

Introduction

One size does not fit all. Human beings are individuals and now, in the twenty-first century, medicine is finally catching up to that reality. Personalized medicine is a rapidly emerging paradigm which is based primarily on the belief that understanding genomic regulation of disease processes will lead to better diagnostics treatment and cures of human disease. However, this book serves to point to a broader genomic-Brainmarkers (cognition, emotions, MRI, fMRI, SPECT, PET brain imaging, EEG and ERP EPS electrophysiology and other yet to be developed direct brain measures) approach to personalized medicine, which should accelerate our progress toward success. Brainmarkers refers to a specific objective measure which directly emanates from the brain and is a functional indicator.

A landmark clinical application of personalized medicine was the gene product, HER2. If a woman with breast cancer tests positive for this biomarker, her tumor is overproducing this gene product, and she is thus a good candidate for treatment with Herceptin, which reduces the chance of recurrence. Another example with widespread possible application is the cytochrome P450 liver enzyme, which can be tested genetically and used to calculate the best dose for each individual based upon the amount of drug metabolized by these liver enzymes. The cytochrome P450 system has been extensively researched by pharmacogenetics researchers. You will find a number of other personalized medicine examples in this book. All of this has been facilitated by the completion of the Human Genome Project in 2003.

No genetic tests have yet been unequivocally proven to be clinically useful in brain disorders. This book catalogues many examples that show promise for clinical application (for additional information on genes, visit the Web site of the National Institutes of Health at http://www.ncbi.nlm.nih.gov/sites/GeneTests/?db=GeneTests).

This volume also argues that, for personalized medicine for brain disorders to become a near future reality, a number of new directions should be explored, and in some cases implemented, in the scientific, clinical and health care arenas.

These new directions include:

- The addition of functional brain measurements including cognition, emotions assessment (along with brain imaging), brain electrophysiology and autonomic body measures, along with personalized medicine gene markers.
- Clinical studies that include representative real-world samples of patients, including those with comorbidities.
- Marker-discover studies designed and implemented from the outset with a replication design.

- Studies designed to analyze new patient sub-groupings, as an intermediate measure to personalization, using groupings by genes and brain function, rather than merely using classical "signs and symptoms" diagnostic criteria.
- Diagnostic sensitivity and specificity should be reported in all personalized medicine studies, not just group results.
- There may be a need to group multiple markers together, in which case specificity of an individual market may be less critical.
- Exploration of a marker may not be specific in one disease, but may bring high specificity to the same disease when combined with other markers.
- Highlighting the multiple applications for personal medicine which the patient can benefit from, including prediction of:
  a. Diagnosis.
  b. Effective treatment prediction.
  c. Measures of treatment responsiveness.
  d. Measures of side effects.
  e. Impending disease onset.
  f. Disease remission or cure.
- Maximization of clinical studies using "standardized" objective methods to collect data, which allows direct comparison among and between different studies and classes of disorders, is additive and confirmatory, and uses the statistical power of pooled subject numbers.
- Databases (including standardized databases) are a key driver to make data available for shedding light on what works and the clinical extent to which new markers are useful, and the extent of their clinical effectiveness, comparative effectiveness, and cost effectiveness.

Other needs are more directly relevant to current clinical care and recent health care reforms and funding stimuli, including:

- Establishing interoperable data connectivity (integrated data sets) with electronic health records (EHRs).
- Treating the need of medical personnel to have available analytical computerized models to interpret a "personal" record and provide recommendations of additional tests, diagnosis, treatment, and care based on the latest available marker information (clinical decision support systems).
- Personalized medicine markers require clear reimbursement criteria and optimized financial models, which include paying for performance outcomes rather than just paying for clinical service delivery.
- Numerous scientific and trade organizations, as well as the federal government, should consider funding and tax subsidies to develop:
  a. New targeted gene-brain markers for personalized therapy.
  b. Standardization and use of appropriate measures.
  c. Enhanced commercial development programs to produce cost-efficient analysis of bona fide markers for everyday use.

As you continue to read this book, we urge you consider these recommendations and how they best address your needs and those of your patients in order to achieve effective personalized medicine for brain disorders. You will gain an appreciation of the current status of personalized medicine, as well as an appreciation of what we need to do to develop successful paradigms for the application of personalized medicine in the future, to the benefit of all.

The Foreword by Herbert Pardes offers a perspective from one of the long-time leaders in mental health, a contemporary leading figure in health care, and the President and Chief Executive Officer of New York-Presbyterian Hospital. The bulk of this book, however, comprises four focused sections discussing the genesis of personalized medicine, mental disorders, other brain disorders, and health care. Each chapter in every section starts with basic explanatory issues and then moves on to relevant subject matter, integrating new information with current approaches, and concluding with suggestions for future opportunities and challenges.

In Chapter 1, "The History of Personalized Medicine," Abrahams and Silver begin with Hippocrates around 2400 years ago and move up to the present day, providing insights into the historical path leading to the development of personalized medicine as we know it today. You will learn classical examples of genetic markers proven to be useful in medicine today, put into the context of why personalized medicine is such a high priority.

In Chapter 2, Schatzberg provides examples of the historical progression of biological markers in psychiatry during a period of great change, when psychiatry ushered in biological and molecular approaches to mental illnesses as a replacement for psychoanalysis. It was during this period that it became clear that, as in other human illnesses, mental illnesses are a manifestation of altered biological mechanisms—in this case, structural, neurochemical, biophysical, genetic, and epigenetic.

In Chapter 3, Williams and Gordon present their perspective on applications of personalized medicine based upon an integrated model of the brain, across disciplines and scale, using Gene-Brain markers (i.e., imaging modalities, and quantitative cognitive testing, including emotional, as well as electrical measures of brain and body), and the recommendation that these measures should be collected with standardized methods and protocols. Their chapter explores potential personalized medicine markers from the first standardized international database of the human brain.

The next section of the book is titled "Personalized Medicine and Mental Disorders." One of the most exciting methods of brain research which emerged in the 1990s was the refinement and application of Magnetic Resonance Imaging (MRI) to functional MRI (fMRI), providing a window into the living functioning brain. In Chapter 4, Fornito and Bullmore discuss using fMRI to examine mental illnesses, and the successes and failures of fMRI investigations in providing insights into diagnosis, prediction, and treatment. They argue that we need to move away from traditional case control studies and employ more sophisticated statistical "connectivity" techniques using brain activity patterns to align individual participants with key clinical outcome variables, such as diagnosis or treatment response.

The profound impact that stress has on brain function and how it may affect personalized medicine as it relates to mood and anxiety disorders is discussed next by Gillespie and colleagues. The critical issues explored here are the impact of early life stress on both brain development and the function of the hypothalamic-pituitary-adrenal (HPA) axis. They report findings from studies which examine endocrine function, genomic regulation and expression, and brain imaging to unravel the impact of stress on brain function as related to anxiety and depression. While there are early genomic markers, most need further examination. Most germane to personalized medicine and epigenetic impact is the potency of developmental stress on major mental disorders.

Schizophrenia, another devastating mental illness, is presented in Chapter 6. Ballon and colleagues focus on personalized medicine markers in schizophrenia. Current researchers consider schizophrenia to be a developmental disorder, with early onset and lifelong positive and negative symptoms. Currently there are well-documented major changes in the structure of the schizophrenic brain; however, there are no replicated clear markers for diagnosis, treatment, or prognosis which have been definitively demonstrated. This chapter presents the current data on genomics, brain imaging, and neurophysiological approaches and how this data supports schizophrenia as a neurodevelopmental brain disorder.

Chapter 7, "Personalized Integrative Markers for Attention Deficit/Hyperactivity Disorder in Children and Adolescents,"takes a broader approach to markers than just genomics and demonstrates the value of combining, Brainmarkers of cognitive function, emotions, feeling, self-regulation, and brain imaging with markers of gene function to provide value in diagnosis and treatment prediction for Attention Deficit/Hyperactivity Disorder (ADHD). Kohn and colleagues focus especially on children and adolescents. They make a salient case for looking at integrative brain function, using multiple measures of function, to provide the insight necessary for understanding, diagnosing, and treating ADHD. In their report they also use the Integrate Model (see chapter 3) to test and extend their hypothesis of dysregulation in ADHD.

Section 3, "Personalized Medicine and Other Brain Disorders" begins with Chapter 8, "The Role of Neuroimaging Biomarkers in Personalized Medicine for Neurodegenerative and Psychiatric Disorders." This chapter presents an excellent overview of imaging methods and insights for biomarkers from brain imaging for dementia, schizophrenia, and bipolar and major depressive disorders. The authors delineate nine criteria which need to be met for a marker to qualify as biomarkers, and conclude that in practice, we have no biomarkers in neuropsychiatry that currently fulfill all of these criteria. However, many identified biomarkers meet at least some of their criteria for specific disorders and utility for personalized medicine. For example, in Alzheimer's disease, MRI has demonstrated whole brain and hippocampal atrophy and ventricular expansion as good Brainmarkers. In addition, Magnetic Resonance Spectroscopy (sMRI) has demonstrated changes in metabolites, which may identify early disease and differentiate between variants. Reports with Positron Emission Tomography are positive when looking at amyloid plaques, but still too early to be definitive. The authors conclude that imaging studies hold much promise and that current studies are moving in the right directions.

The brain is, in essence, the master organ, regulating and controlling all physiology and pathology; it is highly interconnected with the entire body and its functions. The implications of this idea are considered in more depth in Chapter 9, by Richard Gevirtz. This chapter provides another important example of markers beyond genes. Here, the clinical usefulness of Heart Rate Variability (HRV) is summarized, as is its relevance as a marker for personalized medicine. The work presented uses a meditational model, focusing on measurement of the autonomic nervous system, respiratory system, and heart rate as critical variables for monitoring brain–body imbalance and implementing change. Convincing data and arguments are presented to suggest that Heart rate variability offers a unique window into the functioning of the autonomic nervous system, which is critical to overall health and necessary for optimal performance. Examples of the "meditational" model are provided in a number of disorders with heterogeneous symptom clusters, which are difficult to treat in clinical practice.

Sleep is critical for optimal performance. However, sleep suffers in a range of disorders, including both sleep disorders and general medical and psychiatric illnesses. Grunstein presents his views on personalized medicine and sleep health in Chapter 10. Current approaches to individualized management in clinical sleep sciences have predominantly focused on phenotypes, such as anatomical structure, presence of sleepiness, or biomarkers. There are many approaches to studying sleep, which range from studies on circadian rhythms and sleep regulation to studies of the different phases of eye movement in sleep and electroencephalographic (EEG) studies, all offering new insights and treatments. Sleep is an extremely important area for this field and impacts on all functions and disorders. Grunstein argues that, despite the prevalence of, comorbidity with, and range of treatments in sleep disorders, sleep researchers have been slow to bring together the diagnostic markers and treatments into the field of personalized medicine. He provides an outline that could be applied to a standardized clinical and sleep EEG assessment across existing sleep centers around the world, in which markers are identified that help point to the most useful treatment options for each patient.

Chapter 11 focuses on multiple sclerosis. Miller and his colleagues present compelling evidence that research on this disorder has begun to embrace the elements of personalized medicine. This chapter presents an excellent overview of the status of identification of biomarkers for diagnosis, disease course, treatment response, and related factors involved in development and response to treatment for multiple sclerosis. It highlights their goal to give the patient the right drug and dose at the right time, thus treating the patient according to their specific needs and characteristics—in other words, implementing personalized medicine.

In Section 4, we deal with personalized medicine in the health care setting. Whitehouse introduces this issue in his chapter on "Brain-Related Health Care: New Models for Personalized Medicine" in Psychiatry. Whitehouse is in a position which offers a bird's-eye view of this world; he is the chief strategist for Optum Health, with responsibility for determining the best options available for both the patient and the insurance company. Whitehouse argues that applications of personalized medicine must be cost-effective and result in an improvement, if not a cure. He proposes that we start with practical alternatives to expensive and time-consuming assessments, imaging, and laboratory measures. As new biomarkers or Brainmarkers become validated and are deemed to be cost-effective, they can be added to current clinical, psychological, and neuroscientific findings and data. A step-by-step personalized medicine implementation model allows the latest information technology advances to be used by the client and patient.

Eugene Baker offers a practical application of personalized medicine to a health care setting in Chapter 13, "Clinical Decision Support in Employee Assistance Programs: Personalizing the Therapeutic Approach." Depression is the most common complaint seen in Employee Assistance Programs (EAP). What is the treatment? It turns out that the treatment offered depends on the training of the health care provider or administrator who sees the patient, and thus the ideal treatment course is not always presented. Baker then presents a compelling case for using a state-of-the-art, computerized, Web-enabled assessment (a clinical decision support system) program to advise on the best course of treatment–the first steps to accommodate personalized medicine in workplace-based health care programs.

It is inevitable that the pace of investment into discovery of new personalized medical markers will be driven by both federal funding and private, corporate investment. Segal and Williams outline these realities in Chapter 14. A core change is that the patient/consumer is at the center of this revolution, and they will be served by a new health care system which is being transformed by the digital revolution. In this chapter you will find new possible business models for implementing care based on a personalized medical approach.

In the concluding chapter, the editors summarize the key points of personalized medicine, health care, and integrative neuroscience, including actionable recommendations of how the field might best be applied so as to create an effective and efficient health care system of personalized medicine for each and every patient.

Stephen H. Koslow
Evian Gordon

Section 1

Genesis of Personalized Medicine

1 The History of Personalized Medicine

Edward Abrahams, PhD and Mike Silver, PhD

It's far more important to know what person the disease has than what disease the person has.

—*Hippocrates*

Introduction

The history of personalized medicine is not so much about development of the concept of "personalization"—first introduced by Hippocrates (ca. 460 BCE–ca. 370 BCE) around 2400 years ago—but about the evolution and increasing precision of diagnosis and treatment. With each step of medical progress, the knowledge and tools used to describe and diagnose disease have shifted from the metaphysical to the physical, from the cellular to the molecular—and most recently, to a system-level understanding of the interactions between molecular events and higher level phenomena, such as cognition and behavior.

The goal of personalization in medicine has never wavered, but each era has had its own increasingly advanced tools for tailoring treatment to the individual. In ancient times, it was the four humors (sanguis, phlegm, choler, and melancholia). Today, it is the four chemical building blocks of DNA. We have moved from ascribing mental disorders to supernatural causes, to understanding them through brain imaging and the actions of neurotransmitters. As we trace the history of personalized medicine from ancient times to the present, we observe a critical shift in what may now be referred to as the post-genomic era. Using genetic and molecular diagnostics and other markers of functional significance, we are on the verge of accurately predicting whether someone will develop an illness many years in the future, respond positively to treatment, or suffer a serious reaction to a drug. The arrival of such advanced medical forecasting will mean that other elements of the health care system and society have to coevolve with the technology, including laws protecting privacy, systems of payment, regulatory guidelines, physician and patient education, and ethical frameworks. Here, we examine the historical path that has led to personalized medicine as we know it today, in the hope of gaining a better perspective on where it will take us in the not-too-distant future.

Ancient Practice, Modern Thinking

From pre-Socratic times to the Middle Ages, the practice of medicine maintained a close connection to philosophy. Physicians and scientists of the ancient world drew from the wells of mysticism, mythology, and cosmology more than what we consider today to be "hard" evidence, and the distinction between medicine and "natural philosophy" was blurred.

In 400 BCE, Hippocrates created a split in the practice of medicine, starting the movement of medical work away from natural philosophy and establishing the physician as a separate professional. The Hippocratic physician was guided by a rational understanding of disease as rooted in physical processes.

Alcmaeon's (dates of birth and death unknown) pioneering exploration of the human body through anatomical dissection in the fifth century BCE, and his theories on the equilibrium of fluids and "humors" established, at least conceptually if not factually, a connection between a person's physical state and his or her environment, nutrition, and lifestyle.

Hippocrates extended Alcmaeon's doctrine, along with Epedocles's (dates of birth and death unknown) concept of the four elements, to define four basic humors: sanguis/blood (air), phlegm (water), choler (fire), and melancholia (earth). Choler and melancholia were sometimes referred to as yellow and black bile, respectively. Disease was considered to be the result of an imbalance (*dyscrasia*) between these humors, and it was the role of the physician and patient to try to reestablish a proper equilibrium (*eucrasia*). A word we use today, "complexion," was derived from the practice of complexing the four humors to define the unique characteristics of each individual and their state of health, a practice we might now refer to as *phenotyping*. We can still classify people today as being "good-humored" or "ill-humored," as a result of the primary role of the humors in early medical practice.

In the context of today's medicine, diagnosis under these criteria appears to be mere fiction, yet it established an important nexus between observation, prognosis, and tailored treatment, and formed the basis for medical theory from ancient times to the late eighteenth century.

Hippocrates made other foundational contributions that ultimately became central tenets of personalized medicine. Prognosis was central to Hippocratic medicine. "He will manage the cure best," he said, "who foresees what is to happen from the present condition of the patient" (Adams 1849). He was the first to accurately describe and diagnose many diseases, such as pneumonia and epilepsy and, notably, he was the first to point out the differences in the manifestation of these diseases in individuals, their severity, and the ability of patients to respond to treatment and recover.

Hippocrates also made a critical connection between the constitutional and environmental causes of disease, emphasizing lifestyle changes in his prescriptive treatments, including rest, appropriate diet, and cleanliness, as described in his treatise, *On Air, Water and Places* (Adams 1849). In that same document, he noted tendencies toward different diseases in the various populations in Europe and Asia, perpetuated by heredity.

Pausing for a moment in this historical review, one may note that medicine in the fifth century BCE had all the fundamental elements of personalized medicine in place. We can draw a direct line between stratifying by humors and other observations, and stratifying by genetic and molecular profiles; between Hippocrates's clinical documentation and the push for electronic health records; between his emphasis on prognosis and the modern-day shift toward predictive medicine. What was missing was an ability to connect observations, and an understanding of the real underlying causes of disease as being based in chemical and biological factors.

From Metaphysical to Physical Observations of Disease

It was about 2000 years later that chemistry started to play a role in medicine. Paracelsus (born Phillippus Aureolus Theophrastus Bombastus von Hohenheim, 1493–1541) was a

Renaissance man: botanist, alchemist, astrologer and physician (Pagel 1982). He developed theories that broke with traditional medicine. Noting the condition of miners and the illnesses to which they succumbed, Paracelsus held the view that disease was a result of an imbalance of minerals within the body, caused by interaction with the environment. The chemicals and minerals were the poisons that led to disease, but they were also the cure. As he noted, "All substances are poisons; there is none which is not a poison... The right dose differentiates a poison from a remedy."

Paracelsus emphasized the use of chemicals and minerals derived from herbs and other natural sources as treatments, and noted their beneficial effects at the right dose. Not content with the ancient dogmas of medicine and the practices of purging and bloodletting, and acknowledging a greater complexity in the root causes of disease, he introduced the language of chemistry into medicine, and insisted that it be founded on scientific experimentation.

Opening a Window on the Microscopic Origins of Disease

The next major shift in the practice of medicine was effected by someone who was not a physician at all. Antonie Philips van Leeuwenhoek (1632–1723) was a Dutch cloth merchant from Delft, the Netherlands. Initially looking for better tools to inspect his wares, he invented over 400 types of microscope capable of magnification up to 500 times. The sole proprietor of a window into a tiny world, he made many important discoveries: the existence of single-celled organisms (Protista) and bacteria, the cellular structure of human tissue, blood flow in capillaries, and the first hints of subcellular organization (such as nuclei and the bands in human muscle fibers). With these discoveries, van Leeuwenhoek launched medicine into the era of microbiology and histology, and enabled a better understanding of the microscopic origins of disease.

Additional improvements in microscope technology, tissue embedding media, microtomes, and histochemical staining procedures brought discoveries much further by the late nineteenth century. Joseph von Gerlach (1820–1896) published the first staining protocol in 1858, and similar staining methods became the tool of choice in identifying tissues and their changes in state (Clark and Kasten 1983).

The methods of identifying and subclassifying diseases using histochemistry and microscopy were used well into the twentieth century and continue to be valuable today. Rudolf Virchow (1821–1902) employed them to identify leukemia cells for the first time, setting in motion a series of discoveries that led to one of the earliest, detailed subclassifications of disease. Leukemia was first diagnosed by John Hughes Benett in 1845, after physicians in the nineteenth century observed patients with abnormally high levels of white blood cells, and referred to the disease as "weisses blut" or "white blood"—a subclassification of one. By 1913, using cellular pathology techniques pioneered by Virchow, four types of leukemia were identified: chronic lymphocytic leukemia, chronic myelogenous leukemia, acute lymphocytic leukemia, and erythroleukemia. Treatment protocols began to diverge along these separate classifications, and response rates began to improve.

Going Molecular: The Ultimate Precision in Diagnosis

> *The existence of chemical individuality follows of necessity from that of chemical specificity, but we should expect the differences between individuals to be still more subtle and difficult of detection.*
>
> —*Archibald Garrod*

The personalization of medicine gained a solid footing in science in the nineteenth century, and many diseases could be diagnosed and distinguished at the cellular level. Nevertheless, the categories of disease remained broad, and many of the treatments limited.

By the turn of the twentieth century, attention had shifted to the next and final stage of reduction in disease diagnosis and treatment—at the molecular level. In 1900, Hugo de Vries and Carl Correns resurrected the writings of Gregor Mendel (1822–1844) who, thirty-five years prior, had conducted detailed experiments on pea plants illuminating the laws of genetic inheritance (Mendel 1866). His work, largely ignored, had only been cited three times, but from 1900 it caught on rapidly, eventually becoming the foundation of biological science.

In that same year, Sir Archibald Garrod (1857–1936) investigated several families that exhibited a metabolic insufficiency called alkaptonuria (Garrod 1902), which causes a darkening of the urine. Patients eventually developed arthritis, with deposition of brown pigment in joint cartilage and connective tissues. Earlier generations of physicians might have attributed the disease to an excess of black bile. Garrod suggested that the disease and the tendency for certain individuals in a family to develop the condition while others did not was a result of the "chemical individuality" of each patient. Armed with the recently dusted-off works of Mendel, Garrod determined the pattern of inheritance (autosomal recessive), and postulated that the disease was caused by a mutation in a gene encoding an enzyme responsible for metabolizing a class of compounds called alkaptans. He originated the "one gene, one enzyme" hypothesis that held sway well into the 1970s, and suggested that chemical individuality and genetics were at the root of most human diseases (Garrod 1909; 1931)

In 1931, A.L. Fox stumbled upon a demonstration of chemical individuality while investigating non-nutritive sweeteners (Fox 1932). One of the intermediate compounds he was using, phenylthiocarbamide (PTC), escaped into the laboratory ventilation. Fox was unaware of what happened, but several of his colleagues noticed a bitter taste, ranging from mild to very intense. About 30 percent of Caucasians shared PTC taste insensitivity with Fox, who thought it was passed on by simple Mendelian recessive inheritance. It was not until seventy years later that research pointed to multiple alleles (Guo and Reed 2001; Tepper 2008).

Fox's discovery would have been the earliest instance of what might be considered nutrigenomics, but as is often the case, the Greeks seemed to have at least understood the concept first. In 520 BCE, Pythagoras (ca. 570–ca. 495 BCE) observed that eating fava beans could result in a fatal reaction in some individuals (Pimohamed 2001), while for most people, it was just lunch. The genetic basis for the selective toxicity, discovered in 1956, was an inherited mutation in glucose-6-phosphate dehydrogenase (Carson et al. 1956). The same enzymatic deficiency was found to be linked to toxicity of the antimalarial drug primaquine (some soldiers who took the drug during World War II developed anemia), and thus became the first molecular proof of the chemical individuality hypothesis offered by Garrod in 1898.

By the middle of the twentieth century, the pharmaceutical industry had grown significantly, and its products were assuming a central role in health-care strategies. Insulin was introduced in the 1920s, greatly extending the length and quality of life for millions of diabetics. Antibiotics, including sulfonamides and penicillin, were developed in the 1930s and 1940s, transforming once deadly encounters with infection into a common ritual of treatment and recovery. Companies built upon that success and ventured into more complex areas, such as cardiovascular disease, bringing the first antihypertensive medicines to market in the 1950s.

The increased prevalence of pharmaceutical treatments in medicine led to observations of individual differences in response to the drugs, usually due to their metabolism (e.g., suxamethonium, isoniazid and debrisoquine) (Evans et al. 1960). When Arno Motulsky published his seminal article on the topic in 1957, "Drug Reactions, Enzymes and Biochemical Genetics," he helped launch a new field of research, and introduced the concept of tailoring medicine to an individual's genetic makeup (Motulsky 1957). It was two years later that Friedrich Vogel coined the term *pharmacogenetics* (Vogel 1959), and then, in 1963, the first book dedicated to the field, *Pharmacogenetics: Heredity and the Response to Drugs*, was published by Werner Kalow (Kalow 1962).

The nature of DNA, its atomic structure, and the arrangement of the chemical bases were discovered in 1953 by James Watson and Francis Crick. The actual significance of the sequence of bases was deciphered by 1967, when 54 (out of a possible 64) of the 3 base "codons" were shown to direct the addition of specific amino acids (or a "stop" or "start" signal) in the synthesis of proteins (Khorana 1968). Scientists could now make the connection between a variable trait—such as the ability to respond to or metabolize a drug—to a gene, protein, or enzyme, and the molecular mechanisms and pathways within the body.

The idea that genetic differences could affect the response to drug treatment did not catch on quickly, mainly because scientists were aware of so many environmental factors that could cause such variation, and that one more technological advance had to transpire that would enable scientists to separate genetic and environmental influence—the ability to rapidly sequence DNA to obtain a genotype. In 1975, Allan Maxam and Walter Gilbert developed a convenient method to sequence DNA (Maxam and Gilbert 1977), by chemically modifying the strands and cleaving them at specific bases. Another method, developed by Frederick Sanger, used dideoxynucleotide triphosphates (ddNTPs) as DNA chain terminators during synthesis, achieving the same effect as chemical modification and cleavage, but using fewer toxic chemicals and lower amounts of radioactivity (Sanger 1981). Sanger sequencing became the method of choice well into the 1990s.

By the 1970s, researchers had discovered at least two major enzyme families responsible for variation in drug metabolism and response. Variable response to the muscle relaxant suxamethonium chloride, used in surgery, ultimately pointed to the role of N-acetyltransferase (NAT) in chemically modifying and metabolizing the drug (Dorkin 1982). Genetic variation of NAT created "slow acetylators" and "fast acetylators," meaning that some individuals, given the same dose, would experience longer or shorter half-lives of the drug in their blood.

During the 1950s, the drug sparteine was shown in clinical trials to be as effective as oxytocin in inducing labor, but in about 7 percent of patients, the duration and intensity of the effects were dramatically increased, leading to prolonged uterine contractions and abnormally rapid labor, in some cases leading to death of the fetus. Another drug used for treating hypertension, debrisoquine, caused significantly prolonged hypotension in 5 to 10 percent of patients. The culprit in both of these cases was identified in 1977—a polymorphic metabolizing enzyme called cytochrome P450 2D6 (CYP2D6; Eichelbaum, et al 1975; Eichelbaum et al. 1979). Slow-metabolizing variants of CYP2D6 would lead to elevated levels of active drug in the bloodstream at standard doses, leading effectively to an "overdose" and an exaggeration of the drug's effects.

Harkening back to the aphorism from Paracelsus that "the dose makes the poison," scientists and physicians now understood that genes, at least in part, can help determine the dose. The 1980s and the 1990s witnessed a flurry of activity in molecular biology. With the invention of the polymerase chain reaction (PCR) in 1983 (Mullis 1990) to replicate trace amounts of DNA in vitro, the exploration of genetic variation and its relationship to drug metabolism, safety, and effectiveness expanded rapidly.

At this point, the long arc of medical history had seen the definition of disease and its treatment evolve from the mystical and alchemical, to the anatomical and physical, to the histological and cellular, and ultimately to a genetic and molecular understanding. It is a combination of genetic identity, lifestyle, and environment that makes us unique, and that determines our susceptibility to disease or response to treatment. Medicine now had the tools to analyze our individuality and precisely tailor treatment.

The Human Genome: A New Chapter in Personalized Medicine

Examples of genetic variability and its effect on response to drug treatment were accumulating, but by the mid-1980s, an idea was brewing that would revolutionize the field and throw the discovery of genetic variation and the study of its effects into high gear. In 1985, Renato Dulbecco, Nobel Laureate and President of the Salk Institute, asserted that the best way forward in cancer research would be to sequence the entire human genome (Cook-Deegan 1994). Original drafts of an article published in *Science* on March 7, 1986 (Dulbecco 1986) outlining his rationale apparently extended claims that sequence information might help explain variation among breast cancer genes and manifestations of disease.

In 1985, a small group of visionary scientists gathered on the campus of the University of California, Santa Cruz, to consider the feasibility of establishing an institute to sequence the human genome. The discussion generated a great deal of excitement, but led to the conclusion that it simply would not work. The technology had not yet arrived. Nevertheless, the seeds of the idea were planted, and one of the group's participants, Walter Gilbert, along with James Watson, became strong advocates for a human genome project.

With the backing of accomplished scientists, the idea began to gain traction, despite the perceived limitations. To the U. S. Department of Energy (DOE), the genome project seemed to be a natural fit because "big science" (such as the Manhattan Project), was the agency's stock-in-trade. The National Institutes of Health (NIH) supported biological research, but never anything on this scale. Nevertheless, concerns about the DOE's possible role were raised, prompting the NIH to create a special office for genome research headed by Watson. In 1990, both the DOE and the NIH were funding genome programs, with a total budget of $84 million. By 1992, Francis Collins had taken the helm of the effort, and the balance of funding and the ultimate goal of completing the human genome had clearly shifted to the NIH.

In the early years of the Human Genome Project, the idea of "big biology" was anathema to many in the field. Biological research had always been conducted in small labs by individual investigators who sometimes collaborated, but largely followed their own individual paths. That independence, it was argued, led to more innovative approaches to scientific discovery. A massive goal-driven effort, and the enormous price tag of $3 billion, would siphon away talent and funding from other NIH projects. The idea of spending three billion dollars of government money to support a project of such unprecedented scale was considered either visionary or ludicrous, depending on whom you asked.

But even for the visionaries, the scope of the impact of the Human Genome Project, as we know it now, escaped their imagination. The project was launched in 1990, and the first draft of the human genome sequence was published in 2001 (International Human Genome Sequencing Consortium 2001). By 2003, nearly all of the sequence of the roughly 23,000 genes in the human genome was recorded, and the public database (GenBank) became the essential reference for future studies in biology, medicine, and, specifically, in genetic variation. It also demonstrated that "big biology" could be a successful endeavor;

in some cases, pooling resources and collaborating among government, nonprofit, and private entities could do more to advance knowledge than having scientists work independently toward separate or overlapping goals.

In the first large-scale follow-up to the Human Genome Project in 1999, the SNP Consortium set out to identify over 300,000 loci of genetic variation in the human genome (which is 99.9 percent identical among all individuals), and release the information to the public. These single-nucleotide polymorphisms (SNPs), in which one base is deleted, added, or substituted, enabled a better understanding of why some individuals are predisposed to cancer, diabetes, Alzheimer's disease, and cardiovascular and other diseases. Ten of the largest global pharmaceutical companies participated, as did several academic research centers and the Wellcome Trust in the UK. By 2003, they had identified 1.8 million of the 10 million SNPs believed to exist in the human genome. The public SNP database, known as dbSNP http://www.ncbi.nlm.nih.gov/SNP) managed by the NIH, has become a rich source of information on genetic variants.

In 2002, a wide-ranging contingent of international governmental, nonprofit, and private entities formed the International HapMap Consortium, which expanded on the SNP Consortium's efforts by providing a more "usable" map of variation within the human genome. Scientists could identify genetic variants linked to disease or treatment response from among the 10 million SNPs thought to exist in the human genome, but the HapMap offered a shortcut by grouping variants on one chromosome into "haplotype" blocks that are typically inherited together. Correlation of genetic variation to disease or treatment response then becomes a much simpler problem of searching first among a smaller set (300,000 to 600,000) of haplotype "tag" SNPs representing each block.

The first draft of the HapMap was completed in February 2005, seven months ahead of schedule, and researchers subsequently embarked on a "Phase II" effort to make the map five times denser in genetic variants, adding another 4.6 million SNPs. With the HapMap constructed, the third phase of "big biology" gathered momentum, with the emergence of large-scale studies to identify haplotype patterns among groups of people with certain diseases or response to treatment. These are the so-called genome-wide association studies (GWAS). These studies often required thousands of participants, spanning research and clinical institutions and geographic regions. Efforts cosponsored by government, academia, and industry began to generate publicly available data and frameworks to support these large-scale gene-disease association studies.

A Legal History of Personalized Medicine

As the role of genetics in medicine became more prominent, the concept of genetic privacy also came into focus. The knowledge of a person's susceptibility to disease, even before he or she shows signs or symptoms, can be either a powerful tool for improving health and quality of life, or a means to discriminate against people in terms of employment, or to limit access to insurance and other resources. An examination of the legal landscape around genetic privacy shows that it is co-evolving with advances in technology and the acceptance of personalized medicine.

The Supreme Court ruled in *Katz v. United States* (389 US 347 [1967]) that there is no "general constitutional 'right to privacy'... the protection of a person's general right to privacy... is, like the protection of his property and of his very life, left largely to the law of the individual States." However, privacy and confidentiality are essential tenets of medical ethics, and the courts have recognized, to a limit, the expectations of privacy around medical records.

In *Whalen v. Roe* (429 US 589 [1977]), the Supreme Court considered a case in which the State of New York was collecting information, without consent, on the prescription of certain drugs. Did the state have the right to access and use personal medical records? The case was decided in favor of the Commissioner of Health for New York (Whalen), ruling that disclosure of personal health information to representatives of the State responsible for public health does not necessarily constitute an unacceptable invasion of privacy. When there is a legitimate use for the public good, it was determined, the government may access medical records.

A ruling by the United States Court of Appeals of the Ninth Circuit in *Norman-Bloodsaw v. Lawrence Berkeley Laboratories* (135 F.3d 1260, 9th Cir. 1998) established that non-consensual testing of employees for medical conditions was a violation of individual rights to due process guaranteed by the Fourth, Fifth, and Fourteenth Amendments to the U. S. Constitution. In this case, the Lawrence Berkeley Laboratories were using the results of tests for syphilis, the sickle cell trait, and pregnancy as a condition for employment.

Each of these cases helped define the scope and limits of protections for medical information in general. In 2001, a case dealing specifically with genetic discrimination by an employer was brought by the U. S. Equal Employment Opportunity Commission (EEOC) in an Iowa district court (Schafer 2001). The EEOC alleged that the Burlington Northern Santa Fe (BNSF) Railroad had conducted genetic testing on employees without their consent. The test was for a genetic marker for carpal tunnel syndrome, presumably to gain information on their employees' high incidence of repetitive stress injuries. Under the Americans with Disabilities Act of 1990, it would be unlawful to conduct the tests because they were not connected with any business necessity, and any employment action attached to the results of the test would amount to discrimination based on a disability. The EEOC did not find evidence that BNSF intended to use the results of the genetic test as a basis for employment action, but at least one employee was threatened with termination for refusing to take the test. The case was settled out of court, with BNSF compensating the affected employees.

In the absence of uniform federal regulations around genetics privacy and discrimination, many states established their own regulations, resulting in an uneven landscape of protection. The Health Insurance Portability and Accountability Act (HIPAA) of 1996 and other federal regulations controlled access to medical records among federally funded institutions, but left wide gaps in genetic privacy protections with respect to employers and insurance providers. The gaps in protection may have worked against a more rapid adoption of personalized medicine.

In 2008, however, the Genetic Information Nondiscrimination Act (GINA) was signed into law; it explicitly prohibits employers and health insurers from discriminating against individuals on the basis of their genetic risk factors. The federal law remains to be tested (actual court actions involving genetic discrimination, even before GINA, have been rare), but it has established a foundation for genetic privacy and nondiscrimination that is building confidence among the public that genetic information will not be used against them, opening the door to greater participation in research, and acceptance of genetic information as part of the medical record.

Products and Progress

By the 1990s, the idea of tailoring treatment to a patient's genetic makeup was well established, but rarely applied. One biotech drug developed for the treatment of breast cancer shifted the landscape. Herceptin® (generic name trastuzumab) is a monoclonal antibody designed to bind to the human epidermal growth factor receptor 2 (HER2), which is responsible for regulating cell growth. In 1997, Phase III study results showed the drug to be ineffective in the overall population tested, but an analysis of trial results revealed a hidden success story: 20 to 30 percent of the women whose tumors presented with HER2 overexpression had a significantly better response to the drug. In 1998, Genentech presented the clinical data to the FDA (Cobleigh et al. 1999), which approved the application as a drug/diagnostic combination. Without patient selection, Herceptin® would have been added to a long list of cancer drugs that failed to demonstrate effectiveness in clinical trials. Instead, it became the first FDA-sanctioned personalized drug/diagnostic product.

Since Herceptin®, a growing list of personalized medicine diagnostics and therapies have received FDA label acknowledgments, recommendations, or mandates in cancer, cardiovascular disease, and transplantation, as well as in neurological and psychiatric conditions (Personalized Medicine Coalition 2009). In the latter category, a test for *HLA-B**1502* was recommended to avoid serious dermatologic reactions (Stevens-Johnson syndrome) when administering carbamazepine for the treatment of epilepsy or bipolar disorder; and codeine, a prodrug of morphine, was noted for its risk of extensive or poor metabolism linked to genetic variants of *CYP2D6*.

Around 2003, a new type of test emerged that measured and analyzed the expression pattern of multiple genes within a tumor as a tool to predict whether a patient would respond to treatment or suffer side effects. The OncoType DX® multivariate index array, the first in this class to be commercialized, linked sixteen genes to the risk of breast cancer recurrence and the likely benefit of chemotherapy in a patient whose tumor is estrogen- or hormone-receptor positive (Paik et al. 2004). MammaPrint, a 70-gene multivariate index array, similarly determined the risk of tumor recurrence for node-negative breast cancer, and in 2007 was the first such diagnostic to be approved by the FDA as a guide to treatment (Pollack 2007). A 384-SNP diagnostic test, PIMS PhyzioType™ System (still in development by Genomas), also uses a multivariate algorithm to assess a patient's risk of metabolic syndrome side effects when treated with atypical antipsychotics olanzapine, quetiapine, or risperidone.

Pharmacogenetic tests detecting cytochrome P450 variants have been applied to psychiatric indications. CYP2D6 is known to affect the metabolism of antipsychotic and antidepressant drugs and CYP2C19 metabolizes several antidepressants (Kirchheiner et al. 2004; de Leon et al. 2006), accounting for a substantial amount of variation in patient response. Such pharmacogenetic effects were known as early as 1971 (Alexanderson and Sjoqvist 1971). The National Institute of Mental Health STAR*D (Sequenced Treatment Alternatives to Relieve Depression) (McMahon et al. 2006; Paddock et al. 2007) and other studies (Binder et al. 2004; Uhr et al. 2008) put a spotlight on the low success rate of initial therapy for depression and examined the role of pharmacogenetics for antidepressants. The discovery of a growing number of molecular and imaging markers related to psychiatric and neurological conditions assures that clinical neuroscience applications in personalized medicine will continue to be an active area of investigation.

Personalized Medicine Becomes a National Priority

The FDA has struggled to keep up with the rapid developments in personalized medicine after 2003. In an effort to establish some clarity for drug developers, the agency acknowledged its role in encouraging the adoption of pharmacogenetics and personalized medicine through the Critical Path Initiative. The FDA also published a guidance document for voluntary pharmacogenetic data submissions (U. S. Food and Drug Administration 2005a), mitigating the threat that submitting such data might hinder a product's pathway to regulatory approval. The agency also published other draft guidance for genetic tests, including one overcoming multivariate index arrays (U. S. Food and Drug Administration 2007) and a concept paper for the co-development of pharmacogenomic drugs and diagnostics (U. S. Food and Drug Administration 2005b). The FDA clearly stated its openness to permitting "adaptive" clinical trials that genetically "enrich" a study population as a trial proceeds, in order to reduce the time required to establish safety and effectiveness (Vastag 2006). Furthermore, the FDA introduced labeling regulations (21 CFR 201.57), addressing the relationship between genotype and drug response.

In 2005, the FDA followed up its progressive, if slow moving, reform toward personalized medicine by approving the AmpliChip. The newly approved product could have had an impact on a large number of drugs for which the metabolism depends on the detected variants. Yet, at the time, most products did not have a requirement for such testing. The number of pharmaceutical products with labels *recommending* a genetic test to guide selection or dosing stood at about 200 by 2009. One such recommendation garnered much attention in 2007, when the FDA suggested that a patient's profile of two genetic variants (VKORC1 and CYP2C9) be determined as a guide to dosing of the clot-busting drug warfarin. Since more than two million people are started on the drug each year in the U. S., the potential for bringing personalized medicine to a large population was significant (U. S. Food and Drug Administration 2007b).

Such widespread adoption proved to be elusive, however. In 2009, the Center for Medicare and Medicaid Services (CMS) indicated that it would not pay for the warfarin genetic test except in clinical trials to develop evidence. (Pollack 2009). CMS found that pharmacogenomics-guided warfarin dosing had insufficient evidence to prove its benefit to patients. With CMS laying down the gauntlet on backing up products with evidence, private insurers seemed likely to follow suit, and deny coverage for warfarin genetic testing, as well. The series of events from FDA approval to CMS denial illustrated one of the central dilemmas in personalized medicine that has often recurred: a large gap between the standard of evidence required for market approval, versus reimbursement and market acceptance.

Recognizing that the scientific evolution of personalized medicine had reached a critical point and that the regulatory, industrial, and social infrastructure to support it had to be improved, personalized medicine became a priority issue at the highest levels of government. In 2006, the U. S. Department of Health and Human Services (HHS) launched the Personalized Health Care Initiative to support research, clinical adoption, product regulation, and legal protections related to the use of genomics in medicine. Reports released by the HHS Secretary's Advisory Committee on Genetics, Health, and Society (SACGHS) examined and recommended actions related to the integration of genetics into healthcare; the ethical, legal, and social implications of genomics in medicine; the medical education curriculum; and the impact of patent policy, privacy legislation, regulation, and insurance reimbursement. The President's Council on Science and Technology made recommendations to the President in eight major policy areas in its 2008 report, "Priorities for Personalized Medicine" (President's Council of Advisors on Science and Technology 2008). In 2007, then-Senator Barack Obama introduced the Genomics and Personalized Medicine Act to encourage the development and adoption of personalized medicine, with provisions to support research and institute regulatory and other policy changes (U. S. Senate 2006; U. S. Congress 2008).

Beginning around 2008, the push for health care reform led to a keen focus on electronic health records (EHRs). Health information systems with EHRs installed in hospitals and doctors' offices would enable the collection, comparison, and analysis of patient genetic and molecular profiles, medical histories, and clinical outcomes, and thereby facilitate the prediction of disease susceptibility and the patient's response to treatment. In 2004, a presidential executive order established the Office of the National Coordinator for Health Information Technology (ONC), with the goal of implementing EHRs nationwide within 10 years.

By 2008, however, only two (Jha et al. 2009) to eleven percent of hospitals (American Hospital Association 2007), and fewer than five percent of solo physicians (Hsiao et al. 2008), had implemented fully operational EHR systems. Acknowledging the challenges, the Obama administration and the U. S. Congress committed an unprecedented $19 billion in funding for health information technology as part of the Health Information Technology for Economic and Clinical Health, or HITECH, Act, a section of the American Recovery and Reinvestment Act of 2009.

Well before these government initiatives were enacted, a number of health care delivery organizations had already put EHRs into practice to support personalized medicine, affecting millions of patients (U. S. Department of Health and Human Services 2008). For example, the Marshfield Clinic in Wisconsin has been one of the earliest adopters of EHRs, beginning in the 1970s and establishing widespread use by the late 1980s. Currently, their system has data on over 2 million patients, supporting studies on the effects of genetic variation on glaucoma, Alzheimer's disease, hypertension, and psychiatric disorders. In 2009, Massachusetts General Hospital committed to recording the genetic profile of tumors from every cancer patient that they treat, in an effort to make personalized medicine the standard of care (Smith 2009).

Moving Toward a More Personal Genome

Even as the clinical infrastructure for personalized medicine was being put into place, certain trends indicated that aspects of personalized medicine would become important outside the clinic, as well. The 2000s marked the emergence of a new breed of direct-to-consumer genetic testing, bypassing the physician and offering customers an extensive scan or full sequence of their genome for genetic variations, an interpretation of their disease predispositions, and a recommendation for lifestyle adaptations. Start-up companies such as Knome, Navigenics, deCODEme, and 23andMe wrapped their entire business model around this concept of "personal genomics."

The trend raised some concerns regarding the consistency and reliability of risk predictions, the clinical relevance of the testing results, the lack of regulation for consumer-focused testing services, and the ethics of providing medical information directly to consumers without the benefit of professional guidance (Ameer et al. 2009).

Whether on the Internet or in a doctor's office, access to one's personal genome sequence is a likely consequence of prodigious advances in sequencing technology. A human genome that took $3 billion and thirteen years to construct in 1990, cost $50 thousand with a time investment of only two months in 2009 (Wade 2009). The prospect of a full sequence costing one thousand dollars or less appeared to be within sight, prompting both public and private investment into reaching that goal (Service 2006). The Personal Genome Project, launched in 2006, has already sequenced and published the full genomes for a group of ten individuals, and is in the process of expanding that cohort to over one hundred thousand people (Goldberg 2008). With a personal genome, virtually anyone would obtain the opportunity to have their entire genome entered into their medical record. Physicians and patients would be able to use genomic information to craft a more holistic and proactive approach to health care. A host of ethical and health care issues related to our access to this information remain to be worked out in the years to come.

Conclusion

The foundations of anatomy, toxicology, histology, and cellular pathology were established in the centuries after Hippocrates, but it was not until the middle of the twentieth century, when we began to get a deeper *molecular* understanding of disease, and then at the turn of the twenty-first century, with the sequencing of the human genome, that we could develop the tools to truly personalize diagnosis and treatment. This set in motion the transformation of personalized medicine from an *intention* to a *practice*.

Neuroscience by necessity has taken a different and longer path to personalized medicine than other fields such as oncology or cardiovascular disease. There are fewer tissue banks from which to examine the molecular and genetic signatures of disease and how they vary from one individual to the next. Variability in response to treatment can be masked by large placebo effects and ambiguity in the evaluation of outcomes. Few tests have emerged in clinical practice to separate patients into categories for improved safety and response to treatments. As we draw lessons from the historical trajectory in other fields, however, it is evident that personalized medicine is manifesting itself in neurological and psychiatric medicine as we develop a more rational understanding of the molecular and environmental causes of these conditions. The use of molecular and neuroimaging biomarkers, including markers of cognitive function, will greatly improve our ability to effectively diagnose and treat the individual patient correctly the first time treatment is prescribed.

References

Adams, F. 1849. *The genuine works of Hippocrates*. New York: W. Wood and Company.

Alexanderson, B., and swF. Sjöqvist. 1971. Individual differences in the pharmacokinetics of monomethylated tricyclic antidepressants: Role of genetic and environmental factors and clinical importance. *Ann N Y Acad Sci.* 179: 739–51.

Ameer B., and N. Krivoy. 2009. Direct-to-consumer/patient advertising of genetic testing: A position statement of the American College of Clinical Pharmacology. *J. Clin. Pharmacol.* 49: 886.

American Hospital Association (AHA). Continued progress: Hospital use of information technology 2007. American Hospital Association. http://www.aha.org/aha/content/2007/pdf/070227-continuedprogress.pdf (accessed January 25, 2010).

Binder, E. B., D. Salyakina, P. Lichtner, G. M. Wochnik, M. Ising, B. Pütz, S. Papiol, S. Seaman, S. Lucae, M. A. Kohli, T. Nickel, H. E. Künzel, B. Fuchs, M. Majer, A. Pfennig, N. Kern, J. Brunner, S. Modell, T. Baghai, T. Deiml, P. Zill, B. Bondy, R. Rupprecht, T. Messer, O. Köhnlein, H. Dabitz, T. Brückl, N. Müller, H. Pfister, R. Lieb, J. C. Mueller, E. Lõhmussaar, T. M. Strom, T. Bettecken, T. Meitinger, M. Uhr, T. Rein, F. Holsboer, and B. Muller-Myhsok. 2004. Polymorphisms in FKBP5 are associated with increased recurrence of depressive episodes and rapid response to antidepressant treatment. *Nat Genet.* 36: 1319–25.

Carson, P. E., C. L. Flanagan, C. E. Ickes, and A. S. Alvong. 1956. Enzymatic deficiency in primaquine sensitive erythrocytes. *Science.* 124: 484–85.

Clark, G. and R. H. Kasten 1983. *History of staining (3rd ed.)* Baltimore: Williams & Wilkins.

Cobleigh, M. A., C. L. Vogel, D. Tripathy, N. J. Robert, S. Scholl, L. Fehrenbacher, J. M. Wolter, V. Paton, S. Shak,G. Lieberman, and D. J. Slamon. 1999. Multinational study of the efficacy and safety of humanized anti-HER2 monoclonal antibody in women who have HER2-overexpressing metastatic breast cancer that has progressed after chemotherapy for metastatic disease. *J Clin Oncol.* 17(9): 2639–48.

Cook-Deegan, R. M. 1994. *The gene wars: Science, politics and the human genome*. New York: Norton.

de Leon, J, S. C. Armstrong, K. L. and K. L. Cozza. 2006. Clinical guidelines for psychiatrists for the use of pharmacogenetic testing for CYP450 2D6 and CYP450 2C19. *Psychosomatics.* 47: 75–85.

Dorkin, H. R. 1982. Suxamethonium—The development of a modern drug from 1906 to the present day. *Med Hist.* 26(2): 145–68.

Eichelbaum, M., N. Spannbrucker, B. Steincke, and H. J. Dengler. 1979. Defective N-oxidation of sparteine in man: A new pharmacogenetic defect. *Eur J Clin Pharmacol.* 16: 183–87.

Eichelbaum, M., N. Spannbrucker, and H. J. Dengler. 1975. N-oxidation of sparteine in man and its interindividual differences. *Arch. Pharmacol.* 287: R94.

Evans, D. A., K. A. Manley, and V. A. McKusick. 1960. Genetic control of isoniazid metabolism in man. *Br Med J.* 2: 4484–91.

Fox, A. L. 1932. The relationship between chemical constitution and taste. *Proc Natl Acad Sci USA.* 18: 115–20.

Garrod, A. E. 1902. Incidence of alkaptonuria: A study in chemical individuality. *Lancet.* 2: 653–56.

Garrod, A. E. 1909. *Inborn errors of metabolism.* London: Oxford University Press.

Garrod, A. E. 1931. *The inborn factors in disease:* An essay. Oxford: Claredon Press.

Goldberg, C. 2008. Subjects' DNA secrets to be revealed. *Boston Globe,* October 20.

Guo, S. W., and D. R. Reed. 2001. The genetics of phenylthiocarbamide perception. *Ann Hum Biol.* 28: 111–42.

Hsiao, C. J., C. W. Burt, E. Rechtsteiner, E. Hing, D. A. Woodwell, and J. E. Sisk. Preliminary estimates of electronic medical records use by office-based physicians: United States, 2008. National Center for Health Statistics. http://www.cdc.gov/nchs/data/hestat/physicians08/physicians08.pdf. (accessed January 25, 2010).

International Human Genome Sequencing Consortium (IHGSC). 2001. Initial sequencing and analysis of the human genome. *Nature.* 409(6822): 860–921.

Jha, A., C. DesRoches, E. Campbell, K. Donelan, S. Rao, T. Ferris, A. Shields, S. Rosenbaum, and D. Blumenthal. 2009. Use of electronic health records in U.S. hospitals. *N Engl J Med.* 360(16): 1628–38.

Kalow, W. 1962. *Pharmacogenetics. Heredity and the response to drugs.* Philadelphia: W. B. Saunders Co.

Khorana, H. G. 1968. Synthetic nucleic acids and the genetic code. *JAMA.* 206(9): 1978–82.

Kirchheiner, J., K. Nickchen, M. Bauer , M, M.-L. Wong, J. Licinio, I. Roots, and J. Brockmöller. 2004. Pharmacogenetics of antidepressants and antipsychotics: the contribution of allelic variations to the phenotype of drug response. *Mol Psychiatry.* 9: 442–473.

Maxam, A. M., and W. Gilbert. 1977. A new method for sequencing DNA. *Proc Natl Acad Sci USA.* 74(2): 560–64.

Mendel, J. G. 1866. Versuche über Pflanzenhybriden, Verhandlungen des naturforschenden Vereines in Brünn. Bd. IV(1865):3-47. For English translation, see: Druery, C.T and W. Bateson (1901). Experiments in plant hybridization. *Journal of the Royal Horticultural Society.* 26: 1–32.

McMahon, F. J.., S. Buervenich, D. Charney, R. Lipsky, A. J. Rush, A. F. Wilson, A. J. Sorant, G. J. Papanicolaou, G. Laje, M. Fava, M. H. Trivedi, S. R. Wisniewski, and H. Manji. 2006. Variation in the gene encoding the serotonin 2A receptor is associated with outcome of antidepressant treatment. *Am J Hum Genet.* 78: 804–14.

Motulsky, A. G. 1957. Drug reactions, enzymes and biochemical genetics. *JAMA.* 165: 835–37.

Mullis, K. 1990. The unusual origin of the polymerase chain reaction. *Scientific American.* 262(4): 56–61, 64–65.

Paddock, S.., G. Laje, D. Charney, A. J.. Rush, A. F. Wilson, A. J. Sorant, R. Lipsky, S. R. Wisniewski, H. Manji, and F. J. McMahon. 2007. Association of GRIK4 with outcome of antidepressant treatment in the STAR*D cohort. *Am J Psychiatry.* 164(8): 1181–8.

Paik, S., S. Shak, G. Tang, C. Kim, J. Baker, M. Cronin, F. L. Baehner, M. G. Walker, D. Watson, T. Park, W. Hiller, E. R. Fisher, D. L. Wickerham, J. Bryant, and N. Wolmark. 2004. A multigene assay to predict recurrence of tamoxifen-treated, node-negative breast cancer. *N Engl J Med.* 51(27): 2817–26.

Personalized Medicine Coalition (PMC). The case for personalized medicine. May 2009. http://www.personalizedmedicinecoalition.org/communications/TheCaseforPersonalizedMedicine_5_5_09.pdf. (accessed January 25, 2010).

Pirmohamed, M. 2001. Pharmacogenetics and pharmacogenomics. *Br J Clin Pharmacol.* 52: 345–47.

Pollack, A. 2007. Test to predict breast cancer relapse risk is approved. *New York Times*, February 7.

Pollack, A. 2009. Gene test for dosage of warfarin is rebuffed. *New York Times,* May 4.

President's Council of Advisors on Science and Technology (PCAST). Priorities for personalized medicine. September 2008. http://www.ostp.gov/galleries/PCAST/pcast_report_v2.pdf. (accessed January 25, 2010).

Sanger, F. 1981. Determination of nucleotide sequences in DNA. *Science.* 214(4526): 1205–10.

Schafer, S. 2001. EEOC sues railroad on genetic tests. *The Washington Post,* February 10.

Secretary's Advisory Committee on Genetics, Health, and Society (SACGHS). Reports online at: http://oba.od.nih.gov/SACGHS/sacghs_documents.html. (accessed January 25, 2010).

Service, R. F. 2006. Gene sequencing. The race for the $1000 genome. *Science* 311(5767): 1544–46.

Smith, S. 2009. MGH to use genetics to personalize cancer care. *Boston Globe*, March 3.

Tepper, B. J. 2008. Nutritional implications of genetic taste variation: The role of PROP sensitivity and other taste phenotypes. *Annu Rev Nutr.* 28: 367–88.

Uhr, M., A. Tontsch, C. Namendorf, S. Ripke, S. Lucae, M. Ising, T. Dose, M. Ebinger, M. Rosenhagen, M. Kohli, S. Kloiber, D. Salyakina, T. Bettecken, M. Specht, B. Pütz, E. B. Binder, B. Müller-Myhsok, and Holsboer. 2008. Polymorphisms in the drug transporter gene ABCB1 predict antidepressant treatment response in depression. *Neuron.* 57: 203–9.

United States Congress. H.R. 6498: Genomics and personalized medicine act of 2008. http://www.govtrack.us/congress/bill.xpd?bill=h110-6498. (accessed January 25, 2010).

United States Department of Health and Human Services (HHS) . 2008 . Personalized health care: pioneers, partnerships, progress. http://www.hhs.gov/myhealthcare/news/phc_2008_report.pdf. (accessed January 25, 2010).

United States Department of Health and Human Services (HHS). 2008. Personalized health care: pioneers, partnerships, progress.

United States Food and Drug Administration (FDA). 2005a. Guidance for industry on pharmacogenomic data submissions. http://www.ocbn.ca/pdfs/pharmacogenomic_data_submissions_guidance.pdf. (accessed January 25, 2010).

United States Food and Drug Administration (FDA). April 2005b. Drug-diagnostic co-development concept paper. http://www.fda.gov/downloads/Drugs/ScienceResearch/ResearchAreas/Pharmacogenetics/UCM116689.pdf. (accessed January 25, 2010).

United States Food and Drug Administration (FDA). 2007a Guidance for industry and FDA staff: Pharmacogenetic tests and genetic tests for heritable markers. http://www.fda.gov/downloads/MedicalDevices/DeviceRegulationandGuidance/GuidanceDocuments/ucm071075.pdf. (accessed January 25, 2010).

United States Food and Drug Administration (FDA). 2007b. FDA approves updated warfarin (coumadin®) prescribing information. http://www.fda.gov/NewsEvents/Newsroom/PressAnnouncements/2007/ucm108967.htm. (accessed January 25, 2010).

United States Senate. 2006. S.3822: Genomics and personalized medicine act of 2006. http://www.govtrack.us/congress/billtext.xpd?bill=s109-3822. (accessed January 25, 2010).

Vastag, B. 2006. New clinical trials policy at FDA. *Nature Biotechnology.* 24(9): 1043.

Vogel, F. 1959. Moderne problem der humangenetik. *Ergeb. Inn. Med. U. Kinderheilk.* 12: 52–125.

Wade, N. 2009. Cost of decoding a genome is lowered. *New York Times,* August 10.

2

An Applied Context for Personalized Medicine in Psychiatry

Alan F. Schatzberg, MD

Introduction

The movement toward personalized medicine has become increasingly popular across specialties, along with other concepts such as translational research. Although this may seem to be a new trend, we have actually welcomed such efforts in medicine and psychiatry for over three decades. But as is often the case, the past is forgotten in the face of new discoveries, or seemingly new findings.

Efforts at bringing biology into the diagnostic process, and improving our ability to predict the likelihood of the patient's response to treatment, have long been promoted by prominent researchers in our field (particularly in the area of major depression), albeit with limited results. Most recently, interest has been aroused by advances in the study of human genetics; we now understand much about DNA makeup and are beginning to understand the clinical significance of allelic variation. But here, too, we are finding that the leap from alleles to clinical application is more difficult than we had imagined.

In this chapter, I review a number of early hypotheses about the role of neurotransmitters in depression. These hypotheses had some potential, but never had the impact one might have hoped for in the effort to shed light on the problems faced in personalized medicine in psychiatry.

Psychiatry is hampered by its inherent inability to physically access the brain. We do not routinely perform biopsies to study illnesses such as depression, and, until recently, imaging the central nervous system was at best crude and at times painful for the patient. Early work depended on the peripheral assessment of monoamines (which are easily affected by external factors, such as diet), as well as the influence of the peripheral nervous system. Recent advances in understanding genetics and depression, as well as results from brain imaging studies, provide opportunities for developing more targeted tests that we hope are less affected by extraneous events. Still, the earlier experiences do provide us with some context, and lessons that can be used to more effectively develop personalized medicine for psychiatry.

Early Research and Lessons Learned

Catecholamine Physiology

Two seminal papers published in the mid-1960s had a major impact on research into the causes of depression (Bunney and Davis 1965; Schildkraut 1965). These papers argued elegantly that disordered catecholamine physiology played a key role in the pathogenesis of depression. These hypotheses were based on both preclinical and clinical observations on the use of antidepressants, as well as agents that were often associated with precipitating mood changes. This foundation formed a so-called "psychopharmacologic bridge." Shortly thereafter, a European investigator made a similar argument, citing that an imbalance in serotonin levels was the basis for depression (Coppen 1969).

This initial research led to promising and more precise tools for improving diagnosis and treatment. In the late 1970s, Joe Schildkraut and I reported on patterns of urinary 3-methoxy-4-hydroxyphenylglycol (MHPG) in unipolar and bipolar depressive patients. MHPG is a metabolite found in blood and urine that is thought to represent a higher percentage of centrally derived stores than other metabolites or norepinephrine itself. Bipolar I patients appeared to excrete low levels of urinary MHPG in comparison to healthy controls, whereas bipolar II patients often showed high MHPG levels, similar to many of the unipolar patients (Schatzberg et al. 1978; Schildkraut et al. 1978). The precision in diagnosis could be enhanced by applying discriminant function analysis to the wider range of catecholamines and metabolites, along with several ratios, to generate a score on the so-called D-type equation. This provided reasonable sensitivity and specificity for separating Bipolar I depression from the other subtypes (Schatzberg et al. 1978).

Although promising, this methodology was not widely adopted. Variation in laboratory results of amines and metabolites, and inadequate sample size for data replication, were but two of the obstacles faced. Most importantly, it was difficult for the patient to comply with the test preparation. It required a drug-free status for one or more a 24-hour urine collection period, and a modified low catecholamine diet. The test also required a laboratory that could perform the measurements. Therefore, the test received little clinical acceptance, although it was available in the Boston area, where we were working. The test was used mainly to help with diagnosis of depressive subtype, and not to determine if a patient was depressed. But as it was not proposed as a screening instrument, this limitation was not a major factor in its acceptance (or lack thereof).

From these early ventures, we learned that test preparation should be manageable, the test should be easy to perform, and it should be conducted in independent, multiple settings (perhaps internationally) to provide sufficient pooled data for a database.

Urinary catecholamines were also used as predictors of antidepressant response. Several groups reported that patients with low urinary MHPG levels demonstrated more rapid and robust responses to noradrenergic tricyclic antidepressants, such as desipramine, nortriptyline, and maprotiline hydrochloride. At least one paper noted poorer responses to amitriptyline, a relatively weak norepinephrine reuptake blocker that was thought to be more serotonergic in activity (Maas 1978; Rosenbaum et al. 1980; Schatzberg et al. 1980). Although the data were reasonably strong for predicting responses to more noradrenergic agents in low-MHPG patients, the test was not routinely adopted for several reasons. It required the same patient-compliant, drug-free status as the D-type equation. Also, the response rate to amitriptyline in high-MHPG subjects was not as well replicated as it was in low-MHPG subjects. This imbalance created a problem for adoption in routine practice. While we could predict who would respond to the noradrenergic agent, we did not know how to best treat high-MHPG patients. Amitriptyline might or might not be effective, and if it was not, what would be the next step—the addition of lithium? Since we couldn't reliably predict the outcome, and since amitriptyline has considerable side effects, it was prudent to administer a noradrenergic drug to all patients and not conduct the test.

Perhaps we would have been more successful had we been able to conduct a trial similar to the Sequenced Treatment Alternatives to Relieve Depression (STAR*D) study or the I-SPOT (International Study for the Prediction of treatment in Depression) trial, in which biological and treatment data were collected, and the next step could be tested based on pretreatment test results (see Chapter 3). To date, the STAR*D study, conducted by the National Institute of Mental Health (NIMH), of the National Institutes of Health (NIH), is the largest, most comprehensive clinical trial on antidepressant treatments for patients in a community setting. It provided solid evidence regarding specific treatment steps for treatment-resistant depression.

Serotonin Metabolite Levels

In the 1970s and 1980s, research in Europe focused on the role of serotonin in depression. However, since only small amounts of metabolites are found in blood and urine, there were major limitations to assessing serotonin peripherally. Instead, a number of techniques were used to enhance the signal, particularly the application of cerebrospinal fluid (CSF) measurements after administering probenecid to increase relative central nervous system contributions. The test yielded some interesting and potentially useful data on risk for suicide, particularly by violent means (Traskman et al. 1981). However, spinal taps were not particularly attractive to psychiatric practitioners or patients, and these tests were not routinely used. Another lesson learned: ease of use for the application needs to be considered from the outset and throughout the process.

Hypothalamic Pituitary Adrenal Axis Activity

Bernard Carroll and Peter Stokes developed the Dexamethasone Suppression Test (DST) as a potential method for assessing the frequently abnormal hypothalamic-pituitary-adrenal (HPA) axis activity seen in major depression. Carroll published the test results in a seminal paper on the specificity for diagnosing melancholia in 1981(Carroll et al. 1981). This generated a spate of studies that, unfortunately, failed to show either the specificity or sensitivity for melancholia. Instead, these subsequent studies were useful in clarifying that the claims were erroneous, and routine use of the test was discouraged. Data from my group, as well as from others (Schatzberg et al 1983; Nelson and Davis 1997), pointed out a potential use in patients with psychotic depression, but the test was never adopted because of the earlier negative publicity. In the DST test, my colleagues and I had a test that was easy to perform and analyze, and the upper and lower limits for abnormal values were relatively clear. We even had a disorder—psychotic major depression—the treatment of which could be greatly assisted by administering the test, since these patients frequently require the addition of antipsychotics to achieve a response. However, the "hype" surrounding the melancholia application had been too great, and use of the test for general screening or confirmation was dismissed.

What are the lessons to be learned here? We need to maintain a disciplined, iterative approach to data so that we do not "throw the baby out with the bath water." That was not the case here and, even today, the test is rarely used to help with a diagnosis that is difficult to make. Objectivity needs to be maintained by all participants, and expectations need to be moderated to avoid disappointment and overly hasty dismissal of new technology.

Newer Approaches

Genetics of Monoamine Systems

One major advance in pharmacogenetics has been the exploration of the allelic variation associated with monoamine transporters and receptors, with particular emphasis on serotonergic agents and genes. Considerable data are available to support that the serotonin transporter promoter gene (5HTTPr) can predict responses to serotonergic agents, such as fluvoxamine or paroxetine (Smeraldi et al. 1998; Zanardi et al. 2000). This was first noted by the Milan group; patients with the s/s form of the promoter tend to respond poorly to selective serotonin reuptake inhibitors (SSRIs), compared to l/l carriers (which are primarily seen in Caucasian populations). In East Asians, the s/s form is the common variation and predicts good responses (Kim et al. 2006). In a study by our group, we observed that elderly, depressed Caucasian patients with the s/s genotype have slightly poorer outcomes using paroxetine. Dramatic differences in tolerability were observed among s/s patients who demonstrated very high dropout rates due to side effects (Murphy et al. 2004). Other researchers noted similar observations with fluoxetine (Perlis et al. 2003). Thus, it may be easy to confuse poor response with poor tolerability, particularly when an intent-to-treat analysis is employed, where early dropout rates due to side effects can be confused with nonresponsiveness. Of interest, some of the side effects appear to be related to the peripheral effects of the drugs (e.g., gastrointestinal effects), such that, here, peripheral serotonin activity is important clinically, in contrast to early serotonin studies using CSF to predict suicide, or urinary serotonin metabolites levels to assess depressive subtypes, where peripheral serotonin systems often obfuscated relative central activity and clinical relevance.

The relatively weaker predictability of response by the s/s variant is of concern for clinical application. Can the test provide better differentiation? One key question thas not yet been addressed: What is the effect of nonspecific responses on test performance? Could l/l patients, in fact, be better responders to a variety of treatments (including placebo), and are the current data somewhat misleading regarding response specificity? While these tests are all designed to predict responses to antidepressants in clinical populations, it is possible that we would be better served if we knew of the test performance for predicting placebo or nonspecific responses to assess the true added value. It would be beneficial if pharmaceutical companies conducting placebo-controlled trials would pool their data into separate databases or a single database on placebo predictions and provide that information to researchers, who could help resolve this key issue. Having a test that could predict the risk for side effects would be useful in a more limited way, as opposed to an overall test for response or non-response. Currently, the 5HTTPr test for predicting the risk of side effects is available, and I will comment shortly on whether it should be used routinely. (Full disclosure: I have a conflict of interest as a named inventor on intellectual property, namely, Stanford University pharmacogenetic and glucocorticoid antagonist use patents. I am also a cofounder of Corcept Therapeutics).

Earlier, I discussed the limitations of applying a urinary MHPG test to predict responses in patients with high MHPG levels. This is similar to the situation with the transporter promoter, as we have little data to guide us in determining what strategy to use for s/s Caucasians or l/l Asians. It seems reasonable to use more conservative dosing strategies to determine if patients can tolerate the adverse events. Beyond that, do we have a priori selections that would be more efficient? Recently, we reported that, in s/s carrier patients who had failed to respond to a trial of sertraline, the addition of atomoxetine (a noradrenergic agent approved for Attention Deficit Disorder [ADD]) was significantly more effective than placebo (Reimherr et al. 2009). Thus, we have some data to develop a strategy.

However, this is not optimal, because it still involves a trial of sertraline that is likely to be associated with nonresponse.

The lesson learned from the early days of catecholamine research is still applicable today: we need to know what to do to help patients who, predictably, may fail to respond to a particular agent or class of medications.

Another serotonergic approach has been to explore alleles for serotonin receptors. One in particular, the *5HT2a* receptor, has generated some interesting data. Our group has again reported that elderly patients with depression, C/C homozygotes for the 102-T/C allele, failed to tolerate paroxetine and demonstrated high dropout rates due to adverse events (Murphy et al. 2003). (Again, we have potential intellectual property on this test.) In the STAR*D trial, another allelic variant of the same gene was associated with better treatment outcome when citalopram was used (McMahon et al. 2006). However, these are not the same alleles. Such data are likely, given the large numbers of alleles for a particular receptor.

What is the lesson learned here? Different sample sizes and sample characteristics must be taken into consideration, and how to integrate the variables is a challenge. One approach is to accept that the findings point to the importance of a particular gene, and that variations at different loci can have similar effects. One could weigh the relative contribution of each gene separately and test patients for all relevant contenders. Thus, a clinician might be given data on a host of genes and their relative contributions for determining treatment outcome.

Over the past two decades, the treatment of depression has shifted away from the heavily noradrenergic antidepressants and toward the SSRIs (and to some extent, the serotonin-norepinephrine reuptake inhibitors [SNRIs]). Efforts to develop purely noradrenergic agents for major depression have not been very fruitful. There have been a number of notable failures involving genetic tests for these agents, as well as for the noradrenergic system. However, it has been reported that a polymorphism for the norepinephrine transporter predicted nortriptyline response in an East Asian geriatric cohort (Yoshida et al 2004; Kim et al. 2006).

The DEX/CRH Test

Researchers at the Max Planck Institute for Psychiatry have made a number of important observations regarding combining dexamethasone suppression with a corticotropin releasing hormone (CRH) stimulation test (DEX/CRH test) to assess the HPA axis. This test has largely replaced the earlier DST test (Ising et al. 2005). This combination is thought to isolate the central component from that of the pituitary. Adrenocorticotrophic hormone (ACTH) release is measured in response to CRH the next day, following the administration of dexamethasone. The test has yielded some very interesting data on diagnosis, as well as some potentially useful data on the adequacy of treatment. However, a recent paper was somewhat cautious about the test and reported lower sensitivity than previously stated (Schule et al. 2009). These types of findings are not uncommon. Commonly, we all have noted smaller effects over time, in part because early positive trials may be more "positive" than would be expected and, hence, they are more likely to be published. Early failures may not find their way into the literature. Also, the tests or agents tend to be tested more at the margins, with less-typical patients or less severely ill patients. Keeping track of the characteristics of the population being studied—age, gender, ethnicity, severity, comorbidity, etc.—is essential for assessing data across studies.

Brain Imaging

The recent advances in both positron emission tomography (PET) and functional magnetic resonance imaging (fMRI) have resulted in a number of interesting biological observations, which have helped us in understanding function in specific subtypes of depression. These techniques may prove useful for predicting differential treatment response; however, the field is just emerging. Obviously, these studies are more difficult and more expensive to conduct. In addition, it is not clear what degree of sensitivity and specificity is required for individual patients. Still, these approaches offer great potential for not only understanding pathophysiology and function, but for assessing treatment effects and predicting outcomes, including in combination with genomics.

When and How to Apply a Test

We need to have agreed-upon methodologies for deciding when to adopt a test. This could be achieved by convening panels of independent experts, who could meet under the auspices of our professional societies, through consortia, or through the relevant NIH institutes. For now, when should tests be adopted in the absence of such panels? How many trials should be required? How large should the sample size be? What degree of sensitivity and specificity do we want these tests to achieve? Drug and medical device guidance, compliance, and regulatory information are available through the U. S. Food and Drug Administration (FDA). Two positive trials are commonly required for a medication, although technically, only one is needed, and only one positive trial is needed for a device that is equivalent to an existing approved technology. We need to have consensus guidelines, and the published outcome of the October 2009 meeting of the Mayflower Action Group's Initiative, "A Call for Action," provides one stellar exemple in that direction (Koslow et al. 2009; see appendix).

Conclusions

As I have discussed, translational research and personalized medicine are not new to psychiatry but, instead, follow a long-standing tradition. Lessons learned from earlier studies in terms of the need for easily accessible replication samples, the significance of peripheral versus central markers, the prediction of positive versus negative responses, treatment options for patients who are predicted by testing not to respond (i.e., the need for follow up, and positive predictions), decision matrices for adopting tests, and pooling data for independent analysis and insights, are as relevant today as they were twenty-five or thirty years ago.

Our biological tools appear to be potentially more specific and relevant for studying the brain and practicing psychiatry, but only time, hard work, and a more integrative strategy will tell us if this is truly the case.

Acknowledgment

Supported in part by grants from the NIMH (MH 50064 and MH 73914) and the Pritzker Foundation.

References

Bunney, W. E., Jr., and J. M. Davis. 1965. Norepinephrine in depressive reactions. A review. *Arch Gen Psychiatry*. 13(6):483–94.

Carroll, B. J., M. Feinberg, J. F. Greden, J. Tarika, A. A. Albala, R. F. Haskett, N. M. James, Z. Kronfol, N. Lohr, M. Steiner, J. P. de Vigne, and E. Young. 1981. A specific laboratory test for the diagnosis of melancholia. Standardization, validation, and clinical utility. *Arch Gen Psychiatry.* 38(1):15–22.

Coppen, A. 1969. Defects in monoamine metabolism and their possible importance in the pathogenesis of depressive syndromes. *Psychiatr Neurol Neurochir.* 72(2):173–80.

Ising, M., H. E. Kunzel, E. B. Binder, T. Nickel, S. Modell, and F. Holsboer. 2005. The combined dexamethasone/CRH test as a potential surrogate marker in depression. *Prog Neuropsychopharmacol Biol Psychiatry.* 29(6):1085–93.

Kim, H., S. W. Lim, S. Kim, J. W. Kim, Y. H. Chang, B. J. Carroll, and D. K. Kim. 2006. Monoamine transporter gene polymorphisms and antidepressant response in Koreans with late-life depression. *JAMA.* 296(13):1609–18.

Koslow, S. H., L. M. Williams, and E. Gordon. 2010. Personalized medicine for the brain: A call for action. *Molecular Psychiatry.* XX:1–2.

Maas, J. 1978. Clinical and biochemical heterogeneity of depressive disorders. *Ann Intern Med.* 88(4):556–63.

McMahon, F. J., S. Buervenich, D. Charney, R. Lipsky, A. J. Rush, A. F. Wilson, A. J. Sorant, G. J. Papanicolaou, G. Laje, M. Fava, M. H. Trivedi, S. R. Wisniewski, and H. Manji. 2006. Variation in the gene encoding the serotonin 2A receptor is associated with outcome of antidepressant treatment. *Am J Hum Genet.* 78(5):804–14.

Murphy, G. M. Jr., S. B. Hollander, H. E. Rodrigues, C. Kremer, and A. F. Schatzberg. 2004. Effects of the serotonin transporter gene promoter polymorphism on mirtazapine and paroxetine efficacy and adverse events in geriatric major depression. *Arch Gen Psychiatry.* 61(11):1163–69.

Murphy, G. M. Jr., C. Kremer, H. E. Rodrigues, and A. F. Schatzberg. 2003. Pharmacogenetics of antidepressant medication intolerance. *Am J Psychiatry.* 160(10):1830–35.

Nelson, J. C., and J. M. Davis. 1997. DST studies in psychotic depression: A meta-analysis. *Am J Psychiatry.* 154(11):1497–1503.

Perlis, R. H., D. Mischoulon, J. W. Smoller, Y. J. Wan, S. Lamon-Fava, K. M. Lin, J. F. Rosenbaum, and M. Fava. 2003. Serotonin transporter polymorphisms and adverse effects with fluoxetine treatment. *Biol Psychiatry.* 54(9):879–83.

Reimherr, F., J. Amsterdam, D. Dunner, L. Adler, S. Zhang, D. Williams, B. Marchant, D. Michelson, A. Nierenberg, A. Schatzberg, and P. Feldman. 2009. Genetic polymorphisms in the treatment of depression: Speculations from an augmentation study using atomoxetine. *Psychiatry Res.* Epub

Rosenbaum, A. H., A. F. Schatzberg, T. Maruta, P. J. Orsulak, J. O. Cole, E. L. Grab, and J. J. Schildkraut. 1980. MHPG as a predictor of antidepressant response to imipramine and maprotiline. *Am J Psychiatry.* 137(9):1090–92.

Schatzberg, A. F.,P. J. Orsulak, A. H. Rosenbaum, T. Maruta, E. R. Kruger, J. O. Cole, and J. J. Schildkraut. 1980. Toward a biochemical classification of depressive disorders IV: Pretreatment urinary MHPG levels as predictors of antidepressant response to imipramine. *Commun Psychopharmacol.* 4(5):441–45.

Schatzberg, A. F., A. J. Rothschild, J. B. Stahl, T. C. Bond, A. H. Rosenbaum, S. B. Lofgren, R. A. MacLaughlin, M. A. Sullivan, and J. O. Cole. 1983. The dexamethasone suppression test: Identification of subtypes of depression. *Am J Psychiatry.* 140(1):88–91.

Schatzberg, A. F., J. A. Samson, K. L. Bloomingdale, P. J. Orsulak, B. Gerson, P. P. Kizuka, J. O. Cole, and J. J. Schildkraut. 1978. Toward a biochemical classification of depressive disorders. X. Urinary catecholamines, their metabolites, and D-type scores in subgroups of depressive disorders. *Arch Gen Psychiatry* 46(3):260–68.

Schildkraut, J. J. 1965. The catecholamine hypothesis of affective disorders: A review of supporting evidence. *Am J Psychiatry.* 122(5):509–22.

Schildkraut, J. J., P. J. Orsulak, A. F. Schatzberg, J. E. Gudeman, J. O. Cole, W. A. Rohde, and R. A. LaBrie. 1978. Toward a biochemical classification of depressive disorders. I. Differences in urinary excretion of MHPG and other catecholamine metabolites in clinically defined subtypes of depressions. *Arch Gen Psychiatry* 35(12):1427–33.

Schule, C., T. C. Baghai, D. Eser, S. Hafner, C. Born, S. Herrmann, and R. Rupprecht. 2009. The combined dexamethasone/CRH Test (DEX/CRH test) and prediction of acute treatment response in major depression. *PLoS One.* 4(1):e4324.

Smeraldi, E., R. Zanardi, F. Benedetti, D. Di Bella, J. Perez, and M. Catalano. 1998. Polymorphism within the promoter of the serotonin transporter gene and antidepressant efficacy of fluvoxamine. *Mol Psychiatry.* 3(6):508–11.

Traskman, L., M. Asberg, L. Bertilsson, and L. Sjostrand. 1981. Monoamine metabolites in CSF and suicidal behavior. *Arch Gen Psychiatry.* 38(6):631–36.

Yoshida, Y. Takashasi, H. Higuchi, H. Kamata, M. Ito, K. Sato, K. Naito, S. Shimzu, T. Itoh, K. Suzuki, T. and C. B. Nemeroff. 2004. Prediction of antidepressant response to milnacipran by norepinephrine transporter gene polymorphisms. *Am J Psychiatry.* 161(9):1575–80.

Zanardi, R., F. Benedetti, D. Di Bella, M. Catalano, and E. Smeraldi. 2000. Efficacy of paroxetine in depression is influenced by a functional polymorphism within the promoter of the serotonin transporter gene. *J Clin Psychopharmacol.* 20(1):105–07.

3

Personalized Medicine and Integrative Neuroscience

Toward Consensus Markers for Disorders of Brain Health

Leanne M. Williams, PhD and Evian Gordon PhD, MBBCH

Introduction

Major disorders of brain health, such as depression, schizophrenia, and ADHD, are usually treated with pharmacological agents or behavior therapies. Yet despite improvement and refinement in the diagnostic approach in the *Diagnostic and Statistical Manual of Mental Disorders* (*DSM*), and a range of newer drugs that produce fewer side effects than older ones, the harsh reality is that clinicians cannot commonly predict, with a high degree of success, which drugs will work well in any individual patient. The inability to target a specific treatment and management plan for a particular patient remains a significant problem. As a result, an individual patient may be subjected to a disheartening sequence of "trials" to find a treatment that is effective. This trial-and-error process puts the patient at greater risk for acute side effects, reduced compliance, and increased risk of a chronic course of illness. In the emerging new paradigm of personalized medicine (PM), the goal is to develop and apply the knowledge necessary to increasingly tailor treatment to each individual, to move the treatment for brain-related illness from "hit-or-miss trial and error" toward a bull's-eye.

To date, much of the research in PM has focused on genetic markers, which are employed as predictors of individual treatment response. The complexity of the brain is likely to require a shift in focus from a single genetic marker to a more integrated approach, in which a wider scope of molecular and brain-related functional, structural, and cognitive information is brought together in a complementary manner. This approach is consistent with the "endophenotype" approach proposed by Gottesman and Gould (2003).

This chapter provides a brief summary of the proof-of-concept framework for demonstrating the markers that will have the greatest impact on predicting brain-related treatment response in personalized medicine and health care. The framework includes the principles of integrative neuroscience, standardized methods for assessment of large populations, and databasing of information to allow for systematic planned analyses of primary and secondary predictions.

The chapter outlines:

1. ***Current status:*** *The paradigm of PM*. Summarizing examples of the proof-of-concept phase achieved with genetic markers for physical disorders.
2. ***An Integrative PM for the Brain:*** The brain is a highly interconnected system. It requires an integrative approach that goes well beyond genetics alone to predict outcomes. It also requires self-reporting, cognition, brain function, and brain structure as candidate markers.
3. ***Building a Taxonomy:*** *Approaches to identifying integrative PM markers for the brain.* Developing a framework for diagnostic and personalized treatment prediction markers for brain disorders requires a taxonomy that extends beyond the current *DSM* signs and symptoms approach. Moreover, markers that distinguish an individual harboring a disorder from controls are not necessarily the same markers that accurately predict response to treatment. If, however, a marker captures the underlying pathophysiology of a disorder, it is likely it will also be useful in supporting decisions about whether to treat or not. Two sample approaches are summarized:
   i) Traditional reviews and meta-analyses of individual studies. These individual studies typically use their own customized methods and measures.
   ii) Standardized approach using methods that are standardized to build up a large database, with large numbers of subjects and the capacity to interconnect different types of data.

   Key issues that must be addressed within each approach include:

   a) The need for *representative* real-world diversity of samples, including appropriate control subjects.
   b) *Comorbidity* (which may relate to representative samples for some disorders).
   c) *Replication* in independent subject groups and across multiple sites.
   d) Demonstration of the *extent* of usefulness in clinical studies or trials.

4. ***Candidate integrative markers for diagnosis and treatment:*** Examples of diagnostic markers are provided for depression, schizophrenia, and ADHD, as are examplar markers for treatment prediction in depression. (Additional complementary details are also provided in the other chapters in this book that outline markers in depression, schizophrenia, and ADHD).
5. ***Translation in the real world:*** Two key tasks are identifying diagnostic and treatment prediction markers that can be realistically implemented (i.e., considering workflow and cost-effectiveness) in routine clinical practice and health care settings.

We will highlight cost-benefit tradeoffs that will determine which markers get used and which get bogged down (due to replication failure or scale impracticality) in discovery research, regardless of whether it is carried out by academic, biotechnological, or pharmaceutical groups.

Current Status: The Paradigm of PM

An evidence-based approach to medicine is about finding reliable and effective personalized solutions to illnesses. The current medical model is "one size fits all," seeking treatments that can be applied generally across the population. Yet, clinicians are increasingly confronted by the complexities of human biology and the enormous individual variation produced by the interaction of each person's genetic predisposition, biology, and experiences across the life span. The one-size- fits-all approach to treatment is imperfect in dealing with these variations, and does not always provide the best solution to each individual's problem.

The last decade has seen the emergence of a paradigm shift toward PM. This trend focuses on evidence that tailors treatment solutions to each person's biological profile. In this approach, the quality of health care depends upon matching the right treatment, to the right patient at the right time. It also requires a focus on proactive prevention and quality of life.

Advances in biomedical science and information technology are increasingly central to making PM a reality (Gordon 2007). The PM paradigm shift has not been as slow as predicted by the London-based Royal Societies (The Royal Society, 2005), or as fast as hoped for by the US-based Personalized Medicine Coalition (Ginsberg and Angrist 2006). Yet, the first proof-of-concept phase of PM has been achieved.

A prime example of the use of PM is identification of over-expression of HER2 (human epidermal growth factor receptor 2) in breast cancer cells, a marker for those women who will respond to an antibody drug, Herceptin (generic name trastuzumab) (Gordon 2007). Other examples include C-reactive protein (CRP), along with HDL (high-density lipoprotein) and LDL (low-density lipoprotein) cholesterol, commonly used as markers for statin treatment of atherosclerosis in cardiac disease; a molecular marker protein that inhibits the enzyme tyrosine kinase, which indicates patients with non-small-cell lung cancer who will benefit from Iressa (gefitinib); and the breakpoint cluster region-Abelson and tyrosine kinase receptor markers for treatment with Gleevec (imatinib mesylate) for chronic myeloid leukemia and gastrointestinal stromal tumor, respectively (Gordon 2007).

These proof-of-concept PM examples have predominantly focused on identifying genetic markers for physical conditions. The following section outlines the extension of PM to the brain.

An Integrative Personalized Medicine for the Brain

Increasing attention is being paid to the role of PM for diseases based in brain functioning (Koslow, Williams, and Gordon 2010). The identification of genetic markers to diagnose and treat physical conditions provides a starting point for applying the PM paradigm to brain conditions.

A number of candidate genetic markers for predicting and moderating response to brain-related treatments have also been identified and replicated in at least one subsequent study (see Table 3-1). These markers have been found to predict both response and non-response to commonly used medications, including selective serotonin reuptake inhibitors (SSRIs) in major depressive disorder, atypical antipsychotics in schizophrenia, and stimulants in attention deficit hyperactivity disorder (ADHD). Genetic markers may also contribute to identifying those with a greater disposition to side effects, such as weight gain with antipsychotics.

Yet, personalized medicine for the brain requires moving beyond the current focus on genomic markers. The interconnected complexity of the brain is underpinned by multiple gene–gene interactions. It is increasingly recognized that the genes involved in heterogeneous syndromes (such as depression, schizophrenia, and ADHD) are multiple in number, with each gene contributing a small (but meaningful) amount of the variance in these syndromes. It is prudent to consider the relationships between multiple genetic markers and a wider set of brain-related measures (Xu et al. 2001; Berry, Jobanputra, and Pal 2003; Gordon 2007; Shen, Lang, and Nakamoto 2008; Koslow, Williams, and Gordon 2010).

Table 3.1 Summary of Selected Genetic Variants (and References) Implicated in Predicting and Moderating Treatment Response in Major Depressive Disorder, Schizophrenia, and ADHD

Condition	*Genetic markers implicated in predicting treatment response*
Depression	**Response**
	Serotonin receptor 2A (*HTR2A*); rs7997013 AA allele (6,7)
	HTR2A 102T/CC & -1438A GG allele (6,8)
	FK506 binding protein 5 (*FKBP5*) (6)
	CRH receptor 1 (*CRHR1*) homozygous GAG** (6,9)
	GRIK4 rs1954787 (10)
	Non-Response
	Serotonin transporter (*5HTTLPR*) short allele* (6,11)
	Serotonin receptor 1A (*HTR1A*); rs6295 – 1019 G allele (6,12)
	COMT val108/158met Val allele (13)
	BDNF Met66 allele* (6,14,15)
Schizophrenia***	**Response**
	COMT Met homozygosity (16)
	Regulator of G-protein signaling 4 (RGS4)* (17)
	HTR1A 1019 CC genotype (85)
	Non-response
	COMT Val homozygosity (16)
	NOTCH4 (86)
	Neuroregulin (*NRG1*) TT genotype (18)
	HTR2A A-1438G GG and A-T102C CC genotypes (87)
ADHD	**Response (stimulants)**
	Dopamine Transporter (*DAT1*) 10-repeat (20)
	Dopamine Receptor D4 (*DRD4*) 4-repeat (21)
	Non-response (stimulants)
	DAT1 9-repeat (22,23)
	DRD4 7-repeat (23)
	Norepinephrine Transporter Protein 1 (NET) (23)

* The role of these genetic variants may vary with ethnicity.

** The role of these genetic variants may vary with clinical profile (e.g., high anxiety) and ethnicity.

*** For genetic markers associated with side effects of antipsychotics see Chapter 6.

Because of the complexity and polygenetic nature of mental conditions, one-to-one mapping between single genes and clinical endpoints (such as treatment response) is unlikely (except in rare cases, such as Huntington's disease). Attention has focused, instead, on a broader definition of a biomarker (or marker) to refer to any characteristic that is objectively measured and evaluated as an indicator of normal biologic processes, pathogenic processes, or pharmacologic responses to a therapeutic intervention (Biomarkers Definitions Working Group 2001). Markers meeting strict criteria for specificity, stability over time, heritability, and familial association/co-segregation have also been termed "endophenotypes," which capture how our genetic disposition is expressed in brain structure and function (Gottesman and Gould 2003; Bearden 2006). Measures of neuroanatomical circuitry, function, and cognition provide a direct window onto how genetic disposition and its interaction with the environment are expressed in the brain.

Brain-related markers may therefore include genomics, self-reporting, cognitive and electrical brain function (EEG, ERPs), and brain imaging (MRI, fMRI, SPECT, PET) measures. Identification of candidate brain markers (together with genetics) is a vital step toward development of objective tests of response to treatment, and, ultimately, their inclusion in personalized prevention and treatment strategies (Gordon 2007; Koslow, Williams, and Gordon 2010).

Building a Taxonomy: Approaches to Identifying Integrative Personalized Medicine Markers for the Brain

Identifying diagnostic and personalized treatment markers for brain disorders will require a taxonomy that extends the current *DSM* classifications based on signs and symptoms. The value of incorporating brain- and gene-based biological information into the diagnostic taxonomy has been recognized in the plans for the upcoming fifth edition of the *DSM* (Charney et al., 2002). This integrative approach is also being explored independently of the *DSM*.

Two approaches to identifying and achieving consensus on markers are outlined below.

Traditional Reviews and Meta-Analyses of Individual Studies Using Their Own Methods

The traditional approach to medical research into brain-related conditions of mental health has been predominant across psychiatry, psychology, neuroscience, and the allied disciplines. In this approach, each study uses its own unique protocols and subject criteria. In many cases, the experimental methods will be newly developed for the study, and these are specific to the laboratory or site undertaking the study. The focus of studies on unique techniques, as well as differences in protocol design and study methods, means that the relationships between measures cannot be examined directly in the same subjects, nor can data be pooled across studies, because the many small differences confound interpretation of what is cause and what is effect.

Comparability of results across such studies currently relies on descriptive reviews and meta-analyses. These are important approaches to collating information. For instance, meta-analyses have helped to highlight the consistency of reductions in hippocampal gray matter volume reduction associated with depression (Seminowicz et al. 2004). The limitation is that the contribution of the different methods and other confounds in each study included in a review or meta-analysis remains unknown and unquantifiable.

Standardized Approach with Multi-Modal Assessments, Large Numbers of Subjects, and Databasing

An alternative approach is to systematically database multiple sources of information (Koslow 2002; Gordon, 2003). In a standardized approach to databasing, each type of data assessment method is identical across subjects (within each site), across sites, and across study populations, thus eliminating method variability. The use of standardized methods makes it possible to collate data in standardized databases, so that relationships between measures and outcomes can be directly examined. The *specificity* of markers particular to select groups of subjects may also be evaluated, given that all subject groups are assessed with the same protocols. Replicability is derived in parallel, to some extent, since multiple sites in different geographical locations use identical methodology. To successfully establish a standardized assessment, this replicability should be maintained and reinforced by independent groups or new clinical studies designed to test the validity of the marker(s).

Replication of markers in new studies using the same standardized methods would provide a strong evidence-base for progressing personalized medicine markers.

The first standardized and integrative gene-brain-cognition methodology was set up by Brain Resource Ltd (http://www.BrainResource.com). Data from these studies have been used to establish the Brain Resource International Database (Gordon 2003; Gordon et al., 2005; Gordon et al., 2008; Williams et al. 2008). These standardized methods encompass:

- Self-reported clinical and medical history, and demographic measures.[1]
- Cognitive measures of thinking and emotional processes, in touch-screen or web-based platforms.[1]
- Electrical (EEG, ERP) brain and autonomic body function.[1]
- Brain imaging, including MRI, functional MRI (fMRI) and diffusion tensor imaging (DTI).[1]
- Genetic and other molecular measures.[1]

Data are currently available for five thousand healthy controls and two thousand clinical subjects, with major depressive disorder (MDD), schizophrenia (first episode, chronic, early onset), anxiety (post-traumatic stress disorder [PTSD], panic), Alzheimer's disease, mild cognitive impairment, ADHD, anorexia nervosa, obesity, and sleep disorder groups.

Data from the Brain Resource International Database that have consent for use in research are made available to the scientific community for independent publication via the BRAINnet consortium. It is governed independently by the BRAINnet Foundation (http://www.BRAINnet.net). Advantages of the standardized approach and databasing include:

- The capacity to form large populations of data with high statistical power.
- Replication of findings in independent samples that are not confounded by the use of different experimental methods.
- Specificity of findings across disorders, as well as sensitivity within disorders.
- Post-database collation, existing hypotheses may be tested and replicated in sub-groups of the database. The data may also be used for hypothesis generation.
- Collation of data from matched healthy controls using the same standardized methods provides a frame of reference for individual differences, neurodevelopmental changes over the life span, and practice effects in test–retest designs. This frame of reference for the brain is arguably of greater importance than for any other part of the body, given that the brain's networks are so highly interconnected and individualized, which is necessary for supporting complex adaptive functions.
- Collation of data obtained using the same methods across different brain disorders to obtain additional information about specificity and valuable comparative data.

1 The trademarked names for each of the data acquisition methods used to establish the Brain Resource International Database are as follows:
Web-based battery of self-report questionnaires: WebQ™
Brain Resource Inventory of Social Cognitions: BRISC™. The BRISC is a Web-based battery implemented in conjunction with WebQ.
Computerized cognitive test battery operating on a touchscreen platform: IntegNeuro™. The version of 'IntegNeuro' that operates in conjunction with LabNeuro (listed below) has also been called 'Psychometrics'.
Computerized resting and task conditions for recording of EEG, ERPs and autonomic data: LabNeuro™.
Standardized sequences and software for MRI, functional MRI and DTI: MRI-Neuro™.
Standardized protocols for acquiring and transporting DNA samples for genotyping; Molecular-Neuro™.

Importantly, a far greater amount of information can be obtained by using new ways to look at processed data that are collated into large groups (as opposed to small samples of raw data) (Koslow 2000; 2005). This approach also promotes a focus on new techniques to observe and analyze multiple sources of data, from traditional ones (such as structural equation modeling), to new data mining techniques.

An example of a current, global, multisite randomized trial in personalized medicine, the international Study to Predict Optimized Treatment–in Depression (iSPOT-D), is being undertaken with two thousand patients with MDD. It is the first study to include all of the above methods (self-report, cognition, brain and genetic measures). A related trial, iSPOT-A, is the first to employ the same array of standardized methods to study stimulants in over six hundred patients with ADHD.

While the benefits of a standardized approach to PM for the brain are clear, this approach cannot address all needs within the field. For instance, there will always be the need for new discovery studies with small subject numbers and specifically designed methods to uncover new drug and diagnostic mechanisms. But for subsequent phases of research, the use of standardized protocols and methods is likely to be of increasing importance in identifying specific and robust PM markers, which can be replicated across sites and studies.

Candidate Integrative Markers for Diagnosis and Treatment

To date, the focus has been on identifying markers that support diagnostic decisions. Many studies have focused on elucidating the differences between patients and matched controls. What make these markers valuable are their stability, sensitivity, and specificity. An implicit assumption in some of this research is that these markers may be helpful in predicting treatment response. While in some cases they are useful for both diagnosis and treatment, markers for predicting personalized treatment response may not always be the same as those relevant to diagnostic classification and other uses. Some markers may be involved in both diagnostic classification and prediction of response to treatment. However, there are also some specific markers involved only in response to treatment. In both cases, replication is the crucial factor in determining robust markers for predicting treatment response.

An Integrative Approach to Identifying Candidate PM Markers

In this section, we provide an overview of candidate integrative PM markers for supporting diagnostic classification and treatment response decisions in depression, schizophrenia, and ADHD.

A standardized approach to identifying these markers is facilitated by using an integrative theoretical framework. The overview of candidate markers is presented within our integrative theoretical framework, the "INTEGRATE Model" (Figure 3-1) (Williams et al. 2008)). The INTEGRATE Model considers how four core processes may capture the inordinate complexity of the brain at the top level of organization: "emotion," "thinking," "feeling," and "self-regulation." A fundamental organizing principle that underpins these four processes is the core motivation to "minimize danger and maximize reward." Susceptibility to impaired emotion, thinking, feeling, and/or self-regulation is contributed to by one's genetic disposition, coupled with ongoing interactions with one's environment, such as family or life stressors. Gene–environment interactions also determine individual differences in the healthy spectrum of emotion, thinking, feeling, and "self-regulation." Over the course of the life span, the ongoing outcomes of these processes will impact the brain's plasticity and shape one's adaptation abilities.

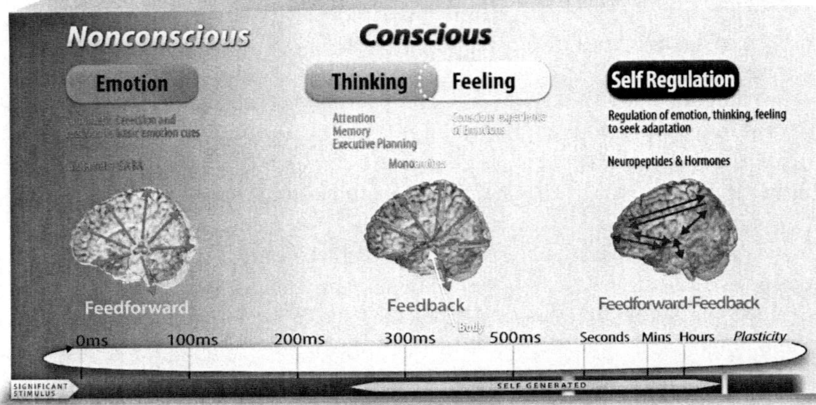

Figure 3.1 A visual summary of the INTEGRATE model framework. A fundamental principle that organizes your brain's processes is: minimize danger and maximize reward. This principle is relevant to the brain's most rapid reactions, as well as longer-term goals. It drives emotion, thinking, feeling, and self-regulation along a continuum of both time and associated modes of brain activity. Brain chemicals that play a key role in modulating feeling and thinking are monoamines, such as dopamine, serotonin, and norepinephrine. At longer time scales of several seconds, several minutes, and longer, the capacity for self-regulation emerges.

The INTEGRATE Model is intended as an organizational framework that may help bring together common concepts that have arisen out of different theoretical or discipline approaches. Three illustrative examples show that it provides a framework for integrating information, and generating new hypotheses:

i) Identifying commonalities across disciplines. For example, the way in which emotion is manifest across different levels of organization (and measurement); from automatic biases in responding to emotion cues (behavioral performance), to approach–avoidance reactions (psychophysiology) to activation of amygdala circuitry that may occur automatically in the absence of stimulus awareness (brain imaging).
ii) To link different theories about the same disorder. How the theories of each disorder, developed from separate traditions, may complement each other. For example, testing the possibility that thinking problems in depression result from an excessive emotionale reactivity, and are not only a sign of incipient dementia in geriatric depression.
iii) To identify what is specific versus common across disorders. For instance, abnormalities in the P300 Event-Related Potential (ERP) in attention tasks are characteristic of depression, schizophrenia, and ADHD. Are these due to a common mechanism in each disorder, associated with different consequences for thinking functions?

In this model, "emotion" refers to the automatic reactions (or action tendencies) generated without our conscious awareness. Emotional reactions are generated by cues relevant to minimizing danger and maximizing reward which are innate or highly conditioned. Facial expressions of emotion are the innate cues we process and react to the most strongly. Our brain and body react to emotional cues before we know that we are reacting to them. New insights from neuroscience show that the brain reacts within one-fifth of a second (200ms) to these cues (Adolphs 2002; Williams et al. 2006; Williams and Gordon 2007). Therefore, we can think of emotional brain systems as "emotional significance detectors." They involve direct connections from brainstem networks to major subcortical systems (limbic, basal ganglia) and their feed-forward rapid connections to medial prefrontal (including anterior cingulate) and other cortical systems (Williams et al. 2006). It is only after a few hundred more milliseconds, with brain and body feedback, that one becomes aware of the emotional experience, which is the basis of of our feelings.

"Feelings" are about how we experience our emotions, and their subsequent influence on our thinking processes (Gordon, 2008; Williams et al. 2008). Changes in the activation of our brain and bodily functions provide the biological basis of our feelings, and how we interpret and label them. For instance, increases in heart rate occur along with the experience of stress, while reductions in heart rate are associated with relaxation. The experience of emotion also gives us the feelings of depression and anxiety that we are able to report on.

"Thinking" processes rely on engagement of conscious awareness and typically commence in half a second to a few seconds. Processes of thinking mostly involve facts, relying on key elements of attention (focus, selective attention), memory (working memory and recall) and executive functions (planning, flexibility). These elements are needed to reflect on the consequences of our actions and to plan ahead. Thinking processes are supported by limbic networks that parallel those involved in emotion processing, such as the hippocampus, and projections with the lateral prefrontal and parietal cortices in particular (Licinio 2004; Choi et al. 2005). The interconnection between these brain systems supports the feedback of thinking processes and also makes us aware of what we are attending to, remembering, and acting upon.

Self-regulation is about how we manage our emotions, thinking processes, and feelings, as well as associated brain processes. If we align these four processes, we may optimize our brain health and how we adapt to our world (Gordon, 2008; Williams et al. 2008). We have a natural bias toward expecting more negative than positive outcomes. Enhancing our positivity bias is associated with a greater capacity for well-being, resilience, effective communication, and productivity. People with a positivity bias tend to be more optimistic. A negativity bias is associated with a more pessimistic outlook. When this negativity bias becomes exaggerated, it may increase one's risk of stress and poor brain health. and depression and anxiety (Gordon 2008).

Contributions to this formulation, or different formulations, and the testing of the details of contributing concepts is undertaken by an international consortium of over 240 scientists, the Brain Research and Integrative Neuroscience Network, or BRAINnet (for more information, vist www.BRAINnet.net).

Candidate PM Markers

Using the INTEGRATE framework, an overview of exemplar candidate personalized markers is presented. These markers bring together evidence from across many categories of brain data. The different types of data include self-reported, cognitive (both "emotion" and "thinking" functions), electroencephalography-derived (EEG), event-related potentials (ERPs), arousal-based (autonomic measures), MRI and fMRI-derived, diffusion tensor imaging (DTI)-derived, genetic (detected by single nucleotide polymorphisms, or SNPs), data gathered from genome-wide association studies (GWAS). Here, we populate the

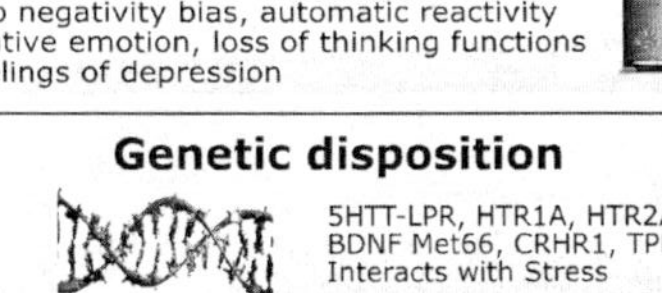
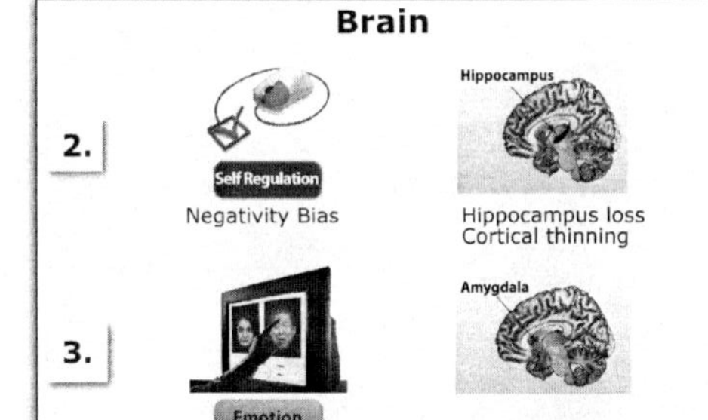
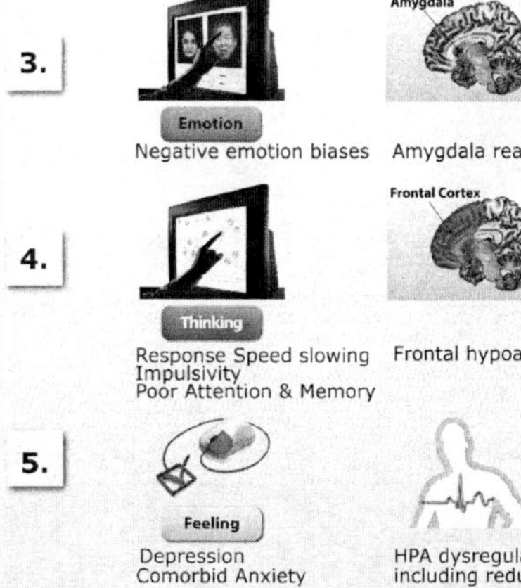

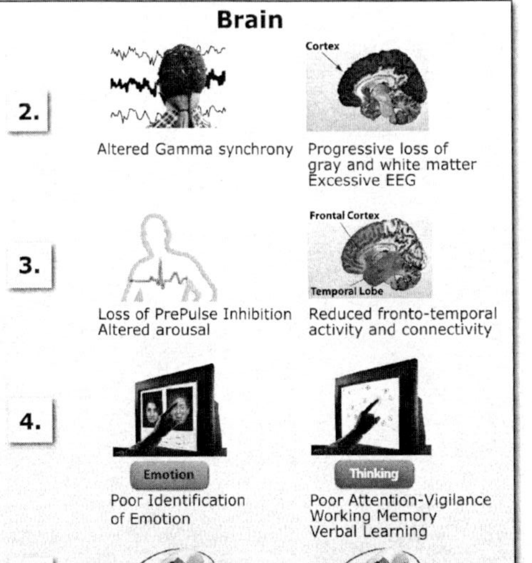

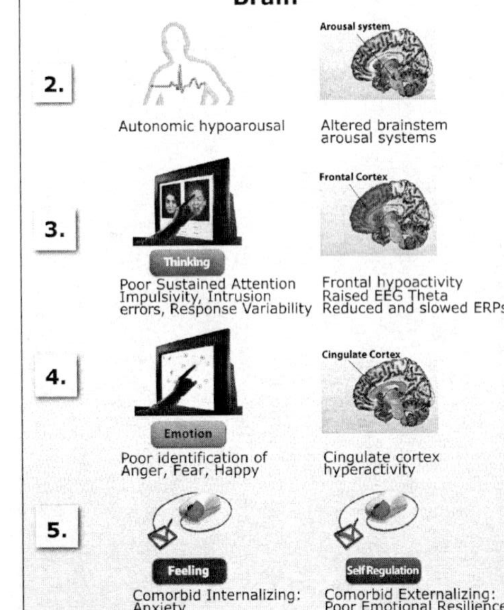

Figure 3.2 Summary of candidate diagnostic markers. These markers are each related to the hypothesized mechanism for each condition: depression, schizophenia, and ADHD. Key: DLPFC: Dorsal Lateral Prefrontal Cortex; VLPFC: Ventral Lateral Prefrontal Cortex; ACC: Anterior Cingulate Cortex; EEG: electroencephalogram.

INTEGRATE framework with those markers that are supported by the most robust evidence to date.

Markers from self-reports and cognitive markers are most readily classified into the processes of emotion, thinking, feeling, and self-regulation. On the other hand, candidate markers from electrical brain recordings (EEG, ERPs), autonomic body recordings of arousal (autonomic measures), brain imaging (structural MRI, fMRI) and genetics may reflect underlying pathophysiological mechanisms that contribute to one or more of these processes. As evidence accumulates for these markers, more definitive testing of their role within the time frames of emotion, thinking, feeling, and self-regulation will become possible.

Figure 3-2 provides a summary of candidate PM markers linking to the *diagnostic* classification of major depressive disorder (MDD), schizophrenia, and ADHD.

The PM markers implicated in treatment response (both positive response and nonresponse) for depression are summarized in Figure 3-3.

Depression and Diagnostic Markers

Markers for depression capture elements of the hypothesized mechanisms. The hypothesis is that genetic vulnerability exacerbated by stress leads to a negativity bias, which produces automatic emotional reactivity. This emotional reactivity detracts from thinking functions relying on higher cortical systems. The consequence is that even normal daily events are perceived as stressful, and there is an inability to self-regulate responses to them. Defining symptoms and feelings of depression are the behavioral products of this process.

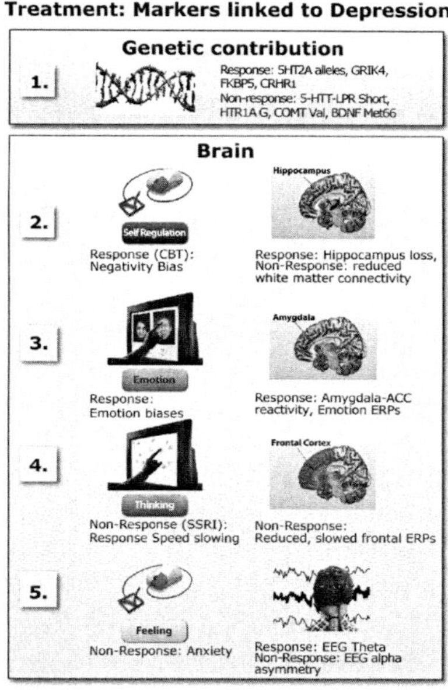

Figure 3.3 Summary of candidate markers for treatment prediction in depression (black text = markers contributing to positive treatment response; blue text = markers contributing to non-response). 1. Genetic markers. 2. Negativity bias in and structural brain change markers. 3. "Emotion" and related brain circuitry markers. 4. "Thinking" and related brain circuitry markers. 5. Self-reported markers of "feeling" (anxiety) and EEG brain arousal markers. Key: SSRI: Selective Serotonin Reuptake Inhibitor; SNRI: Selective Norepinephrine Reuptake Inhibitor.

Candidate markers are organized below in relation to how they may capture each aspect of these mechanisms. This organization correlates with the recent formulation by Beck (2008). The candidate markers summarized below have been identified in more than one study.

1. Genetic Disposition to Depression, Exacerbated by Stress.
   Genetic markers: Meta-analyses suggest that the *5-HTTLPR* short allele is a marker of risk for depression (Kiyohara and Yoshimasu, 2010). Other genetic markers related to monoamine neurotransmission are polymorphisms of serotonin 2A and 1A receptors, and the Met allele of catechol-o-methyltransferase *(COMT)* Val108/158Met (Levinson, 2006). The COMT Met allele is associated with risk for depression and with emotional brain markers (Williams et al., 2010). Another set of genetic markers reflects alterations in brain plasticity (the Met allele of BDNF Val66Met polymorphism) and overactivity of the hypothalamic-pituitary-adrenal axis (HPA) (tryptophan hydroxylase, or TPH). The viability of these candidate genetic markers requires replication in large independent studies.
   Stress markers: Risk for depression is highest in those with both the short allele and exposure to stress (Caspi et al. 2003; Williams et al. 2009). The BDNF Met allele also interacts with stress in early life to heighten risk for depression (Gatt et al. 2009). (For further details on stress markers, see Chapter 5.)
2. Negativity Bias: Genetic Disposition Interacting With Stress Manifests As A Negativity Bias In Attributions And Structural Brain Changes.
   Self-report markers: Negativity bias is the loss of Self Regulation. It is reflected in the tendency to expect negative outcomes and focus on negative events. It is greater in those with both gene and stress markers, such as both the *5-HTTLPR* short allele and exposure to early life stress (Williams et al. 2009; 2010). Negativity bias captures trait-like markers of likelihood for developing depression (and related mental health problems), supported by longitudinal research (Alloy, Abramson, and Francis 1999; Wicker et al. 2007).
   Brain imaging (MRI) markers: The gene–stress vulnerability to depression is also manifest in structural brain changes. A meta-analysis of brain imaging studies highlights reduced hippocampal volume (8 to 10 percent) as the most consistent candidate marker for depression (Sheline, Gado, and Kraemer, 2003). Loss of hippocampal gray matter is associated with both genetic risk factors and a history of early life stress (Gatt et al. 2009). A risk marker for depression is a thinning of the cortex that is particularly apparent in the right hemisphere (Peterson et al. 2009). Cortical thinning is also associated with the emotion, thinking, and arousal disturbances (Peterson et al. 2009). Reduced white matter connectivity, assessed by DTI, has been observed in depression across a number of studies, particularly in later-life patients (Shimony et al. 2009).
3. Emotional Reactivity: The Combination Of The First Two Points Above Produces An Automatic Emotional Reactivity Associated With Amygdala Circuitry.
   Cognitive "emotion" markers: On cognitive tasks of emotion processing, there is a reduction in mood-congruent processing of negative emotion and an inability to identify positive emotion. These markers are sensitive to change in serotonin and norepinephrine systems.
   Brain imaging (fMRI) markers: In fMRI, markers for depression include excessive limbic (amygdala) activation in response to negative emotional stimuli (Sheline et al. 2001; Chen et al. 2007). Excessive paralimbic activity (anterior cingulate) has also been observed.

Electrical brain (ERP) markers: ERPs elicited around 200 milliseconds post-stimulus during an emotion-processing task, and have also been implicated in depression (Kerestes et al. 2009). These emotion ERPs may reflect amygdala reactivity, and the impact of cognitive performance on emotion-based tasks.

4. Thinking Functions: Emotional Reactivity Within Amygdala Circuitry Detracts From Thinking Functions Associated With Higher-Level Cortical Circuitry.
   Cognitive "thinking" markers: While depression has traditionally been considered an "emotion" disorder, it is commonly defined by a cluster of difficulties in the domains of attention and concentration, memory, executive function, and speed of information processing (Hasler et al. 2004). It has been postulated that excessive negativity bias and sensitivity to negative emotion in cognitive tasks may disrupt these "thinking" functions, which are subserved by higher-order cortical systems. The form of depression traditionally referred to as melancholic is characterized by slowed response speed, also known as psychomotor slowing (Pier, Hulstijn, and Sabbe 2004). On the other hand, the traditional subclassification of atypical depression with anxiety is defined by impulsivity in cognitive tasks that assess the capacity to inhibit behavior and automatically generated responses (Posternak, 2003; Langenecker et al. 2007).
   Electrical brain (ERP) markers: ERPs elicited during "thinking" tasks suggest hypoactivation of the frontal cortex as a candidate marker for depression. A reduced and delayed P300 ERP during attention tasks, and reduced ERPs elicited by an error-processing task, equivalent to a "go-no-go" task, have been observed in depression (Kalayam and Alexpoulous 1999; 2003; Kemp et al. 2009). These candidate ERP markers elicited in "thinking" tasks have been identified in older as well as younger depressed patients. They are most apparent over the frontal cortex, which is consistently implicated in "thinking" functions.
5. Feelings Of Depression: This Cycle Of Negativity Produces The Feelings Associated With Depression Which Are Its Defining Symptoms. It Is Maintained By Dysregulation Of The HPA Axis And Arousal Systems.
   Self-report markers: The two key clinical diagnostic features of depression are self-reported experience of prolonged depressed mood and the inability to experience pleasure. Feelings of anxiety and stress are commonly comorbid.
   Electrical brain (EEG) markers: In EEG measures, depressed states are associated with asymmetric brain activity, especially in the alpha power band (Allen et al. 2004). Level of depression is also indicated by a heightened slow wave theta (and delta) and reduced alpha activity (Gatt et al. 2008) in EEG recordings.
   Arousal (autonomic) markers: A well-replicated marker of the loss of regulation in arousal systems is reduced heart rate variability (HRV) (Nahsoni et al. 2004). For further details on heart rate markers see Chapter 9.

Future studies are required to replicate these candidate markers in large samples. It would be valuable to include all candidates within a single study, so as to compare the relative sensitivity of each marker, and the relationships between them. Future studies may also consider markers for anxiety comorbidity in depression. The iSPOT-D study uses a standardized battery of self-reported, cognitive, brain, and genetic measures to evaluate these markers in two thousand patients with depression.

Depression and Treatment Markers

Treatment response in depression has been linked to the following markers in more than one study (Figure 3-3).

1. Genetics and Treatment Response.
   Genetic markers: The search for genetic predictors of treatment response has largely been focused on a group of key models of depression and its treatment. These include the monoaminergic system, neurotrophic factors relevant to neurogenesis, and the HPA. The majority of studies that have sought to identify genetic markers for response have focused on the monoaminergic pathways. Yet, it is clear that alteration of the monoaminergic system alone may not be sufficient for amelioration of depressive symptoms or understanding the action of currently available antidepressants. The neurotrophin hypothesis of depression may be at least as important in identifying predictors of personalized treatment response. This hypothesis is based on observations that depression involves alterations in the stress response, with loss of neuronal connectivity and reduction of BDNF in the hippocampus. Other strategies for selection of candidate genes have been based on prior evidence of association with MDD, or antidepressant response in MDD patients (Kemp et al. 2008).

   Replicated genetic variants predicting positive response to treatment include the *5HT2A* rs7997013 AA allele, *5HT2A* 102T/CC and -1438A/G G alleles, *GRIK4* rs1954787gene, and *FKBP5* and *CRHR1* alleles (Binder et al. 2004; Licinion et al. 2004; Choi et al. 2005; Binder and Holsboer 2006; McMahon et al. 2006;Paddock et al. 2007; Lee and Ham 2008; Lekman et al. 2008; Drago, Ronchi, and Serretti 2009). Of these, the *5HT2A* rs7997013 AA allele has shown the most robust effect size and consistency across studies, including the most sensitive markers in the U. S. National Institute of Mental Health (NIMH)-funded STAR*D (Sequenced Treatment Alternatives to Relieve Depression) study. It contributes 16 to 18 percent of the effect size, which is substantial for a single allelic variant. The *5HTTLPR* short allele has been implicated as a biomarker for nonresponse to treatment (Zanardi et al. 2000). Other studies indicate that the short allele may predict better response in patients of Asian ethnicity, highlighting the importance of replication in large samples to confirm markers that may be personalized in their application. Other candidate genetic markers for non-response include the *5HTTLPR* short allele, *HTR1A* (rs6295)G allele, *COMT* (val108/158met) Val allele, and the *BDNF* Met66 allele (which may also vary with ethnicity) (Baune et al., 2008; Choi et al., 2006; Tsai et al., 2003; Villafuerte et al., 2009). It has been observed that differences in tolerability of medication might also contribute to the association between nonresponse and the *5HTTLPR* short allele (see Chapter 2 and Chapter 5).
2. Negativity Bias, Structural Brain Changes, and Treatment Response.
   Self-report markers: Cognitive-behavioral therapy (CBT) was designed in large part to "re-attribute" the negative attributional biases in depression (Beck, 2008). To date, trials have not been designed to evaluate whether this bias predicts response to medications. iSPOT-D is the the the first randomized trial underway to evaluate self-reported negativity bias as a potential predictor of response to antidepressant medications.

   Brain imaging (MRI) markers: Antidepressant treatment has been reported to block or even reverse hippocampal atrophy via neurogenesis, indicating its role as a candidate marker of treatment response (Sheline, Gado, and Kraemer 2003). Nonresponse to treatment has been predicted by lower frontal white matter connectivity from DTI data (Alexopoulos et al. 2002), (Kiesappä et al. 2009). These findings are from studies of geriatric depression, and need replication in younger patients to determine if they can be generalized to a wider population.

3. Emotional Reactivity, Amygdala Circuitry, and Treatment Response.
Cognitive "emotion" markers: Alterations in identifying negative versus positive emotion on cognitive tasks improve with acute antidepressant treatment (Harmer et al. 2003; Venn et al. 2005; Harmer et al. 2009). Accumulating evidence points to the value of objective measures of emotion identification and recognition as sensitive markers of change in serotonin and norepinephrine systems, and response to antidepressant treatments (Venn et al. 2005). Indeed, these emotion cognition markers are proposed to be more sensitive than traditional endpoints, such as the Hamilton Depression Rating Scale.
Brain imaging (fMRI) markers: In fMRI studies, hyperreactivity of the amygdala to negative emotion words predicts future response to CBT in depressed patients (Siegle, Carter, and Thase 2006). The response of patients with enhanced activation of the paralimbic anterior cingulate cortex to negative emotion stimuli at pre-treatment baseline shows the most robust response to medication treatment (Chen, Ridler, and Suckling 2007; Mulert et al. 2007). These markers may capture the neural circuitry underlying emotional cognition markers of treatment response. Convergent evidence from positron emission tomography (PET) suggests that distinct limbic and paralimbic systems may distinguish response to behavioral versus medication treatments (Martin et al. 2001; Goldapple et al. 2004; Kennedy et al. 2007).
Electrical brain function (ERP) markers: Emotion-task elicited ERPs peaking around 200 ms post-stimulus are indicators of a positive response to antidepressants (Kerestes et al. 2009). ERP markers also show promise as moderators of response to specific treatments. For instance, a heightened baseline P100/N200 ERP elicited by noise bursts (also known as the Loudness-Dependent Evoked Potential) predicts response to SSRI, but not to SNRI, treatment (Paige et al. 1994; Linka et al. 2005; Mulert et al. 2007).
4. "Thinking" Functions, Cortical Circuitry, and Treatment Response.
Cognition "thinking" markers: A slowing of response speed (psychomotor), a primary feature of melancholia, has been found to predict nonresponse to the SSRI fluoxetine (Caligiuri et al. 2003; Taylor et al. 2006). Response speed is, therefore, a specific candidate marker for predicting response to alternative treatments that target multiple neurotransmitter systems, such as SNRIs.
Electrical brain function (ERP) markers: A reduced and delayed P300 ERP elicited by attention tasks, particularly over the frontal cortex, is a candidate marker for poor treatment response. Frontal ERPs elicited by an error-processing task, capturing a "go-no-go" response, have also been linked to poorer treatment response (Bruder et al. 1999; Kalayam and Alexpoulous 1999; 2003). These candidate ERP markers of poor response have been identified particularly in older depressed patients.
5. "Feelings" of Depression, Symptoms, Arousal Systems, and Treatment Response.
Self-report markers: There is evidence for anxiety and depression severity as predictors of poorer response to SSRIs in major depressive disorder, or MDD (Fava et al. 2008). More moderate depression severity has commonly been associated with a non-melancholic or atypical form of depression, with good response to SSRIs. By contrast, more severe depression has been associated with melancholic features and poorer response to SSRIs, which may be more pronounced when comorbid with severe anxiety. It has been suggested that the latter patients may respond better to SNRIs, which modulate both serotonin and norepinephrine systems (Kemp et al. 2008).

Electrical brain (EEG) markers: A positive response to medication has also been associated with heightened baseline EEG Theta power, localized to the anterior cingulate cortex (Pizzagalli et al. 2001). By contrast, patients with greater EEG Alpha asymmetry (reflecting greater right than left hemisphere activity during resting conditions) respond poorly to fluoxetine (Linka et al. 2005).

Arousal (autonomic) markers: Reduced heart rate variability (HRV) is a predictive moderator of positive responses to CBT. Positive responses are related to reduced severity of depression in patients with both PTSD and heart disease (Carney et al. 2000; Nishith et al. 2003). These findings highlight the importance of arousal markers for moderating responses to non-pharmaceutical treatments.

Schizophrenia and Diagnostic Markers

Markers for schizophrenia capture elements of the hypothesized mechanisms for this condition, with a focus on the markers present from the first schizophrenic episode. It is proposed that genetic disposition interacts with insults during critical periods of development to trigger schizophrenia. The mechanism by which schizophrenia manifests is a loss of normal neural integration (or "dis-integration"). It is related to progressive loss of gray and white matter. The loss of neural integration and gray and white matter produces functional alterations in brain and body arousal systems. Behaviorally, the unfolding of neural disintegration is reflected in problems with emotion (social cognition), and thinking (general cognition). An ongoing cycle of poor self-regulation develops, reflected in associated feelings of depression and a negativity bias.

Candidate markers are organized below in relation to the hypothesized manner in which they unfold from genetic risk to behavior (see Figure 3-2, part B). These candidate markers have been identified in more than one study (see Chapter 6 a detailed review).

1. Genetic Disposition To Schizophrenia: Interacting with Critical Periods Of Development.
   Genetic markers: While there is variation in findings from genetic studies of schizophrenia, replicated markers include variants on chromosomes 1q, 6p/6q, 8p, 10p, 13q, 15q, and 22q. Candidate genetic marker variants on these chromosomes are *NSG4* and *DISC1* (chromosome 1), *dysbindin (DTNBP1)* and *NOTCH4* (chromosome 6), neuregulin (chromosome 8), phosphatidylinositol-4-phosphate-5-kinase type-II alpha (*PIP4K2A*) (chromosome 10), *BDNF* Met66 (intersection of chromosomes 11 and 13), the cholinergic alpha7-like nicotinic receptor (*CHRNA7*, chromosome 15), the *COMT* Val108/158Met Val allele, and *PRODH2* (chromosome 22) (Egan et al. 2001; Xu et al. 2001; Berry, Jobanputra and Pal 2003; Chen, Lipksa, and Weinberger 2006; Gur et al., 2007; Ho et al. 2006; Kampman et al., 2004; Shen, Lang, and Nakamoto 2008; Weickert et al., 2004).

   Genetic markers may have a cumulative impact during successive critical "windows of vulnerability" during brain development and the early course of the illness.
2. Neural "Dis-integration" and Associated Gray and White Matter Loss.
   Electrical brain (EEG neural synchrony) markers: With brain insults, disposition to schizophrenia may manifest as a loss of neural integration (or "dis-integration").
   A promising marker for schizophrenia is provided by EEG measures of neural synchrony, particularly in the 40Hz gamma band. Gamma synchrony is altered in the first schizophrenic episode and in chronic patients during "thinking" and "emotion" tasks. These patients show an excess of synchrony that is unrelated to the

task demands at hand, and present as first onset of schizophrenia (Flynn et al. 2008; Williams et al. 2009). This excess of synchrony suggests excessive integration of "random" input. During a task, schizophrenia patients are unable to increase their synchrony in response to a specific stimulus requirement. This lack of stimulus-elicited synchrony is most apparent over the left fronto-temporal cortices (Haig et al. 2000; Lee et al. 2001;Spencer et al. 2004; Uhlhass et al. 2006). By contrast, healthy people show a comparatively lower baseline synchrony, and a more pronounced increase in synchrony in response to task stimuli. The pattern of synchrony in schizophrenia is consistent with the clinical phenotype—namely, an inability to increase synchrony for task demands, within a background of general over-association of information.

Electrical brain (EEG) markers: A complementary candidate marker for schizophrenia is the reversal of the normal developmental pattern of EEG activity. Schizophrenia is characterized by a preservation or even increase in slow-wave EEG Theta from the time of the first episode to several years later, while healthy individuals show a reduction in theta. Since slow-wave EEG captures neurodevelopmental changes, this marker may index the abnormal developmental trajectory in schizophrenia. An abnormal increase in EEG accords with excessive synchrony.

Brain imaging (MRI) markers: A promising brain imaging marker for progression from risk to f rst-episode to chronic schizophrenia is reduction in gray and white matter. First-episode and high-risk individuals show gray matter loss in frontal brain regions which distinguishes them from those exhibiting bipolar psychosis, and which progresses into the temporal and parietal-occipital regions over time (Farrow et al. 2005; Whitford et al. 2006a,b; Williams 2008). Corresponding progressive loss of white matter in fronto-temporal regions is apparent from the first episode of schizophrenia (Whitford et al. 2007). Chronic patients show loss of gray matter in the superior temporal cortex, with extensive loss in additional frontal and parietal-occipital regions (Williams 2008). The progression of gray matter loss is tracked by corresponding gamma synchrony alterations (Williams et al. 2009).

3. Functional Alterations in Brain and Body Arousal Systems.

Arousal (autonomic) markers: A well-replicated marker for schizophrenia is elicited by the prepulse inhibition (PPI) task, which assesses automatic filtering of relevant from irrelevant input (Parwani et al. 2000; Ludewig, Geyer, and Vollenweider 2003). Similarly, excessive phasic skin conductance is present in both early and first-episode-onset schizophrenia, and serves as a reliable marker of this illness and its progression (Zahn et al. 1997; Williams et al. 2004).

Brain imaging (fMRI) markers: Candidate fMRI markers capture reduced activation in fronto-temporal circuitry during attention tasks of thinking-based processes (Kiehl et al. 2005; Wolf et al. 2008). Both first-episode and chronic patients also show reduced fronto-temporal activation (including amygdala-medial prefrontal regions) in emotion-based tasks (Williams et al. 2004; Das et al. 2007; Hall et al. 2008; Russell et al. 2008). A complementary marker is the reversal of normal functional connectivity in these networks (Das et al. 2007), corresponding to the "dis-integration" seen in Gamma synchrony markers (see Chapter 4).

Electrical brain function (ERP) markers: A consistent candidate marker for schizophrenia is reduction of the P300 ERP, elicited around 300 ms, during the same attention tasks. This marker is sensitive to both chronic and first-episode schizophrenia (Mathalon, Ford, and Pfefferbaum 2000; Brown et al. 2002; Salisbury et al., 1998; Strik et al., 1993), indicating it may qualify for trait-like status. ERPs are also altered for emotion tasks in schizophrenia, within 200 ms post-stimulus (Horley et al. 2001; Streit et al. 2001).

4. Reflected in Emotion (Social Cognition) and Thinking (General Cognition) Problems.
   Cognitive "emotion" markers: Schizophrenia patients generally perform poorly on emotion identification tasks, compared to healthy controls (Mandal, Pandey, and Prasad 1998; Hooker and Park 2002; Kohler et al. 2009; Russell et al., 2007). Meta-analysis reports a large overall effect size of -0.91 for 86 of these studies (Kohler et al. 2009). Some of these studies suggest the impairment is most pronounced for negative emotions.
   Cognitive "thinking" markers: The consortium known as 'Measurement And Treatment Reserach to Improve Cognition in Schizophrenia' (MATRICS) has recommended core domains for assessment of cognition in schizophrenia. These domains encompass both emotion (social cognition) and thinking (general cognition). Within these domains, both chronic and first-onset patients show impairments that are most pronounced for working memory, vigilance-attention, verbal learning, and social cognition domains (Symond et al. 2005; Harvey, Patterson, and Potter 2006). Future studies might examine the direct relationship between cognitive performance on measures of MATRICS domains and brain structure and function, using brain-imaging techniques.
5. A Cycle of Poor Self Regulation Develops, Reflected in Feelings of Depression and Development of Negativity Bias
   Self-report markers: A negativity bias, as noted earlier, refers to the tendency to expect negative outcomes and focus on negative events (Williams et al. 2008). This bias may also be linked to emotional intelligence problems that are candidate markers for schizophrenia (Neuchterlein et al. 2004).
   Treatment response in schizophrenia has been linked to a number of markers, several of which have also been implicated in diagnostic classification (as outlined above). (For further details on these markers and other cognition-brain markers linked to antipsychotic treatment response and side effects, see Chapter 6).

ADHD and Diagnostic Markers

Markers for ADHD capture how the hypothesized mechanism for this condition develops from genetic risk to behavior (see Figure 3-2, part C). The hypothesis is that genetic disposition manifests as hypoarousal in brain and body systems. This disposition produces difficulty with higher-order thinking functions, that involve sustaining attention on relevant output. ADHD symptoms reflect an attempt to boost arousal through hyperactivity, for instance. There is also associated difficulty in early appraisal of significant emotional cues. These functional difficulties are coupled with experiences of anxious feelings that define comorbid internalizing conditions and/or poor emotional resilience characterizing comorbid externalizing conditions. These candidate markers have been identified in more than one study. (See Chapter 7 for a detailed review and references.)

1. Genetic Disposition to ADHD.
   Genetic markers: Across studies of genetic variants, the *DAT1* (10R allele) and *DRD4* (7R allele) have been most consistently implicated in disposition to ADHD, These genetic markers are implicated in the pathophysiology of ADHD, and are candidates for indicating who is at elevated risk for developing the disorder.

2. Manifests as Hypoarousal in Brain and Body Systems.
   Arousal (autonomic) markers: The key mechanism for ADHD is a loss of arousal ability for processing significant input, which is reflected in candidate markers of reduced autonomic arousal. These include lack of skin conductance responses and reduced heart rate, implicating the brainstem and ascending reticular activating system.
3. Produces Difficulty with Higher-Order Thinking Functions.
   Cognitive "thinking" markers: Thinking markers for ADHD encompass poor performance on tasks of sustained attention, inhibition and impulsivity, and errors of intrusion from irrelevant information and response variability.
   Electrical brain (EEG, ERP) mrkers: "Thinking markers are associated with brain alterations, including raised EEG Theta, and slowed and reduced ERPs during the same cognitive tasks.
   Brain imaging (fMRI) markers: During the same cognitive thinking tasks, the following candidate brain imaging markers have been observed: frontal hypoactivation in the dorsal and ventral prefrontal cortex, and related anterior cingulate cortex. The parietal cortex and basal ganglia circuitry have also been implicated.
4. Associated Difficulty in Early Appraisal of Significant Emotion Cues.
   Cognitive "emotion" markers: Impairments in identifying facial expressions of emotion characterize ADHD, and are associated with comorbid internalizing features, such as anxiety.
   Electrical brain (ERP) markers: Emotion-processing difficulties in ADHD are captured by a reduction in early electrical brain activity (the P120 ERP), which is most apparent over the occipital cortex.
   Brain imaging (fMRI) markers: In processing facial expressions of anger, ADHD is distinguished by hyperreactivity of the frontal and posterior cingulate cortex. Neural hyperreactivity to emotion may interfere with the capacity to accurately identify emotion cues.
5. Coupled with Experiences of Anxious Feelings and/or Poor Emotional Resilience.
   Self-report markers: Self-reported anxiety commonly co-occurs with ADHD, and may identify those at most risk of later anxiety and depressive disorders (internalizing conditions). Markers of poor Self-Regulation in ADHD include self-report measures of emotional resilience and social function (including empathy), behavioural measures of theory of mind, and management of motivational goal-directed behavior. These markers may be most associated with conduct and oppositional-defiance behaviors (externalizing conditions).
   A number of these markers have also been implicated in predicting response and nonresponse to treatment for ADHD. A review of the evidence for these treatment markers is provided in Chapter 7, which also reviews the roles of these markers in identifying comorbidity in ADHD.

Translation in the Real World

A crucial determinant of which markers will be most actionable in the real world of clinical practice and health care are the potential cost-benefit tradeoffs of those markers. This will significantly influence which markers get used, regardless of whether they emerge from research done in academic or corporate settings.

The ultimate goal of personalized medicine is the discovery of markers that can be used in clinical practice to improve the individualization of patient care, and the overall improvement of treatment outcome. Methods and procedures that may facilitate the ready transfer of markers into routine clinical care include:

- The core assessment battery is made available in a scalable platform (such as Web-based, or tablet), making its use in everyday clinical practice simple and inexpensive.
- The vast amounts of data collected using standardized approaches (such as with www.BRAINnet.net and www.BrainResource.com) provide an existing base of evidence for integrating measures of brain structure and function with cognitive measures. This will allow, in many cases, the use of proxy measures from cognition; they may be coupled with inevitably cheaper genomics for a first-phase use, with more expensive neuroimaging procedures reserved for more complex and treatment-resistant cases.
- Implementation in a Web-based or other widely accessible platform is necessary for scalability. Delivering markers via a Web-based structure and electronic health records (EHR) makes it relatively easy to rapidly inform clinicians of marker research findings that should be considered in clinical care. The same platform can be populated with clinical care consideration messages to the clinician, based on algorithms that analyze an individual patient's assessment results, and link those findings to established evidence regarding the clinical and treatment implications of the particular marker profile a patient presents. Thus, a smooth, uninterrupted flow can be established from assessment to markers and their application in clinical care. This represents a paradigmatic advance that bridges the "knowledge gap" between new scientific discovery and clinical care.
- Brain imaging markers may be of most value for informing clinicians about complex cases, rather than for routine use in personalized medicine in practice. A progressive set of markers may be considered. For instance, routine assessment with Web-implemented markers may offer a first step ("triage") which identifies more complex or treatment-resistant cases for follow-up with brain-imaging assessments.

The adoption of personalized medicine for the brain will require consensus about candidate markers, and their cost-benefit ratio in the real world of clinical care (Gordon 2007). Identifying those markers with high predictive validity and replicability, and that potentially are actionable now needs to be distinguished from identifying markers that have failed to be replicated or still need more research.

The speed with which the adoption of marker-based PM progresses to a widespread uptake in mainstream clinical care will be determined by the availability of unambiguous and cost-effective replicated markers, which can be readily implemented into the workflow of clinical practices and reimbursed by the government or private health insurers.

Conclusion

This chapter offers an integration of candidate PM markers for disorders of brain health as a step toward achieving consensus about these markers. The expanding scope of variables that could be considered, and the exponentially large interactions among these variables, will be best resolved by standardized methods and large, standardized databases that allow knowledge to be shared, and the largest correlations to be systematically elucidated and replicated.

Personalized medicine that can be translated to Web-based and her-embedded applications will have particular ease of use for clinicians. The Web-based delivery of objective PM markers will revolutionize the clinician's ability to prescribe brain-related treatments to individual patients, thus improving treatment success rates.

To achieve a consensus in PM markers for the brain, there are practical considerations that are likely to accelerate the success of this endeavor, including the following factors:

1. Markers that integrate genetics with brain and cognitive measures will need to be identified in trials that capture a representative sample of outpatients.
2. Standardized assessment methodology is a practical approach to reducing variance in measurement among multiple distributed data-gathering sites, and providing value over a long period.
3. With standardization, databasing of results is facilitated. This enables replication across subjects and sites, which is key to the both the research and clinical adoption of consensus markers.
4. Ultimately, for clinical translation of markers, the experimental methods should be packed and bundled in a way that makes them easily understood and learned, and implemented and distributed, along with being inexpensive and reimbursable. That is, more use should progressively decrease the unit cost of assessment, and improve the usefulness and accessibility for the clinician.

References

Adolphs, R. 2002. Neural systems for recognizing emotion. *Current Opinion in Neurobiology* 12(2): 169–77.

Alexopoulos, G., D. N. Kiosses, S. F. Choi, C. F. Murphy, and K. O. Lim. 2002. Frontal white matter microstructure and treatment response of late-life depression: A preliminary study. *American Journal of Psychiatry* 159(11): 1929–32.

Allen, J. J. B., H. Urry, S. K. Hitt, and J. A. Coan. 2004. Stability of resting frontal EEG asymmetry across different clinical states of depression. *Psychophysiology* 41: 269–80.

Alloy, L. B., L. Y. Abramson, and E. L. Fancis. 1999. Do negative cognitive styles confer vulnerability to depression? *Current Directions in Psychological Science* 4: 128, 132.

Baune, B. T., C. Hohoff, K. Berger, A. Neumann, S. Mortensen, T. Roehrs, J. Deckert, V. Arolt, and K. Domschke. 2008. Association of the COMT val158met variant with antidepressant treatment response in major depression. *Neuropsychopharmacology* 33: 924–32.

Bearden, C. E., and N. B. Freimer. 2006. Endophenotypes for psychiatric disorders: Ready for primetime? *Trends in Genetics* 22: 306–13.

Beck, A.T., 2008. The evolution of the cognitive model of depression and its neurobiological correlates. *American Journal of Psychiatry* 165: 969–977.

Berry, N., V. Jobanputra V, and H. Pal. 2003. Molecular genetics of schizophrenia: A critical review. *Journal of Psychiatry and Neuroscience* 28(6): 415–29.

Binder, E. B., and F. Holsboer. 2006. Pharmacogenomics and antidepressant drugs. *Annals of Medicine* 38: 82–94.

Binder, E. B., D. Salyakina, P. Lichtner, G. M. Wochnik, M. Ising, B. Putz, S., Seaman, S. Lucae, M.A. Kohli, T. Nickel, H.E. Künzel, B. Fuchs, M. Majer, A. Pfennig, N. Kern, J. Brunner, S. Modell, T. Baghai, T. Deiml, P. Zill, B. Bondy, R. Rupprecht, T. Messer, O, Köhnlein, H. Dabitz, T. Brückl, N. Müller, H. Pfister, R. Lieb, J. Mueller, E. LÕhmussaar, T.M. Strom, T. Bettecken, T., Meitinger, M. Uhr, T. Rein, F. Holsboer, B. Muller-Myhsok. 2004. Polymorphisms in FKBP5 are associated with increased recurrence of depressive episodes and rapid response to antidepressant treatment. *Nature Genetics* 36: 1319–25.

Biomarkers Definitions Working Group. 2001. Biomarkers and surrogate endpoints: Preferred definitions and conceptual framework. *Clinical Pharmacology and Therapeutics* 69: 89–95.

Brown, K., A. W. F. Harris, C. Gonsalves, L. M. Williams, and E. Gordon. 2002. Target and non-target ERP disturbances in first episode vs chronic schizophrenia. *Clinical Neurophysiology* 113: 1754–63.

Bruder, G. E., C. E. Tenke, J. W. Stewart, P. J. McGrath, and F. M. Quitkin. 1999. Predictors of therapeutic response to treatments for depression: A review of electrophysiologic and dichotic listening studies. *CNS Spectrums* 4: 30–36.

Caligiuri, M. P., V. Gentili, S. Eberson, J. Kelsoe, M. Rapaport, and J. C. Gillin. 2003. A quantitative neuromotor predictor of antidepressant non-response in patients with major depression. *Journal of Affective Disorder* 77: 135–41.

Carney, R. M., K. E. Freedland, P. K. Stein, J. A. Skala, P. Hoffman, and A. S. Jaffe. 2000. Change in heart rate and heart rate variability during treatment for depression in patients with coronary heart disease. *Psychosomatic Medicine* 62(5): 639–47.

Caspi, A., K. Sugden, T. E. Moffitt, A. Taylor, I. W. Craig, H. L. Harrington, J. McClay, J. Mill, J. Martin, A. Braithwaite, and R. Poulton. 2003. Influence of life stress on depression: Moderation by a polymorphism in the 5-HTT gene. *Science* 301(5631): 386–89.

Charney, D. S., D. H. Barlow, K. Botteron, J.D. Cohen, D. Goldman, R.E. Gur, K.M. Lin, J.H. Meador-Woodruff, S.O. Moldin, E.J. Nestler, S.J. Watson, S.J., Zalcman. 2002. Neuroscience research agenda to guide development of a pathophysiologically based classification system. Research agenda for DSM-V (pp. 31–83). Kupfer DJ, First MB, Regier DA (Ed.). The American Psychiatric Association, Washington D.C., US

Chen, C. H., K. Ridler, J. Suckling, S. Williams, C.H. Fu, E. Merlo-Pich, E. Bullmore. 2007. Brain imaging correlates of depressive symptom severity and predictors of symptom improvement after antidepressant treatment. *Biological Psychiatry* 62: 407–41.

Chen, J., B. K. Lipksa, and D. R. Weinberger. 2006. Genetic mouse models of schizophrenia: From hypothesis-based to susceptibility gene-based models. *Biological Psychiatry* 59: 1180–88.

Choi, M. J., R. H. Kang, B. J. Ham, H. Y. Jeong, and M. S. Lee. 2005. Serotonin receptor 2A gene polymorphism (-1438A/G) and short-term treatment response to citalopram. *Neuropsychobiology* 52(3): 155–62.

Choi, M. J., R. H. Kang, S. W. Lim, K. S. Oh, and M. S. Lee. 2006. Brain-derived neurotrophic factor gene polymorphism (Val66Met) and citalopram response in major depressive disorder. *Brain Research* 1118: 176–82.

Das, P., A. H. Kemp, G. Flynn, A. W. F. Harris, B. J. Liddell, T. J. Whitford, A. Peduto, E. Gordon, L.M. Williams. 2007. Functional disconnections in the direct and indirect amygdala pathways for fear processing in schizophrenia. *Schizophrenia Research* 90(1–3): 284–94.

Drago, A., D. Ronchi, and A. Serretti. 2009. Pharmacogenetics of antidepressant response: An update. *Human Genomics* 3(3): 257–74.

Egan, M. F., T. E. Goldberg, B. S. Kolachana, J. H. Callicott, C. M. Mazzanti, R. E. Straub, D. Goldman, and D. R. Weinberger. 2001. Effect of COMT Val108/158 Met genotype on frontal lobe function and risk for schizophrenia. *Proc Natl Acad Sci USA*. 98: 6917–22.

Farrow, T. F. D., T. J. Whitford, L. M. Williams, L. Gomes, and A. W. F. Harris. 2005. Gray matter deficits and symptom profile in first episode schizophrenia. *Biological Psychiatry* 58: 713–23.

Fava, M., A. J. Rush, J. E. Alper, G. K. Balasubramani, S. R. Wisniewski, C.N. Carmin, M.M. Biggs, S. Zisook, A. Leuchter, R. Howland, D. Warden, M.H. Trivedi. 2008. Difference in treatment outcome in outpatients with anxious versus nonanxious depression: A STAR*D report. *American Journal of Psychiatry* 165: 342–51.

Flynn, G., D. Alexander, A. Harris, T. Whitford, W. Wong, C. Galletly, Silverstein S, Gordon E, Williams LM. 2008. Increased absolute magnitude of gamma synchrony in first-episode psychosis. *Schizophrenia Research* 105(1–3): 262–71.

Gatt, J. M., S. A. Kuan, C. Dobson-Stone, R. H. Paul, R. T. Joffe, A. H. Kemp, E. Gordon, P. R. Schofield, and L. M. Williams. 2008. Association between BDNF Val66Met polymorphism and trait depression is mediated via resting EEG alpha band activity. *Biological Psychology* 79: 275–84.

Gatt, J. M., C. B. Nemeroff, C. Dobson-Stone, R. H. Paul, R. A. Bryant, P. R. Schofield, E. Gordon, A. H. Kemp, and L. M. Williams. 2009. Interactions between BDNF Val66Met polymorphism and early life stress predict brain and arousal pathways to syndromal depression and anxiety. *Molecular Psychiatry* 14(7): 681–95.

Ginsberg, G. S., and M. Angrist. 2006. The future may be closer than you think: A response from the Personalized Medicine Coalition to the Royal Society's report on personalized medicine. *Personalized Med* 3(2): 119–23.

Goldapple, K., Z. Segal, C. Garson, M. Lau, P. Bieling, S. Kennedy, H. Mayberg . 2004. Modulation of cortical-limbic pathways in major depression: Treatment-specific effects of cognitive behavior therapy. *Archives of General Psychiatry* 61: 34–41.

Gordon, E. 2003. Integrative neuroscience. *Neuropsychopharmacology* 28: S2–S8.

Gordon, E. 2007. Genomic genetics and neuromarkers are both required for the era of brain-related personalized medicine. *Personalized Medicine* 4(2): 201–15.

Gordon, E., N. Cooper, C. Rennie, D. Hermens, and L. M. Williams. 2005. Integrative neuroscience: The role of a standardized database. *Clin EEG Neuroscience* 36: 64–75.

Gordon, E., K. J. Barnett, N. J. Cooper, N. Tran, and L. M. Williams. 2008. An "'integrative neuroscience'" platform: Application to profiles of negativity and positivity bias. *J Integr Neuroscience* 7: 345–66.

Gottesman, I., and T. D. Gould. 2003. The endophenotype concept in psychiatry: Etymology and strategic intentions. *American Journal of Psychiatry* 160: 636–45.

Gur, R. E., M. E. Calkins, R. C. Gur, W. P. Horan, K. H. Nuechterlein, L. J. Seidman, and W. S. Stone. 2007. The consortium on the genetics of schizophrenia: Neurocognitive endophenotypes. *Schizophrenia Bulletin* 33: 49–68.

Haig, A. R., E. Gordon, V. De Pascalis, R. A. Meares, H. Bahramali, and A. Harris. 2000. Gamma activity in schizophrenia: Evidence of impaired network binding? *Clinical Neurophysiology* 111(8): 1461–68.

Hall, J., H. C. Whalley, J. W. McKirdy, L. Romaniuk, D. McGonigle, A. M. McIntosh, B.J. Baiga, V-E. Gountounaa, D.E. Joba, D.I. Donaldson, R. Sprengelmeyer, A.W. Young, E.C. Johnston, S.M. Lawrie. 2008. Overactivation of fear systems to neutral faces in schizophrenia. *Biological Psychiatry* 64(1): 70–73.

Harmer, C. J., Z. Bhagwagar, D. I. Perrett, B. A. Vollm, P. J. Cowen, and G. M. Goodwin. 2003. Acute SSRI administration affects emotional processing in healthy volunteers. *Neuropsychopharmacology* 28: 148–52.

Harmer, C. J., U. O'Sullivan, E. Favaron, R. Massey-Chase, R. Ayres,R, Reinecke, A, Goodwin, GM, Cowen, PJ. 2009. Effect of acute antidepressant administration on negative affective bias in depressed patients. *American Journal of Psychiatry* 166: 1178–84.

Harvey, P. D., T. L. Patterson, and L. S. Potter. 2006. Improvement in social competence with short-term atypical antipsychotic treatment: A randomized, double-blind comparison of quetiapine versus risperidone for social competence, social cognition, and neuropsychological functioning. *American Journal of Psychiatry* 163: 1918–25.

Hasler, G., W. C. Drevets, H. K. Manji, and D. S. Charney. 2004. Discovering endophenotypes for major depression. *Neuropsychopharmacology* 29: 1765–81.

Ho, B. C., N. C. Andreasen, J. D. Daswon, T.H. Wassink. 2006. Association between brain-derived neurotrophic factor Val66Met gene polymorphism and progressive brain volume changes in schizophrenia. *American Journal of Psychiatry* 164: 1890–99.

Hooker, C., and S. Park. 2002. Emotion processing and its relationship to social functioning in schizophrenia patients. *Psychiatry Research* 112(1): 41–50.

Horley, K., C. Gonsalvez, L. Williams, I. Lazzaro, H. Bahramali, and E. Gordon. 2001. Event-related potentialsto threat-related faces in schizophrenia. *International Journal of Neuroscience* 107(1–2): 113–30.

Kalayam, B., and G. S. Alexopoulos. 1999. Prefrontal dysfunction and treatment response in geriatric depression. *Arch Gen Psychiatry,* 56: 713–18.

Kalayam, B., and G. S. Alexpoulos. 2003. A preliminary study of left frontal region error negativity and symptom improvement in geriatric depression. *Am J Psychiatry* 160: 2054–56.

Kampman, O., S. Anttila, A. Illi, M. Saarela, R. Rontu, K. M. Mattila, E. Leinonen, and T. Lehtimäki. 2004. Neuregulin genotype and medication response in Finnish patients with schizophrenia. *NeuroReport* 15(16): 2517–20.

Kemp, A. H., E. Gordon, A. J. Rush, and L. M. Williams. 2008. Improving the prediction of treatment response in depression: Integration of clinical, cognitive, psychophysiological, neuroimaging and genetic measures. *CNS Spectrums* 13(12): 1066–86.

Kemp, A. H., P. J. Hopkinson, D. F. Hermens, D. L. Rowe, A. L. Sumich, C. R. Clark, W. Drinkenburg, N. Abdi, R. Penrose, A. McFarlane, P. Boyce, E. Gordon, and L. M. Williams. 2009. Fronto-temporal alterations within the first 200 ms during an attentional task distinguish major depression, non-clinical participants with depressed mood and healthy controls: A potential biomarker? *Human Brain Mapping* 30(2): 602–14.

Kennedy, S. H., J. Z. Konarski, Z. V. Segal, M.A. Lau , P.J. Bieling, R.S. McIntyre, H.S. Mayberg. 2007. Differences in brain glucose metabolism between responders to CBT and venlafaxine in a 16-week randomized controlled trial. *Am J Psychiatry* 164: 778–88.

Kerestes, R., I. Labuschagne, R. J. Croft, B. V. O'Neill, Z. Bhagwagar, et al. 2009. Evidence for modulation of facial emotional processing bias during emotional expression decoding by serotonergic and noradrenergic antidepressants: An event-related potential (ERP) study. *Psychopharmacology* 202(4): 621–34.

Kiehl, K. A., M. C. Stevens, K. Celone, M. Kurtz, and J. H. Krystal. 2005. Abnormal hemodynamics in schizophrenia during an auditory oddball task. *Biol Psychiatry* 57(9): 1029–40.

Kieseppä, T., M. Eerola, R. Mäntylä, T. Neuvonen, .V. Poutanen, K. Luoma, A. Tuulio-Henriksson, P. Jylhä, O. Mantere, T. Melartin, H. Rytsälä, M. Vuorilehto, E. Isometsä. 2009. Major depressive disorder and white matter abnormalities: A diffusion tensor imaging study with tract-based spatial statistics. *Journal of Affective Disorders* 120(1): 240–44.

Kiyohara, C. and K. Yoshimasu. I. 2010. Association between major depressive disorder and a functional polymorphism of the 5-hydroxytrptamine (serotonin) transporter gene: A meta analysis. *Psychiatric Genetics* 20(2): 49–58.

Kohler, C. G., J. B. Walker, E. A. Martin, K. M. Healey, and P. J. Moberg. 2009. Facial emotion perception in schizophrenia: A meta-analytic review. *Schizophrenia Bulletin*. doi:10.1093/schbul/sbn192.

Koslow, S. H. 2000. Should the neuroscience community make a paradigm shift to sharing primary data? *Nature Neuroscience* 3(9): 863–65.

Koslow, S. H. 2002. Sharing primary data: A threat or asset to discovery? *Nature Reviews Neuroscience* 3(4): 311–13.

Koslow, S. H. 2005. Discovery and integrative neuroscience. *Clinical EEG and Neuroscience* 36(2): 55–63.

Koslow, S. H., L. M. Williams, and E. Gordon. 2010. Personalized medicine for the brain: A call for action. News and Commentary. *Molecular Psychiatry* 15: 229–30.

Langenecker, S. A., S. E. Kennedy, L. M. Guidotti, E. M. Briceno, L. S. Own, T. Hooven, E. A. Young, H. Akil, D. C. Noll, and J-K Zubieta. 2007. Frontal and limbic activation during inhibitory control predicts treatment response in major depressive disorder. *Biological Psychiatry* 62(11): 1272–80.

Lee, B. T., and B. J. Ham. 2008. Serotonergic genes and amygdala activity in response to negative affective facial stimuli in Korean women. *Genes Brain Behavior* 7: 899–905.

Lee, K. H., L. M. Williams, A. Haig, E. Goldberg, and E. Gordon. 2001. An integration of 40 Hz gamma and phasic arousal: Novelty and routinization processing in schizophrenia. *Clinical Neurophysiology* 112(8): 1499–507.

Lekman, M., G. Laje, D. Charney, A. J. Rush, A.F. Wilson, A.J. Sorant, R. Lipsky, S.R. Wisniewski, H. Manji, F.J. McMahon, S. Paddock. 2008. The FKBP5-gene in depression and treatment response–an association study in the STAR*D-cohort. *Biological Psychiatry* 63(12): 1103–10.

Levinson, D. F. 2006. The genetics of depression: A review. *Biological Psychiatry* 60(2): 84–92.

Licinio, J., F. O'Kirwan, K. Irizarry, B. Merriman, S. Thakur, R. Jepson, S. Lake, K. G. Tantisira, S. T. Weiss, and M. L. Wong. 2004. Association of a corticotropin-releasing hormone receptor 1 haplotype and antidepressant treatment response in Mexican-Americans. *Molecular Psychiatry* 9(12): 1075–82.

Linka, T., B. W. Muller, S. Bender, G. Sartory, and M. Gastpar. 2005. The intensity dependence of auditory evoked ERP components predicts responsiveness to reboxetine treatment in major depression. *Pharmacopsychiatry* 38: 139–43.

Ludewig, K., M. A. Geyer, and F. X. Vollenweider. 2003. Deficits in prepulse inhibition and habituation in never-medicated, first-episode schizophrenia. *Biological Psychiatry* 54(2): 121–28.

Mandal, M. K., R. Pandey, and A. B. Prasad. 1998. Facial expressions of emotions and schizophrenia: A review. *Schizophrenia Bulletin* 24(3): 399–412.

Martin, S. D., E. Martin, S. S. Rai, M. A. Richardson, and R. Royall. 2001. Brain blood flow changes in depressed patients treated with interpersonal psychotherapy or venlafaxine hydrochloride: Preliminary findings. *Archives of General Psychiatry* 58: 641–48.

Mathalon, D. H., J. M. Ford, and A. Pfefferbaum. 2000. Trait and state aspects of P300 amplitude reduction in schizophrenia: A retrospective longitudinal study. *Biological Psychiatry* 47(5): 434–49.

McMahon, F. J., S. Buervenich, D. Charney, R. Lipsky, A.J. Rush, A.F. Wilson, A.J. Sorant, G.J. Papanicolaou, G. Laje, M. Fava, M.H. Trivedi, S.R. Wisniewski, H. Manji. 2006. Variation in the gene encoding the serotonin 2A receptor is associated with outcome of antidepressant treatment. *American Journal of Human Genetics* 78(5): 804–14.

Mulert, C., G. Juckel, M. Brunnmeier, S. Karch, G. Leicht, R. Mergl, H.J. Möller, U. Hegerl, O. Pogarell. 2007. Prediction of treatment response in major depression: Integration of concepts. *Journal of Affective Disorders* 98: 215–25.

Nahshoni, E., D. Aravot, D. Aizenberg, M. Sigler, G. Zalsman, B. Strasberg, S. Imbar, E. Adler, A. Weizman. 2004. Heart rate variability in patients with major depression. *Psychosomatics* 45: 129–34.

Neuchterlein, K. H., D. M. Barch, J. M. Gold, T. E. Goldberg, M. F. Green, and K. Heaton. 2004. Identification of separable cognitive factors in schizophrenia. *Schizophrenia Research* 72: 29–39.

Nishith, P., S. P. Duntley, P. P. Domitrovich, M. L. Uhles, B. J. Cook, and P. K. Stein. 2003. Effect of cognitive behavioral therapy on heart rate variability during REM sleep in female rape victims with PTSD. *Journal of Traumatic Stress* 16(3): 247–50.

Paddock S., G. Laje, D. Charney, A. J. Rush, A. F. Wilson, A.J. Sorant, R. Lipsky, S.R. Wisniewski, H. Manji, F.J. McMahon. 2007. Association of GRIK4 With outcome of antidepressant treatment in the STAR*D cohort. *American Journal of Psychiatry* 164: 1181–88.

Paige, S. R., D. F. Fitzpatrick, J. P. Kline, S. E. Balogh, and S. E. Hendricks. 1994. Event-related potential amplitude/intensity slopes predict response to antidepressants. *Neuropsychobiology* 30: 197–201.

Parwani, A., E. J. Duncan, E. Bartlett, S. H. Madonick, T. R. Efferen, R. Rajan, M. Sanfilipo, P. B. Chappell, S. Chakravorty, S. Gonzenbach,G.N. Ko, J.P. Rotrosen. 2000. Impaired prepulse inhibition of acoustic startle in schizophrenia. *Biological Psychiatry* 47(7): 662–69.

Peterson, B. S., V. Warner, R. Bansal, H. Zhu, X. Hao, J. Liu, K. Durkin, P. B. Adams, P. Wickramaratne, and M. M. Weissman. 2009. Cortical thinning in persons at increased familial risk for major depression. *Proceedings of the National Academy of Sciences* 106 (15): 6273–78.

Pier, M. P. B. I., W. Hulstijn, and B. G. C. Sabbe. 2004. Differential patterns of psychomotor functioning in unmedicated melancholic and nonmelancholic depressed patients. *Journal of Psychiatric Research* 38(4): 425–35.

Pizzagalli, D., R. D. Pascual-Marqui, J. B. Nitschke, T. R. Oakes, C. L. Larson, H.C. Abercrombie, S.M. Schaefer, J.V. Koger, R.M. Benca, R.J. Davidson. 2001. Anterior cingulate activity as a predictor of degree of treatment response in major depression: Evidence from brain electrical tomography analysis. *American Journal of Psychiatry* 158: 405–15.

Posternak, M. A. 2003. Biological markers of atypical depression. *Harvard Review of Psychiatry* 11 (1): 1–7.

The Royal Society. 2005. *Personalised medicine: Hopes and realities*. London: OML.

Russell, T. A., E. Reynaud, K. Kucharska-Pietura, C. Ecker, P. J. Benson, F. Zelaya, et al. 2007. Neural responses to dynamic expressions of fear in schizophrenia. *Neuropsychologia* 45(1): 107–23.

Salisbury, D. F., M. E. Shenton, A. R. Sherwood, I. A. Fischer, D. A. Yurgelun-Todd, M. Tohen, et al. 1998. First-episode schizophrenic psychosis differs from first-episode affective psychosis and controls in P300 amplitude over left temporal lobe. *Arch Gen Psychiatry* 55(2): 173–80.

Seminowicz, D. A., H. S. Mayberg, A. R. McIntosh, K. Goldapple, S. Kennedy, Z. Segal, and S. Rafi-Tari. 2004. Limbic-frontal circuitry in major depression: A path modeling meta analysis. *Neuroimage* 22: 409–18.

Sheline, Y. L., D. M. Barch, J. M. Donnelly, J. M. Ollinger, A. Z. Snyder, and M. A. Mintun. 2001. Increased amygdala response to masked emotional faces in depressed subjects resolves with antidepressant treatment: An fMRI study. *Biological Psychiatry* 50: 651–58.

Sheline, Y. I., M. H. Gado, and H. C. Kraemer. 2003. Untreated depression and hippocampal volume loss. *American Journal of Psychiatry* 160: 1516–18.

Shen, S., G. Lang, and C. Nakamoto. 2008. Schizophrenia-related neural and behavioral phenotypes in transgenic mice expressing truncated Disc1. *Journal of Neuroscience* 28: 10893–904.

Shimony, J. S., Y. I. Sheline, G. D'Angelo, A. A. Epstein, T. L. Benzinger, M. A. Mintun, R. C. McKinstry, and A. Z. Snyder. 2009. Diffuse microstructural abnormalities of normal-appearing white matter in late life depression: a diffusion tensor imaging study. *Biological Psychiatry* 66(3): 245–52.

Siegle, G. J., C. S. Carter, and M. E. Thase. 2006. Use of FMRI to predict recovery from unipolar depression with cognitive behavior therapy. *American Journal of Psychiatry* 163: 735–38.

Spencer, K. M., P. G. Nestor, R. Perlmutter, M. A. Niznikiewicz, M. C. Klump, M. Frumin, M.E. Shenton, R.W. McCarley . 2004. Neural synchrony indexes disordered perception and cognition in schizophrenia. *Proc Natl Acad Sci USA*. 101(49): 17288–93.

Streit, M., W. Wolwer, J. Brinkmeyer, R. Ihl, and W. Gaebel. 2001. EEG-correlates of facial affect recognition and categorisation of blurred faces in schizophrenic patients and healthy volunteers. *Schizophrenia Research* 49(1–2): 145–55.

Strik, W. K., T. Dierks, and K. Maurer. 1993. Amplitudes of auditory P300 in remitted and residual schizophrenics: correlations with clinical features. *Neuropsychobiology* 27(1): 54–60.

Symond, M. P., A. W. Harris, E. Gordon, and L. M. Williams. 2005. "Gamma synchrony" in first-episode schizophrenia: A disorder of temporal connectivity? *American Journal of Psychiatry* 162(3): 459–65.

Taylor, B. P., G. E. Bruder, J. W. Stewart, P.J. McGrath, J. Halperin, H. Ehrlichman, F.M. Quitkin. 2006. Psychomotor slowing as a predictor of fluoxetine nonresponse in depressed outpatients. *American Journal of Psychiatry* 163: 73–78.

Tsai, S. J., C. Y. Cheng, Y. W. Yu, T. J. Chen, and C. J. Hong. 2003. Association study of a brain-derived neurotrophic factor genetic polymorphism and major depressive disorders, symptomatology, and antidepressant response. *American Journal of Medical Genetics B; Neuropsychiatric Genetics*, 123: 19–22.

Uhlhaas, P. J., D. E. J. Linden, W. Singer, C. Haenschel, M. Lindner, K. Maurer, et al. 2006. Dysfunctional long-range coordination of neural activity during gestalt perception in schizophrenia. *Journal of Neuroscience* 26(31): 8168–75.

Venn, H. R., S. Watson, P. Gallagher, and A. H. Young. 2005. Facial expression perception: An objective outcome measure for treatment studies in mood disorders? *International Journal of Neuropsychopharmacology* 9: 1–17.

Villafuerte, S. M., K. Vallabhaneni, E. Sliwerska, F. J. McMahon, E. A. Young, and M. Burmeister. 2009. SSRI response in depression may be influenced by SNPs in HTR1B and HTR1A. *Psychiatric Genetics* 19(6): 281–91.

Weickert, T. W., T. E. Goldberg, A. Mishara, J. A. Apuda, B. S. Kolachana, M. F. Egan, and D. F. Weinberger. 2004. Catechol-O-methyltransferase val108/158met genotype predicts working memory response to antipsychotic medications. Biological Psychiatry 56(9): 677–682.

Whitford, T. J., T. F. Farrow, C. J. Rennie, S. M. Grieve, L. Gomes, J. Brennan, A. W. Harris, and L. M. Williams. 2006a. Longitudinal changes in neuroanatomy and neural activity in early schizophrenia. *Neuroreport* 18: 435–39.

Whitford, T. J., S. M. Grieve, T. F. D. Farrow, L. Gomes, J. Brennan, A. W. F. Harris, E. Gordon, and L. M. Williams. 2006b. Progressive gray matter atrophy over the first 2–3 years of illness in first-episode schizophrenia: A tensor-based morphometry study. *NeuroImage* 32: 511–19.

Whitford, T. J., S. M. Grieve, T. F. D. Farrow, L. Gomes, J. Brennan, A. W. F. Harris, E. Gordon, and L. M. Williams. 2007. Volumetric white matter abnormalities in first-episode schizophrenia: a longitudinal, tensor-based morphometry study. *American Journal of Psychiatry* 164: 1082–89.

Wicker, M., N. Jacobs, F. Peeters, C. Derom, P. Delespaul, and J. Van Os. 2007. Genetic risk of depression and stress-induced negative affect in daily life. *The British Journal of Psychiatry* 191: 218–23.

Williams, L. M. 2008. Voxel-based morphometry in schizophrenia: Implications for neurodevelopmental connectivity models, cognition and affect. *Expert Reviews in Neurotherapeutics* 8(7): 1049–65.

Williams, L. M., P. Das, A. W. Harris, B. B. Liddell, M. J. Brammer, G. Olivieri, D. Skerrett, M.L. Phillips, A.S. David, A. Peduto, E. Gordon. 2004. Dysregulation of arousal and amygdala-prefrontal systems in paranoid schizophrenia. *American Journal of Psychiatry* 161(3): 480–89.

Williams, L. M., P. Das, B. J. Liddell, A. H. Kemp, C. J. Rennie, and E. Gordon. 2006. Mode of functional connectivity in amygdala pathways dissociates level of awareness for signals of fear. *Journal of Neuroscience* 26: 9264–71.

Williams, L. M., J. M. Gatt, S. M. Grieve, C. Dobson-Stone, R. H. Paul, E. Gordon, and P. R. Schofield. 2010. COMTVal108/158Met polymorphism effects on emotional brain function are modulated by level of awareness and associated with negativity biases. *Neuroimage* (online early view; doi:10.1016/j.neuroimage.2010.01.084).

Williams, L. M., J. M. Gatt, A. Hatch, D.M. Palmer, M. Nagy, C. Rennie, N.J. Cooper, C. Morris, S. Grieve, C. Dobson-Stone, P. Schofield, C.R. Clark, E. Arns, R.H. Paul, E. Gordon. 2008. The INTEGRATE model of emotion, thinking and self regulation: Application to the "'paradox of aging.'" *J Integrative Neuroscience* 7: 367–404.

Williams, L. M., J. M. Gatt, P. R. P. R. Schofield, G. Olivieri, A. Peduto, E. Gordon. 2009. "Negativity bias" in risk for depression and anxiety: Brain-body fear circuitry correlates, 5-HTT-LPR and early life stress. *Neuroimage* 47(3): 804–14.

Williams, L. M., D. Palmer, B. J. Liddell, L. Song, and E. Gordon. 2006. The "when" and "where" of perceiving signals of threat versus non-threat. *NeuroImage* 31: 458–67.

Williams, L. M., A. J. Rush, S. H. Koslow, S. R. Wisniewski, N. Cooper, C. B. Nemeroff, A. Schatzberg, and E. Gordon. International Study to Predict Optimized Treatment for Depression (iSPOT-D), a Randomized Controlled Trial: Rationale and Design. *Trials Submitted.* (in submission)

Williams, L. M., T. J. Whitford, G. Flynn, W. Wong, B. J. Liddell, S. Silverstein, C. Galletly , A.W. Harris, E. Gordon. 2008. General and social cognition in first episode schizophrenia: Identification of separable factors and prediction of functional outcome using the IntegNeuro test battery. *Schizophrenia Research* 99(1–3): 182–91.

Williams, L. M., T. J. Whitford, E. Gordon, L. Gomes, K. J. Brown, and A. W. Harris. 2009. Neural synchrony in patients with a first episode of schizophrenia: Tracking relations with gray matter and symptom profile. *Journal of Psychiatry and Neuroscience* 34(1): 21–29.

Williams, L. M., T. J. Whitford, M. Nagy, G. Flynn, A. W. Harris, S. M. Silverstein, E. Gordon. 2009. Emotion-elicited gamma synchrony in patients with first-episode schizophrenia: A neural correlate of social cognition outcomes. *Journal of Psychiatry and Neuroscience* 34(4): 303–13.

Wolf, D. H., B. I. Turetsky, J. Loughead, M. A. Elliott, R. Pratiwadi, R. E. Gur, R.C. Gur. 2008. Auditory oddball fMRI in schizophrenia: Association of negative symptoms with regional hypoactivation to novel distractors. *Brain Imaging Behav* 2(2): 132–45.

Xu, J., M. T. Pato, C. D. Torre, H. Medeiros, C. Carvalho, V. S. Basile, A. Bauer, A. Dourado, J. Valente. M. J. Soares, A. A. Macedo, I. Coelho, C. P. Ferreira, M. H. Azevedo, F. Macciardi, J. L. Kennedy, and C. N. Pato. 2001. Evidence for linkage disequilibrium between the alpha 7-nicotinic receptor gene (CHRNA7) locus and schizophrenia in Azorean families. *American Journal of Medical Genetics* 105(8): 669–74.

Zahn, T. P., L. K. Jacobsen, C. T. Gordon, K. McKenna, J. A. Frazier, and J. L. Rapoport. 1997. Autonomic nervous system markers of psychopathology in childhood-onset schizophrenia. *Archives of General Psychiatry* 54(10): 904–12.

Zanardi, R., F. Benedetti, D. Di Bella, M. Catalano, and E. Smeraldi. 2000. Efficacy of paroxetine in depression is influenced by a functional polymorphism within the promoter of the serotonin transporter gene. *Journal of Clinical Psychopharmacology* 20: 105–07.

Section 2

Personalized Medicine and Mental Disorders

4

Does fMRI Have A Role in Personalized Health Care for Psychiatric Patients?

Alex Fornito, MD and Edward T. Bullmore, MD

Introduction

The development and widespread application of functional magnetic resonance imaging (fMRI) was met with great enthusiasm in clinical and academic psychiatry. The ability to correlate patterns of brain activity with different cognitive and emotional states in vivo, with a spatial resolution on the order of a few millimeters, offered an unprecedented capacity to characterize the neural correlates of psychiatric dysfunction and response to treatment interventions.

Nevertheless, after nearly two decades of fMRI research, many of these lofty hopes remain unfulfilled and fMRI is far from being a routinely used clinical tool in psychiatry. However, recent conceptual and methodological advances provide a sufficient basis for cautious optimism concerning the future clinical applicability of fMRI, and suggest that it may yet prove to be a powerful tool, as medical practice moves toward a more personalized model of care. In this chapter, we focus on some areas of clinical practice in which we think fMRI has the greatest potential to make a significant impact. We begin with a brief overview of the fundamentals of fMRI acquisition, experimental design, and analysis. We then focus on recent developments with the potential to advance personalized treatment in three key clinical domains: clinical diagnosis, prediction of illness, and treatment monitoring. Finally, we speculate on the future role that fMRI may play in personalized psychiatric treatment, and outline a research strategy that might see these speculations realized.

Basic Principles of fMRI

The Biophysical Basis of fMRI Measurements

Functional MRI measures cerebral haemodynamics as an indirect marker of neuronal function. The blood-oxygenation-level-dependent (BOLD) contrast, the most widely used fMRI measure, has its physical basis in the differing magnetic properties of oxyhaemoglobin and deoxyhaemoglobin. Neuronal activity increases oxygen metabolism, provoking a concomitant increase in local oxygenated blood flow and volume to supply metabolic demands. The increase in blood supply typically exceeds that required purely for neuronal oxygen consumption. Since deoxyhaemoglobin is paramagnetic (i.e., it distorts a magnetic field it is placed in) and oxyhaemoglobin is diamagnetic (i.e., it has no effect on a surrounding magnetic field, and correspondingly, tissue relaxation parameters), the ratio of oxygenated to de-oxygenated haemoglobin can be used as a marker of neuronal activity in the region (Ogawa et al. 1990). The highly localized nature of these responses allows cerebral haemodynamic changes to be mapped with a spatial resolution of a few millimeters in whole-brain acquisition protocols.

The precise relationship between neuronal signaling and corresponding haemodynamic responses, termed *neurovascular coupling*, is complex and remains a topic of active investigation. Combined electrophysiological and BOLD studies in monkeys have suggested that, while measures of both local field potentials and multiunit activity are correlated with the BOLD response, the association with the former is stronger and more robust (Logothetis et al. 2001; Goense and Logothetis 2008). This suggests that the BOLD response may predominantly index local input or processing, rather than spiking output, although the relationship may be context-specific (Maier et al. 2008). For most purposes, fMRI results are interpreted in terms of regional activation increases or decreases, and the underlying neural mechanisms are often overlooked. However, BOLD and electrophysiological measures can sometimes produce discrepant results (Bartels et al 2008), largely because they index different aspects of brain physiological function (see Logothetis 2008 for a detailed discussion). As such, any inferences regarding underlying neural mechanisms drawn from fMRI findings must be made with care, and cross-validated using other techniques. In general, however, fMRI has proved to be a useful tool for mapping brain activity changes in health and disease, and remains the most powerful tool available for identifying spatially localized activation differences in living humans.

fMRI Design and Analysis

Broadly, fMRI studies can be divided into two categories: those examining task-related activation, and those examining brain activity fluctuations occurring in the absence of a specific task. Task-related activation studies represent the bulk of fMRI research, and represent the basis of cognitive neuroscience. In this work, study participants perform a task in the scanner, during which time repeated scans are acquired (usually every 2 to 3 seconds) to measure the BOLD signal changes induced by different task conditions. The result is a time series of activity fluctuations for each voxel in the measured brain volume. Following acquisition, the images pass through one of a variety of pre-processing pipelines. These commonly involve correction for various scanner- and measurement-related artifacts (such as distortions in the scanner's magnetic field, or differences in the time at which each slice making up a given functional volume is acquired); correction for participant movement during the acquisition; spatial realignment of each individual's images to a standard template space to enable comparison of datasets across participants; and some degree of spatial smoothing to increase the signal-to-noise ratio. While the precise steps implemented, and their order of implementation, may vary across studies, the motivation behind such pipelines is to eliminate as many known artifacts and boost the signal-to-noise ratio as much as possible, given that the signal changes measured using fMRI are typically small (i.e., 3 to 5 percent).

Once pre-processing is complete, a model is fit to the imaging data that attempts to identify brain activity related to the task conditions of interest. Typically, this involves regressing the time course of stimulus presentations against each voxel's measured time course. Voxels in which this regression is significant, after appropriate correction for multiple comparisons (e.g., Worsley and Friston 1995; Poline et al. 1997; Bullmore et al. 1999; Genovese et al. 2002), are deemed to be activated by the task condition (i.e., they show activity fluctuations that are time-locked to that particular task manipulation). As such, determination of whether a brain region is active or not in response to a given stimulus is based on statistical criteria, meaning that the results are susceptible to the usual array of factors affecting statistical inferences, such as differences in analysis parameters, sample size, and power. The design can then be extended to examine differences between groups (e.g., patients versus controls) in regional activation.

The second category of fMRI studies involves measuring BOLD signal fluctuations while participants lie quietly in the scanner without performing any explicit task, a technique often referred to as resting-state fMRI (rs-fMRI). Interest in this approach arose from the initial observation that activity fluctuations in regions of the left-motor cortex were correlated with homologous regions in the right hemisphere, in addition to medial motor areas, when no overt motor behavior was being performed (Biswal et al. 1995). Subsequently, it was discovered that spontaneous BOLD signal fluctuations (i.e., fluctuations occurring in the absence of any specific task) are highly organized across the brain, recapitulating well-known functional networks, such as the dorsal and ventral attentional systems, sensorimotor, visual and cerebellar networks, the cingulo-opercular system, and the so-called default mode network (Fox et al. 2005; Damoiseaux et al. 2006; De Luca et al. 2006; Fox et al. 2006). Typically, the strongest correlations (i.e., functional connectivity) are observed at relatively low frequencies (i.e., <0.1 Hz). These so-called resting-state networks (RSNs) are stable and robust across time and individuals (Damoiseaux et al. 2006; De Luca et al. 2006; Shehzad et al. 2009), suggesting they do not merely reflect a neural correlate of unconstrained mind-wandering (e.g., Mason et al. 2007). Rather, these correlated fluctuations have been interpreted as reflecting an intrinsic property of brain organization (Fox and Raichle 2007), potentially related to more enduring aspects of brain physiology.

Both task-based and task-free fMRI have strengths and weaknesses in terms of their potential clinical applications. In task-based fMRI, experimental manipulations are tightly controlled. This allows specific hypotheses to be tested about the relationship between cognitive and emotional processes and evoked brain activity. However, in case-control studies, there is no guarantee that patients will perform a given task in the same manner as controls; they may use different strategies or experience different levels of motivation, each of which can cause differences in brain activity that are not necessarily pathophysiologically informative (e.g., see discussion in Manoach 2003). Additionally, the task used will determine the type of brain changes detected, as different tasks vary in the degree to which they probe the function of different brain regions. Thus, brain dysfunction in some regions may go undetected simply because those regions were not sufficiently engaged by the experimental paradigm employed.

One strength of rs-fMRI is that confounds associated with group differences in cognitive strategy and motivation are minimized, as no explicit task is performed. Additionally, multivariate, whole-brain techniques afford investigation of both local and global brain dynamics that are unconstrained by a particular experimental paradigm. However, the psychology and phenomenology of the so-called resting-state is ill-defined meaning that any conclusions from such data regarding underlying cognitive or emotional processes must be drawn cautiously (Morcom and Fletcher 2007).

This limitation, to a large extent, is mitigated if the primary goal is to identify biomarkers with discriminative or predictive utility, which is the main concern of research attempting to exploit the potential clinical applications of fMRI. In such work, the phenomenological correlates of changes in brain activity are less important than the efficacy of a given brain measure in differentiating between patients and controls (or different patient groups), or offering some other clinically useful information. Indeed, the possibility that resting-state networks reflect an enduring and intrinsic property of brain organization suggests they may provide more reliable biomarkers for psychiatric disorders than task-based fMRI. Some suggestive evidence in favor of this view was provided by a recent study of Alzheimer's disease (Fleisher et al. 2009), where resting-state measures provided better discrimination between healthy controls and people at elevated risk for the disease than group activation differences elicited by a memory task. A more detailed discussion of some of the issues associated with using resting-state fMRI in case-control designs can be found in Fornito and Bullmore (2010).

fMRI Applications in Psychiatry

fMRI and Psychiatric Diagnosis

The application of fMRI to the study of psychiatric disorders was initially met with hopes that it could be used to identify patterns of brain dysfunction that could provide diagnostically useful information; i.e., that it could be used to identify pathognomonic markers of distinct categories of psychiatric illness. Much of the work since then has suggested that fMRI-measured brain activation differences between patient and control groups are quantitative rather than qualitative, and few diagnostically useful markers have been identified. However, recent methodological advances have renewed interest in fMRI's potential to identify such characteristics.

One factor limiting the identification of useful diagnostic markers has been an overreliance on the case-control design. In its simplest and most widely used form, this approach involves comparing mean differences between patient and control groups. Any differences between the groups are assumed to reflect some pathophysiological characteristic of the illness, after appropriate controls for various confounds are made. Functional MRI studies using this design have been useful in identifying the key brain regions affected by psychiatric disease, and recent meta-analyses of this work have begun to identify the areas that are most consistently implicated in each disorder. For example, one meta-analysis of forty-one fMRI studies examining executive functions in schizophrenia found that patients showed reduced activation in lateral and medial prefrontal, lateral parietal, thalamic and striatal regions (Minzenberg et al. 2009). Meta-analyses of fMRI studies of patients with obsessive-compulsive disorder (OCD) have confirmed involvement of fronto-striatal pathology in the disorder, in addition to dysfunction in several additional regions, such as parietal and temporal areas, that are worthy of further investigation (Menzies et al. 2008; Rotge et al. 2008). Meta-analysis of fMRI and tomographic studies in major depression found that two networks showed consistent activation differences between patients and controls during rest, following pharmacological treatment and when viewing happy or sad faces (Fitzgerald, Srithiran et al. 2008). The first comprised the dorsal and pregenual cingulate, bilateral middle frontal, insula, and superior temporal regions, and was associated with decreased activity at rest, a relative lack of activation in response to viewing sad faces, and increased activation following selective serotonin reuptake inhibitor (SSRI) therapy. The second network, primarily comprising inferior frontal cortex and basal ganglia, showed increased activity at rest, hyperactivity during viewing of sad faces, and reduced activity following SSRI treatment.

Interpretation of the meta-analytic findings must be tempered, however, because the results of case-control studies are contingent on the type of cognitive paradigm used. For example, one meta-analysis found that the activation deficits in ventrolateral prefrontal areas shown by schizophrenia patients during encoding conditions of episodic memory paradigms were not observed in studies where patients were provided with encoding strategies (Ragland et al. 2009). Thus, inferences regarding specific loci of brain dysfunction in patients will depend on the cognitive context of the experiment, even within the same cognitive domain. Similarly, whether patients show an increase or decrease in task-related activity relative to controls can depend on whether or not there are also concomitant differences in behavioral measures of task performance (Manoach 2003; Van Snellenberg, Torres, and Thornton 2006). As previously stated, rs-fMRI can minimize some of these confounds, and early findings suggest that there are differences between patients and controls in various resting-state network parameters. These include the degree of connectivity between regions (Cherkassky et al. 2006; Liang et al. 2006; Greicius et al. 2007; Zhou et al. 2007; Church et al. 2009), the amplitude of low-frequency BOLD signal fluctuations (Zang et al. 2007), the degree of regional homogeneity (spatial autocorrelation) in signal fluctuations (Shi et al. 2007; Yao et al. 2009) and the topological organization of resting-state connectivity networks (Liu et al. 2008). Correlations among these measures and clinical indices have been reported (Greicius et al. 2007; Liu et al. 2008; Yao et al. 2009), supporting the clinical relevance of rs-fMRI indices, although further work will be required to better characterize the potential clinical utility of measures derived from rs-fMRI.

The primary factor limiting the clinical applicability of findings obtained using the case-control design is that this work is primarily concerned with characterizing mean differences between populations. This makes the case-control design well-suited to identifying brain regions showing abnormal activity patterns in a patient group, or how dysfunctional brain activity in certain regions relates to specific patient deficits. However, these issues are distinct from those concerned with establishing direct clinical utility, which usually determine whether fMRI can help in the diagnosis or treatment of a given disorder. As diagnostic and treatment decisions are made on a case-by-case basis, clinically relevant fMRI research must seek to understand individual differences within and among populations, rather than mean population differences. Indeed, a better understanding of individual differences in any biomedical context represents a critical foundation for the enhancement of personalized medicine.

Recent methodological developments have provided a quantitative framework within which fMRI has been used to understand diagnostically relevant individual differences in various patient populations. Principal among these developments has been the application of machine-learning algorithms and classification statistics to fMRI datasets (Haynes and Rees 2006; Demirci et al. 2008; Formisano et al. 2008). Typically, this approach involves the use of an algorithm that attempts to learn the relationship between changes in BOLD activity and specific cognitive, perceptual, or emotional states, such as the brain activity associated with viewing faces versus houses. This learned relationship is then used to predict a given state induced by a new set of stimuli: for example, based on a particular pattern of BOLD signal changes, the algorithm will predict whether a participant is viewing a face or a house. Importantly, these predictions are made at the individual level, which is of utmost importance in any clinical application, and the methods are easily extended to classify groups of individuals rather than specific cognitive states. In this case, the algorithm learns the patterns of brain activity associated with either patients or controls, and then attempts to predict whether a new participant is a patient or control based on the measured brain activity.

The application of these techniques to the classification of psychiatric disorders is a nascent field, and various methodological issues need to be addressed (see Demirci et al. 2008 for a review). Nonetheless, the results published to date have been promising. Fu and colleagues (2008) used machine-learning techniques to classify healthy controls and patients with major depression, using brain activity measured while participants viewed emotional faces. They were able to classify patients with 84 percent sensitivity and 89 percent specificity, corresponding to an overall accuracy rate of 86 percent. Using activation during an auditory oddball task, Demirci and colleagues (2008) were able to classify schizophrenia patients and healthy controls with 80 to 90 percent accuracy following a multivariate decomposition (independent component analysis) of the key brain networks subserving task performance. A separate study by the same group reported comparable accuracy rates, also reporting that classification accuracy was consistent across different scanning sites, but varied somewhat depending on the specific cognitive paradigm used (Demirci et al. 2008). While this finding reiterates concerns about the dependence of classification statistics on the specific task used, preliminary work suggests that classification accuracy can be improved by appropriately combining activation information from different tasks (Michael et al. 2008), which may prove useful in reducing the biases associated with reliance on one specific experimental paradigm. Notably, group classification based on rs-fMRI data appears to be just as accurate as task-based fMRI. Jafri and Calhoun (2006) reported a classification accuracy of approximately 76 percent for discriminating between schizophrenia patients and controls based on resting-state connectivity measures, while Shi and colleagues (2007) reported 80 percent accuracy using a measure of the regional homogeneity of spontaneous BOLD signal fluctuations. A similar approach applied to children with attention-deficit hyperactivity disorder (ADHD) yielded a correct classification rate of 85 percent (Zhu et al. 2008).

One limitation of this work is that only one patient group has been classified as different from controls. Such distinctions tend to be easier than those faced by practitioners in a psychiatric clinic, where patients can present with one or more of a variety of different disorders. As such, a more appropriate test of the clinical applicability of fMRI-based classification methods would involve making distinctions between several different, yet partially overlapping, syndromes. To our knowledge, only one fMRI study has examined classification accuracy with two patient groups. Calhoun and colleagues (2008) attempted to classify schizophrenia patients, bipolar patients, and healthy controls based on functional network connectivity measured during performance of an oddball task. On average, sensitivity and specificity rates were 90 percent and 95 percent, respectively, indicating that high accuracy rates are obtainable even when multiple patient groups are investigated.

To summarize, the high accuracy rates achieved thus far in fMRI studies of diagnostic classification are encouraging, although the results are preliminary. Further work is required to test the robustness of the findings to the classification algorithm, experimental design, and scanner used, as well as to ascertain the accuracy of these methods in more realistic clinical settings, where diagnostic distinctions must be made in diffuse and heterogeneous samples presenting with a variety of overlapping syndromes. The ultimate goal of this work would be to identify fMRI biomarkers that can be used to diagnose, or help diagnose, a presenting patient. One factor limiting progress toward this goal is the uncertain validity of current diagnostic constructs, which do not provide biologically valid gold standards against which these methods can be evaluated. Consequently, more refined diagnostic systems may be necessary before the full clinical potential of these techniques can be realized. Indeed, fMRI-based classification research may play an important role in identifying more biologically valid diagnostic constructs, given that it can be used to quantitatively cluster patients based on key clinical indicators, such as outcome or response to specific treatments. This, combined with other measures derived from molecular genetics or peripheral biomarkers, may promote a new era of diagnostic accuracy in psychiatry.

fMRI and Illness Prediction

An important goal for neurobiological researchers in psychiatry is to identify markers that can predict who is likely to become affected before the transition to illness occurs. This is based on accumulating evidence that early intervention, at or before illness onset, can ameliorate outcomes or perhaps even prevent psychiatric morbidity (Yung et al. 1996; 1998; Berk, Hallam, and McGorry 2007; McGorry et al. 2007; Hetrick et al. 2008; McGorry et al. 2009). Typically, these vulnerability markers are studied in so-called high-risk samples: samples of individuals deemed to be at elevated risk for a disorder either by virtue of a familial liability, or the expression of sub-threshold symptoms known to precede illness onset.

The high-risk design has been most widely applied to predicting the onset of psychosis. In this literature, two types of high-risk strategies have been used. In one, termed the genetic high-risk strategy, unaffected family members of patients are investigated for evidence of any abnormalities resembling those seen in their affected relatives. If such abnormalities are apparent, it is assumed that they are caused by a genetic susceptibility to the illness shared between the affected and unaffected participants (Gottesman and Gould 2003). In the second, termed the symptomatic high-risk design, individuals displaying sub-threshold symptoms known to increase risk for illness onset are recruited, regardless of their family history. This approach is based on the idea that many psychiatric disorders are preceded by a prodromal period characterized by mild, sub-threshold symptomatology and/or a change in daily functioning (Yung and McGorry 1996). The genetic high-risk design is useful for providing information that might facilitate the identification of risk genes, but is limited because transition rates to frank illness in unaffected relatives are relatively low (e.g., ~10 percent for a first-degree family member of a schizophrenia proband). Transition rates in symptomatic high-risk samples tend to be higher, but these samples often comprise individuals with and without family histories, making the genetic basis of any putative risk markers difficult to ascertain unless large samples are employed. Nonetheless, both techniques have shown that fMRI may provide biomarkers useful for identifying individuals at risk of illness onset.

The disorder most widely studied using the high-risk design is schizophrenia. Genetic high-risk studies have found that unaffected relatives of patients show altered activity in a number of brain regions in a manner that is similar to that seen in patients. For example, one systematic review of twenty-four studies found that, regardless of whether genetic or symptomatic high-risk samples were studied, medium-to-large effect sizes were consistently observed in contrasts of prefrontal activation with healthy controls (Fusar-Poli et al. 2007). This finding, however, may have reflected a bias towards most of the included studies using tests of executive function designed to tap prefrontal processes. A separate review of twenty studies found evidence for consistent increases in activation of ventrolateral prefrontal and right parietal cortices, with two-thirds of studies showing activation differences in unaffected relatives compared to controls in cerebellar, thalamic, and striatal and lateral prefrontal and temporal regions (Macdonald et al. 2008). Such findings are consistent with evidence that schizophrenia is primarily characterized by abnormalities in a network of frontal, striatal, limbic, temporal, and cerebellar regions (Ellison-Wright et al. 2008; Fornito et al. 2009), although considerable variability across studies has been observed. Part of the reason for this variability is that diagnostic outcome information has not been obtained in most of this research, meaning that the proportion of high-risk individuals that subsequently developed an illness is unknown. It is therefore unclear whether the abnormalities identified represent risk for incipient illness onset, or characteristics of a more generalized at-risk state.

To our knowledge, the Edinburgh High Risk Study is the only genetic high-risk study to have obtained diagnostic outcomes. An early attempt to predict schizophrenia onset using fMRI was encouraging, with a combination of activity changes in parieto-occipital regions during performance of a sentence completion task predicting schizophrenia onset with very high positive and negative predictive power (.80 and 1.0, respectively) (Whalley et al. 2006). However, only four participants made the transition to schizophrenia in this study. In a more recent longitudinal study, the same group found that high-risk individuals who developed psychotic symptoms during an 18-month follow-up period showed increased left-temporal activation during the sentence completion task, relative to high-risk individuals who did not develop symptoms during the follow-up period (Whalley et al. 2009). This suggests that brain changes occurring during the transition to illness may also be used in the prediction of which specific individuals will make the transition to illness, an idea that is consistent with structural MRI evidence for dynamic and progressive brain abnormalities before, during, and after the onset of psychosis (Pantelis et al. 2003; 2007; Sun et al. 2008; 2009).

Several studies of symptomatic high-risk samples have been published. The majority have focused on structural MRI measures, which have generally shown predictive utility, including some diagnostic specificity (Garner et al. 2005; Fornito et al. 2008). One recent study employing machine-learning algorithms and classification statistics reported accuracies exceeding 85 percent when classifying either controls, high-risk individuals who made the transition to psychosis, or high-risk individuals who did not make the transition, based on neuroanatomical indicators (Koutsouleris et al. 2009). To our knowledge, there have been no fMRI studies of symptomatic high-risk individuals in which the diagnostic outcomes of the sample were ascertained, although several group comparisons of baseline characteristics seen in these high-risk individuals and healthy controls have been published. In general, the results have varied depending on the task used. Studies using attentional or working memory paradigms have generally found differences in fronto-temporal or fronto-parietal systems (Morey et al. 2005; Benetti et al. 2009; Broome et al. 2009; Crossley et al. 2009), while one study of face processing found activation differences in posterior cortical regions (Seiferth et al. 2008). Such differences illustrate how task selection can greatly affect the results and consequent conclusions regarding pathophysiological mechanisms and useful biomarkers. Accordingly, it may be necessary to combine information from different tasks to help improve prediction accuracy (e.g., Michael et al. 2008). To our knowledge, no rs-fMRI studies have been conducted in high-risk samples, so it is unclear whether resting-state measures may further enhance prediction.

One relatively recent extension of the high-risk design has involved determining whether baseline measures of fMRI can predict the expression of psychiatric symptoms under pharmacological challenge. This design is predicated on the assumption that the neurobiological mechanisms mediating individual variability in vulnerability to the effects of various psychotropic agents may parallel those mediating individual variability in susceptibility to psychiatric disease. The validity of the results crucially depends on the validity of the pharmacological model of the disorder of interest. Initial results have been encouraging. For example, one study of healthy volunteers found that individual differences in baseline levels of right prefrontal activation during an associative learning task correlated with the severity of delusions and perceptual aberrations experienced following administration of ketamine, an antagonist of the *N*-methyl-*D*-aspartate receptor (NMDAR) and a potent psychotomimetic agent (Corlett et al. 2006). A similar pattern of results was observed in first-episode delusional patients (Corlett et al. 2007), supporting the validity of the pharmacological model. In a subsequent study, the same group found that individual differences in baseline activation during performance of three different tasks correlated with the expression of different symptoms under ketamine in a specific manner. Negative symptoms were correlated with fronto-thalamic responses elicited by a working memory task and bilateral prefrontal responses evoked by an attentional task, while fronto-temporal activation during a language comprehension task predicted thought disorder and auditory illusions under ketamine

(Honey et al. 2008). While these findings again illustrate how fMRI findings can be sensitive to the precise experimental paradigm used, they also highlight the potential utility of fMRI for uncovering differing vulnerabilities to the varied symptoms associated with schizophrenia. Pharmacological challenges using different agents to investigate other disorders with fMRI have also been published (Van der Veen et al. 2007), although none have attempted to predict the effects of the drug based on individual differences in baseline activity levels.

To summarize, the available findings suggest that individual differences in brain activity levels, as measured with fMRI, can help to predict which specific individuals are at heightened risk of developing an illness. To date, the majority of the work has focused on psychosis, but the high-risk design could theoretically be extended to other disorders in a similar manner. The major difficulty in executing such studies is that large samples are required to ensure that adequate numbers of people make the transition to illness, to provide sufficient reliable data. Indeed, relatively small sample sizes have been the major methodological limitation of most high-risk research to date, and, as such, the findings discussed above must be viewed as tentative until replicated in larger samples. Nonetheless, the relatively large effects reported thus far (e.g., Whalley et al. 2006) suggest that this is an area worthy of further investigation.

fMRI and Treatment Monitoring

An emerging literature suggests that fMRI may be a powerful tool in facilitating more effective psychiatric treatment planning and management. As yet, there are few guidelines to advise clinicians on which specific treatment should be prescribed to a given patient, and patients will vary in the degree to which they respond to such treatments. Consequently, the efficacy of various treatments is often explored in a trial-and-error manner, which can prolong morbidity and cause considerable distress for patients and their caregivers. The considerable delay between initiation and response for many psychiatric treatments (usually three to four weeks) underscores the need to identify objective markers that suggest a limited range of specific treatments to which a given individual might respond. This represents a core goal of the personalized health care approach.

Functional MRI has shown promise as a predictor of treatment response and clinical outcome across a range of psychiatric disorders. In general, studies attempting to identify fMRI predictors of treatment response employ a repeated-measures design, where a baseline pre- or early-treatment assessment is undertaken before a subsequent assessment is made some time (usually two to four weeks) after treatment has commenced. Baseline brain activation measures are then tested to see whether they can predict symptom improvement.

Perhaps the disorder most studied using this design has been unipolar depression (see Kemp et al. 2008 for a detailed review). This emphasis followed early positron mission tomography (PET) research suggesting that pre-treatment activity in the anterior cingulate cortex, particularly in the pre- and sub-genual regions, was predictive of treatment response (Mayberg et al. 1997). The fMRI findings have been consistent with this work, with most studies converging to suggest that baseline anterior cingulate activation is indeed predictive of symptom amelioration following treatment with either fluoxetine (Chen et al. 2007), venlafaxine (Davidson et al. 2003), generic antidepressant therapy (Keedwell et al. 2009a; Keedwell et al. 2009b), or cognitive-behavioral therapy (CBT) (Siegleet al 2006). Moreover, these treatments typically normalize baseline activity differences in the region (Fu et al. 2004).

Functional MRI studies of treatment-response predictors in depression have generally contrasted brain activity elicited by viewing sad or negative emotional stimuli with that evoked by neutral emotional conditions. It is therefore unclear whether the specificity of the findings to the anterior cingulate cortex is due to a reliance on this particular emotional paradigm. For example, one recent study of schizophrenia patients found that baseline activity in fronto-striatal and cerebellar regions during performance of a working memory task predicted response to CBT (Kumari et al. 2009). Working memory tasks are known to probe fronto-striatal function, whereas affective tasks are known to activate limbic regions, such as the anterior cingulate cortex and amygdala. As such, it is unclear whether fronto-striatal activity during working memory performance in depressed patients would be as good a predictor of treatment response as cingulate activation during affective stimulation, or vice versa for schizophrenia patients. In depression, the convergence between task-based fMRI findings and resting PET and SPECT studies suggest that the predictive utility of cingulate activity may indeed generalize across various experimental manipulations, but further work is required to assess the specificity of the findings to the task used and disorder under investigation. One recent study of people with post-traumatic stress disorder (PTSD) found that baseline anterior cingulate and amygdala activity during an affective task similar to that commonly used in studies of depression was predictive of outcome following CBT, an effect that could not be explained simply by amelioration of depressive symptoms in this group. Thus, activity in these regions may represent a generic marker of one's emotional adaptability or capacity to respond to treatment, regardless of the particular diagnosis.

One potential criticism is that fMRI-based predictors of treatment response may be secondary to differences in clinical or psychosocial parameters that distinguish responders from nonresponders, such as baseline differences in symptom severity or cognitive ability. This is an important criticism that is yet to be comprehensively addressed, although several studies in depressed patients that have attempted to control for such confounds indicate that fMRI offers predictive power over and above that provided by other variables (reviewed in Kemp et al. 2008). This suggests that fMRI may indeed provide valuable information for developing appropriate treatment strategies. However, much of the work to date has been correlational, and several methodological shortcomings must be overcome before firm conclusions can be drawn regarding the clinical utility of the technique. First, as previously stated, the dependency of the findings on the specific task used, and disorder studied, must be better understood. Second, the sensitivity, specificity, and positive and negative predictive power of these measures must be quantified. To this end, a powerful way forward would involve using machine learning and multivariate classification techniques to quantitatively classify treatment responders and nonresponders using baseline characteristics. Third, all studies to date have only examined treatment response to one specific intervention. A comparison of predictors of response to different types of treatments (e.g., CBT vs. fluoxetine vs. venlafaxine, etc.) will bear the greatest clinical relevance, as clinicians are typically faced with deciding on one specific treatment from a pool of several possible candidates. Such work will undoubtedly require large samples for sufficient power to make these distinctions, but will ultimately be necessary if fMRI is to become routinely used for clinical investigation in psychiatry. Table 4-1 provides a summary of the status of fMRI brainmarkers for diagnosis and treatment.

The Future Role of fMRI in Personalized Psychiatric Care

In recent years, the initial exuberance surrounding the potential clinical applications of fMRI in psychiatry has been replaced by guarded skepticism. In this chapter, we have considered some of the reasons why the full clinical potential of fMRI has not been realized,

Table 4.1a Markers for Diagnosis: Status and Future

Marker	*Disorder*	*Specificity*	*Sensitivity*	*First study*	*Replications*	*Next steps*
Brain activation during affective task performance	MDD	89%	84%	Fu et al. (2008)	n/a	Replication with a prospective sample; specificity relative to other patient groups.
Brain network functional connectivity during oddball task	SZ	Up to 100% for some parameter combinations	Up to 100% for some parameter combinations	Demirci, Clark & Calhoun (2008)	n/a	As above.
Brain network functional connectivity during a range of tasks	SZ	Up to 100% for some parameter combinations	>70% for some parameter combinations	Demirci et al. (2008)	n/a	As above.
Brain activation during a range of tasks	SZ	n/a	n/a	Michael et al. (2008)	n/a	Need diagnostic accuracy statistics.
Resting-state functional brain connectivity	SZ	n/a	n/a	Jafri & Calhoun (2006)	n/a	As above.
Regional homogeneity of resting-state functional connectivity	SZ	n/a	n/a	Shi et al. (2007)	n/a	As above.
Regional homogeneity of resting-state functional connectivity	ADHD	91%	78%	Zhu et al. (2008)	n/a	Replication with a prospective sample; specificity relative to other patient groups.
Brain network functional connectivity during oddball task	SZ & BD	95%	90%	Calhoun et al. (2008)	n/a	Replication with a prospective sample; specificity relative to other patient groups.

MDD=Major depressive disorder; SZ=Schizophrenia; ADHD=Attention deficit hyperactivity disorder; BD=Bipolar disorder.

Table 4.1b Markers for Treatment: Status and Future

Marker	*Disorder*	*Specificity*	*Sensitivity*	*First study*	*Replications*	*Next steps*
Brain activation during affective task performance	MDD	n/a	n/a	Davidson et al. (2003)	Chen et al. (2007); Keedwell et al. (2009a; 2009b); Siegle, Carter, and Thase (2006)	Need diagnostic accuracy statistics; characterize how results depend on task used, disorder studied and treatment administered.
Brain activation during working memory performance	SZ	n/a	n/a	Kumari et al. (2009)	n/a	As above.
Brain activation during affective task performance	PTSD	n/a	n/a	Bryant et al. (2007)	n/a	As above.

MDD=Major depressive disorder; SZ=Schizophrenia; PTSD=Post-traumatic stress disorder.

and have highlighted recent research that offers reasons for optimism about the clinical role fMRI may play in the future.

We propose that a major reason why the full clinical potential of fMRI has not been exploited is due to conceptual rather than technological reasons. The persistent focus on mean patterns of brain activation and mean case-control differences, while facilitating the development of theoretical models of neurobiological dysfunction in psychiatric disorders, has limited the clinical applicability of the findings. In part, this is because studies of mean effects typically treat inter-individual variability as a source of noise that needs to be minimized. In clinical settings: however, inter-individual variability is one of the most prominent features of a patient's presentation, and represents perhaps the greatest challenge to be overcome for accurate diagnosis and effective treatment. Thus, a greater understanding of individual differences is a necessary precondition, not only for realizing the full clinical potential of fMRI, but also for developing personally tailored health-care programs.

Important steps toward these goals have already been initiated. Statistically, this is reflected by a shift away from traditional comparisons of group mean differences, toward an emphasis on classification statistics and correlational analyses, as exemplified by the work discussed in this chapter. Experimentally, there is greater emphasis on asking more clinically relevant questions. That is, rather than asking, "In brain region Y, does patient group X differ from healthy individuals?" researchers are now asking questions such as, "Can fMRI markers be used to predict illness onset or treatment response?" These classification-based experimental designs and methods will play a central role in the further validation of fMRI as a clinically useful tool, although several relevant issues will need to be addressed.

First, samples much larger than those currently studied will be necessary, to better characterize the true variability of patient populations. Second, the samples studied will need to include people with a range of different diagnoses, to better emulate actual clinical settings. Third, a concerted effort is required to determine the degree to which variations in experimental paradigms affect findings. Researchers have understandably focused most of their efforts on key brain regions or experimental tasks thought to be relevant to the core psychological dysfunction of the disorder under investigation. So, for example, tasks requiring processing of emotional facial expressions have been widely used to probe affective function in depression, and functional abnormalities in cortico-limbic systems have emerged as a potential neural basis for such deficits. However, there is also mounting evidence for dysfunction of prefrontal regions during performance of working memory and cognitive control tasks in patients with depression (e.g., Fitzgerald et al. 2008), with such changes showing some diagnostic utility in preliminary classification studies (Marquand et al. 2008). These findings suggest that brain dysfunction in psychiatric disorders may not be confined to a circumscribed set of regions. Rather, functional brain changes may be expressed in different ways, depending on the cognitive and emotional contexts. As such, further work is required to identify which brain changes and which experimental contexts provide the highest classification accuracy, either for diagnostic, predictive, or treatment planning purposes. An important goal for such work will also involve comparing the relative efficacy of task-based versus resting-state measures of brain function, and how the two interact with each other (Fox, et al. 2007; Barnes et al 2009). It will also be important to determine whether the addition of information from other imaging modalities (e.g., structural or diffusion MRI, spectroscopy) improves classification accuracy.

The endgame for all fMRI research in psychiatry is to identify a biological phenotype or phenotypes that can be used to aid in some kinds of clinical activity, whether they are diagnostic, predictive, or therapeutic. Ultimately, however, this work will be limited by the conceptual shortcomings of current nosological schemes. Even the application of classification-based statistics requires some kind of gold standard against which to evaluate the accuracy of the method, and current psychiatric diagnoses fall short of providing such a standard for biological investigations. Thus, for example, while a particular method may show high accuracy in distinguishing unipolar from bipolar depressed patients, it may be possible that accuracy could be enhanced using some other criterion (e.g., treatment response, cognitive impairment, presence of certain risk factors) not captured by the diagnostic scheme used. Indeed, further developments in classification-based fMRI may facilitate the development of more valid diagnostic constructs by examining which specific clinical phenotypes can be predicted with the highest accuracy.

Conclusions

In this chapter, we have tried to consider reasons why fMRI has failed to live up to its early promise as a revolutionary tool for psychiatric practice, and have highlighted recent advances suggesting that fMRI may still play a role in addressing key clinical questions. We have argued that an increasing move away from mean group comparisons, and toward classification-based paradigms, represents a conceptual shift that has the potential to greatly increase the clinical applications of fMRI. However, similar arguments could be made for nearly any other investigative modality, ranging from the neuropsychological to the molecular. Thus, any potential clinical application of fMRI will need to be evaluated against these other methods to determine whether it provides a cost-effective solution to current clinical problems. At a minimum: however, the evidence considered here provides sufficient cause for optimism regarding the utility of fMRI for better understanding psychiatric illness phenotypes in a more clinically oriented fashion, which will be crucial for enhancing evidence-based, personalized care in psychiatry.

References

Barnes, A., E. T. Bullmore, and J. Suckling. 2009. Endogenous human brain dynamics recover slowly following cognitive effort. *PLoS One* 4: e6626.

Bartels, A., N. K. Logothetis, and K. Moutoussis. 2008. fMRI and its interpretations: An illustration on directional selectivity in area V5/MT. *Trends in Neurosciences* 31: 444–53.

Benetti, S., A. Mechelli, M. Picchioni, M. Broome, S. Williams, and P. McGuire. 2009. Functional integration between the posterior hippocampus and prefrontal cortex is impaired in both first episode schizophrenia and the at risk mental state. *Brain.* 132: 2426–36.

Berk, M., K. T. Hallam, and P. D. McGorry. 2007. The potential utility of a staging model as a course specifier: a bipolar disorder perspective. *J Affect Disord.* 100: 279–81.

Biswal, B., F. Z. Yetkin, V. M. Haughton, and J. S. Hyde. 1995. Functional connectivity in the motor cortex of resting human brain using echo-planar MRI. *Magn Reson Med.* 34: 537–41.

Broome, M. R., P. Matthiasson, P. Fusar-Poli, J. B. Woolley, L. C. Johns, P. Tabraham, E. Bramon, L. Valmaggia, S. C. Williams, M. J. Brammer, X. Chitnis, and P. K. McGuire. 2009. Neural correlates of executive function and working memory in the "at-risk mental state." *Br J Psychiatry* 194: 25–33.

Bullmore, E. T., J. Suckling, S. Overmeyer, S. Rabe-Hesketh, E. Taylor, and M. J. Brammer. 1999. Global, voxel, and cluster tests, by theory and permutation, for a difference between two groups of structural MR images of the brain. *IEEE Trans Med Imaging* 18: 32–42.

Calhoun, V. D., P. K. Maciejewski, G. D. Pearlson, and K. A. Kiehl. 2008. Temporal lobe and "default" hemodynamic brain modes discriminate between schizophrenia and bipolar disorder. *Hum Brain Mapp* 29: 1265–75.

Chen, C. H., K. Ridler, J. Suckling, S. Williams, C. H. Fu, E. Merlo-Pich, and E. Bullmore. 2007. Brain imaging correlates of depressive symptom severity and predictors of symptom improvement after antidepressant treatment. *Biol Psychiatry* 62: 407–14.

Cherkassky, V. L., R. K. Kana, T. A. Keller, and M. A. Just. 2006. Functional connectivity in a baseline resting-state network in autism. *Neuroreport* 17: 1687–90.

Church, J. A., D. A. Fair, N. U. Dosenbach, A. L. Cohen, F. M. Miezin, S. E. Petersen, and B. L. Schlaggar. 2009. Control networks in paediatric Tourette syndrome show immature and anomalous patterns of functional connectivity. *Brain* 132: 225–38.

Corlett, P. R., G. D. Honey, M. R. Aitken, A. Dickinson, D. R. Shanks, A. R. Absalom, M. Lee, E. Pomarol-Clotet, G. K. Murray, P. J. McKenna, T. W. Robbins, E. T. Bullmore, and P. C. Fletcher. 2006. Frontal responses during learning predict vulnerability to the psychotogenic effects of ketamine: Linking cognition, brain activity, and psychosis. *Arch Gen Psychiatry* 63: 611–21.

Corlett, P. R., G. K. Murray, G. D. Honey, M. R. Aitken, D. R. Shanks, T. W. Robbins, E. T. Bullmore, A. Dickinson, and P. C. Fletcher. 2007. Disrupted prediction-error signal in psychosis: Evidence for an associative account of delusions. *Brain* 130: 2387–400.

Crossley, N. A., A. Mechelli, P. Fusar-Poli, M. R. Broome, P. Matthiasson, L. C. Johns, E. Bramon, L. Valmaggia, S. C. Williams, and P. K. McGuire. 2009. Superior temporal lobe dysfunction and frontotemporal dysconnectivity in subjects at risk of psychosis and in first-episode psychosis. *Hum Brain Mapp.* 30: 4129–37.

Damoiseaux, J. S., S. A. Rombouts, F. Barkhof, P. Scheltens, C. J. Stam, S. M. Smith, and C. F. Beckmann. 2006. Consistent resting-state networks across healthy subjects. *Proc Natl Acad Sci USA.* 103: 13848–853.

Davidson, R. J., W. Irwin, M. J. Anderle, and N. H. Kalin. 2003. The neural substrates of affective processing in depressed patients treated with venlafaxine. *Am J Psychiatry* 160: 64–75.

De Luca, M., C. F. Beckmann, N. De Stefano, P. M. Matthews, and S. M. Smith. 2006. fMRI resting state networks define distinct modes of long-distance interactions in the human brain. *Neuroimage* 29: 1359–67.

Demirci, O., V. P. Clark, and V. D. Calhoun. 2008. A projection pursuit algorithm to classify individuals using fMRI data: Application to schizophrenia. *Neuroimage* 39: 1774–82.

Demirci, O., V. P. Clark, V. A. Magnotta, N. C. Andreasen, J. Lauriello, K. A. Kiehl, G. D. Pearlson, and V. D. Calhoun. 2008. A review of challenges in the use of fMRI for disease classification/characterization and A projection pursuit application from multi-site fMRI schizophrenia study. *Brain Imaging Behav*. 2: 147–226.

Ellison-Wright, I., D. C. Glahn, A. R. Laird, S. M. Thelen, and E. Bullmore. 2008. The anatomy of first-episode and chronic schizophrenia: an anatomical likelihood estimation meta-analysis. *Am J Psychiatry* 165: 1015–23.

Fitzgerald, P. B., A. R. Laird, J. Maller, and Z. J. Daskalakis. 2008. A meta-analytic study of changes in brain activation in depression. *Hum Brain Mapp*. 29: 683–95.

Fitzgerald, P. B., A. Srithiran, J. Benitez, Z. Z. Daskalakis, T. J. Oxley, J. Kulkarni, and G. F. Egan. 2008. An fMRI study of prefrontal brain activation during multiple tasks in patients with major depressive disorder. *Hum Brain Mapp*. 29: 490–501.

Fleisher, A .S., A. Sherzai, C. Taylor, J. B. Langbaum, K. Chen, and R. B. Buxton. 2009. Resting-state BOLD networks versus task-associated functional MRI for distinguishing Alzheimer's disease risk groups. *Neuroimage* 47: 1678–90.

Formisano, E., F. De Martino, and G. Valente. 2008. Multivariate analysis of fMRI time series: Classification and regression of brain responses using machine learning. *Magn Reson Imaging* 26: 921–34.

Fornito, A., and Bullmore, E. T. 2007. What can spontaneous fluctuations of the blood oxygenation-level-dependent signal tell us about psychiatric disorders. *Curr Op Psychiatry* 23: 239-249.

Fornito, A., M. Yucel, J. Patti, S. J. Wood, and C. Pantelis. 2009. Mapping grey matter reductions in schizophrenia: An anatomical likelihood estimation analysis of voxel-based morphometry studies. *Schizophr Res* 108: 104–113.

Fornito, A., A. R. Yung, S. J. Wood, L. J. Phillips, B. Nelson, S. Cotton, D. Velakoulis, P. D. McGorry, C. Pantelis, and M. Yucel. 2008. Anatomic abnormalities of the anterior cingulate cortex before psychosis onset: An MRI study of ultra-high-risk individuals. *Biol Psychiatry* 64: 758–65.

Fox, M. D., M. Corbetta, A. Z. Snyder, J. L. Vincent, and M. E. Raichle. 2006. Spontaneous neuronal activity distinguishes human dorsal and ventral attention systems. *Proc Natl Acad Sci USA* 103: 10046–51.

Fox, M. D., and M. E. Raichle. 2007. Spontaneous fluctuations in brain activity observed with functional magnetic resonance imaging. *Nat Rev Neurosci* 8: 700–11.

Fox, M. D., A. Z. Snyder, J. L. Vincent, M. Corbetta, D. C. Van Essen, and M. E. Raichle. 2005. The human brain is intrinsically organized into dynamic, anticorrelated functional networks. *Proc Natl Acad Sci USA* 102: 9673–78.

Fox, M. D., A. Z. Snyder, J. L. Vincent, and M. E. Raichle. 2007. Intrinsic fluctuations within cortical systems account for intertrial variability in human behavior. *Neuron* 56: 171–84.

Fu, C. H., J. Mourao-Miranda, S. G. Costafreda, A. Khanna, A. F. Marquand, S. C. Williams, and M .J. Brammer. 2008. Pattern classification of sad facial processing: Toward the development of neurobiological markers in depression. *Biol Psychiatry* 63: 656–62.

Fu, C. H., S. C. Williams, A. J. Cleare, M. J. Brammer, N. D. Walsh, J. Kim, C. M. Andrew, E. M. Pich, P. M. Williams, L. J. Reed, M. T. Mitterschiffthaler, J. Suckling, and E. T. Bullmore. 2004. Attenuation of the neural response to sad faces in major depression by antidepressant treatment: A prospective, event-related functional magnetic resonance imaging study. *Arch Gen Psychiatry* 61: 877–89.

Fusar-Poli, P., J. Perez, M. Broome, S. Borgwardt, A. Placentino, E. Caverzasi, M. Cortesi, P. Veggiotti, P. Politi, F. Barale, and P. McGuire. 2007. Neurofunctional correlates of vulnerability to psychosis: A systematic review and meta-analysis. *Neurosci Biobehav Rev* 31: 465–84.

Garner, B., C. M. Pariante, S. J. Wood, D. Velakoulis, L. Phillips, B. Soulsby, W. J. Brewer, D. J. Smith, P. Dazzan, G. E. Berger, A. R. Yung, M. van den Buuse, R. Murray, P. D. McGorry, and C. Pantelis. 2005. Pituitary volume predicts future transition to psychosis in individuals at ultra-high risk of developing psychosis. *Biol Psychiatry* 58: 417–23.

Genovese, C. R., N. A. Lazar, and T. Nichols. 2002. Thresholding of statistical maps in functional neuroimaging using the false discovery rate. *NeuroImage* 15: 870–78.

Goense, J. B., and N. K. Logothetis. 2008. Neurophysiology of the BOLD fMRI signal in awake monkeys. *Curr Biol.* 18: 631–40.

Gottesman, I. I., and T. D. Gould. 2003. The endophenotype concept in psychiatry: Etymology and strategic intentions. *American Journal of Psychiatry* 160: 1–10.

Greicius, M. D., B. H. Flores, V. Menon, G. H. Glover, H. B. Solvason, H. Kenna, A. L. Reiss, and A. F. Schatzberg. 2007. Resting-state functional connectivity in major depression: Abnormally increased contributions from subgenual cingulate cortex and thalamus. *Biol Psychiatry* 62: 429–37.

Haynes, J. D., and G. Rees. 2006. Decoding mental states from brain activity in humans. *Nat Rev Neurosci.* 7: 523–34.

Hetrick, S. E., A. G. Parker, I. B. Hickie, R. Purcell, A. R. Yung, and P. D. McGorry. 2008. Early identification and intervention in depressive disorders: Towards a clinical staging model. *Psychother Psychosom.* 77: 263–70.

Honey, G. D., P. R. Corlett, A. R. Absalom, M. Lee, E. Pomarol-Clotet, G. K. Murray, P. J. McKenna, E. T. Bullmore, D. K. Menon, and P. C. Fletcher. 2008. Individual differences in psychotic effects of ketamine are predicted by brain function measured under placebo. *J Neurosci* 28: 6295–303.

Jafri, M. J., and V. D. Calhoun. 2006. Functional classification of schizophrenia using feed forward neural networks. *Conf Proc IEEE Eng Med Biol Soc.* Suppl: 6631–34.

Keedwell, P., D. Drapier, S. Surguladze, V. Giampietro, M. Brammer, and M. Phillips. 2009a. Neural markers of symptomatic improvement during antidepressant therapy in severe depression: Subgenual cingulate and visual cortical responses to sad, but not happy, facial stimuli are correlated with changes in symptom score. *J Psychopharmacol.* 23: 775–88.

Keedwell, P. A., D. Drapier, S. Surguladze, V. Giampietro, M. Brammer, and M. Phillips. 2009b. Subgenual cingulate and visual cortex responses to sad faces predict clinical outcome during antidepressant treatment for depression. *J Affect Disord.* 120: 120–25.

Kemp, A. H., E. Gordon, A. J. Rush, and L. M. Williams. 2008. Improving the prediction of treatment response in depression: Integration of clinical, cognitive, psychophysiological, neuroimaging, and genetic measures. *CNS Spectr.* 13: 1066–86; quiz 1087–1068.

Koutsouleris, N., E. M. Meisenzahl, C. Davatzikos, R. Bottlender, T. Frodl, J. Scheuerecker, G. Schmitt, T. Zetzsche, P. Decker, M. Reiser, H. J. Moller, and C. Gaser. 2009. Use of neuroanatomical pattern classification to identify subjects in at-risk mental states of psychosis and predict disease transition. *Arch Gen Psychiatry* 66: 700–12.

Kumari, V., E. R. Peters, D. Fannon, E. Antonova, P. Premkumar, A. P. Anilkumar, S. C. Williams, and E. Kuipers. 2009. Dorsolateral prefrontal cortex activity predicts responsiveness to cognitive-behavioral therapy in schizophrenia. *Biol Psychiatry* 66: 594–602.

Liang, M., Y. Zhou, T. Jiang, Z. Liu, L. Tian, H. Liu, and Y. Hao. 2006. Widespread functional disconnectivity in schizophrenia with resting-state functional magnetic resonance imaging. *Neuroreport* 17: 209–13.

Liu, Y., M. Liang, Y. Zhou, Y. He, Y. Hao, M. Song, C. Yu, H. Liu, Z. Liu, and T. Jiang. 2008. Disrupted small-world networks in schizophrenia. *Brain* 131: 945–61.

Logothetis, N. K. 2008. What we can do and what we cannot do with fMRI. *Nature* 453: 869–78.

Logothetis, N. K., J. Pauls, M. Augath, T. Trinath, and A. Oeltermann. 2001. Neurophysiological investigation of the basis of the fMRI signal. *Nature* 412: 150–57.

Macdonald, A. W., III, H. W. Thermenos, D. M. Barch, and L. J. Seidman. 2008. Imaging genetic liability to schizophrenia: Systematic review of fMRI studies of patients' nonpsychotic relatives. *Schizophr Bull.* 35: 1142–162.

Maier, A., M. Wilke, C. Aura, C. Zhu, F. Q. Ye, and D. A. Leopold. 2008. Divergence of fMRI and neural signals in V1 during perceptual suppression in the awake monkey. *Nat Neurosci* 11: 1193–1200.

Manoach, D. S. 2003. Prefrontal cortex dysfunction during working memory performance in schizophrenia: Reconciling discrepant findings. *Schizophrenia Research* 60: 285–98.

Marquand, A. F., J. Mourao-Miranda, M. J. Brammer, A. J. Cleare, and C. H. Fu. 2008. Neuroanatomy of verbal working memory as a diagnostic biomarker for depression. *Neuroreport* 19: 1507–11.

Mason, M. F., M. I. Norton, J. D. Van Horn, D. M. Wegner, S. T. Grafton, C. N. Macrae. 2007. Wandering minds: The default network and stimulus-independent thought. *Science New York, NY* 315: 393–95.

Mayberg, H. S., S. K. Brannan, R. K. Mahurin, P. A. Jerabek, J. S. Brickman, J. L. Tekell, J. A. Silva, S. McGinnis, T. G. Glass, C. C. Martin, and P. T. Fox. 1997. Cingulate function in depression: A potential predictor of treatment response. *Neuroreport* 8: 1057–61.

McGorry, P. D., B. Nelson, G. P. Amminger, A. Bechdolf, S. M. Francey, G. Berger, A. Riecher-Rossler, J. Klosterkotter, S. Ruhrmann, F. Schultze-Lutter, M. Nordentoft, I. Hickie, P. McGuire, M. Berk, E. Y. Chen, M. S. Keshavan, and A. R. Yung. 2009. Intervention in individuals at ultra high risk for psychosis: A review and future directions. *J Clin Psychiatry* 70: 1206–12.

McGorry, P. D., R. Purcell, I. B. Hickie, A. R. Yung, C. Pantelis, and H. J. Jackson. 2007. Clinical staging: A heuristic model for psychiatry and youth mental health. *Med J Aust.* 187: S40–42.

Menzies, L., S. R. Chamberlain, A. R. Laird, S. M. Thelen, B. J. Sahakian, and E. T. Bullmore. 2008. Integrating evidence from neuroimaging and neuropsychological studies of obsessive-compulsive disorder: The orbitofronto-striatal model revisited. *Neurosci Biobehav Rev.* 32: 525–49.

Michael, A. M., V. D. Calhoun, N. C. Andreasen, and S. A. Baum. 2008. A method to classify schizophrenia using inter-task spatial correlations of functional brain images. *Conf Proc IEEE Eng Med Biol Soc.* 2008: 5510–13.

Minzenberg, M. J., A. R. Laird, S. Thelen, C. S. Carter, and D. C. Glahn. 2009. Meta-analysis of 41 functional neuroimaging studies of executive function in schizophrenia. *Arch Gen Psychiatry* 66: 811–22.

Morcom,A. M., and P. C. Fletcher. 2007. Does the brain have a baseline? Why we should be resisting a rest. *Neuroimage* 37: 1073–82.

Morey, R. A., S. Inan, T. V. Mitchell, D. O. Perkins, J. A. Lieberman, and A. Belger. 2005. Imaging frontostriatal function in ultra-high-risk, early, and chronic schizophrenia during executive processing. *Arch Gen Psychiatry* 62: 254–62.

Ogawa, S., T. M. Lee, A. R. Kay, and D. W. Tank. 1990. Brain magnetic resonance imaging with contrast dependent on blood oxygenation. *Proc Natl Acad Sci USA* 87: 9868–72.

Pantelis, C., D. Velakoulis, P. D. McGorry, S. J. Wood, J. Suckling, L. J. Phillips, A. R. Yung, E. T. Bullmore, W. Brewer, B. Soulsby, P. Desmond, and P. K. McGuire. 2003. Neuroanatomical abnormalities before and after onset of psychosis: A cross-sectional and longitudinal MRI comparison. *The Lancet* 361: 281–88.

Pantelis, C., D. Velakoulis, S. J. Wood, M. Yucel, A. R. Yung, L. J. Phillips, D. Q. Sun, and P. D. McGorry. 2007. Neuroimaging and emerging psychotic disorders: the Melbourne ultra-high risk studies. *Int Rev Psychiatry* 19: 371–81.

Poline, J. B., K. J. Worsley, A. C. Evans, and K. J. Friston. 1997. Combining spatial extent and peak intensity to test for activations in functional imaging. *Neuroimage* 5: 83–96.

Ragland, J. D., A. R. Laird, C. Ranganath, R. S. Blumenfeld, S. M. Gonzales, and D. C. Glahn. 2009. Prefrontal activation deficits during episodic memory in schizophrenia. *Am J Psychiatry* 166: 863–74.

Rotge, J. Y., D. Guehl, B. Dilharreguy, E. Cuny, J. Tignol, B. Bioulac, M. Allard, P. Burbaud, and B. Aouizerate. 2008. Provocation of obsessive-compulsive symptoms: A quantitative voxel-based meta-analysis of functional neuroimaging studies. *J Psychiatry Neurosci.* 33: 405–12.

Seiferth, N. Y., K. Pauly, U. Habel, T. Kellermann, N. J. Shah, S. Ruhrmann, J. Klosterkotter, F. Schneider, and T. Kircher. 2008. Increased neural response related to neutral faces in individuals at risk for psychosis. *Neuroimage* 40: 289–97.

Shehzad, Z., A. M. Kelly, P. T. Reiss, D. G. Gee, K. Gotimer, L. Q. Uddin, S. H. Lee, D. S. Margulies, A. K. Roy, B. B. Biswal, E. Petkova, F. X. Castellanos, M. P. Milham. 2009. The resting brain: Unconstrained yet reliable. *Cereb Cortex.* 19: 2209–29.

Shi, F., Y. Liu, T. Jiang, Y. Zhou, W. Zhu, J. Jiang, H. Liu, and Z. Liu. 2007. Regional homogeneity and anatomical parcellation for fMRI image classification: Application to schizophrenia and normal controls. *Med Image Comput Comput Assist Interv Int Conf Med Image Comput Comput Assist Interv*. 10: 136–43.

Siegle, G. J., C. S. Carter, and M. E. Thase. 2006. Use of FMRI to predict recovery from unipolar depression with cognitive behavior therapy. *Am J Psychiatry* 163: 735–38.

Sun, D., L. Phillips, D. Velakoulis, A. Yung, P. D. McGorry, S. J. Wood, T. G. van Erp, P. M. Thompson, A. W. Toga, T. D. Cannon, and C. Pantelis. 2009. Progressive brain structural changes mapped as psychosis develops in "at risk" individuals. *Schizophr Res*. 108: 85–92.

Sun, D., G. W. Stuart, M. Jenkinson, S. J. Wood, P. D. McGorry, D. Velakoulis, T G. van Erp, P. M. Thompson, A. W. Toga, D. J. Smith, T. D. Cannon, and C. Pantelis. 2009. Brain surface contraction mapped in first-episode schizophrenia: a longitudinal magnetic resonance imaging study. *Mol Psychiatry*. 14: 976–86.

Van der Veen, F. M., E. A. Evers, N. E. Deutz, and J. A. Schmitt. 2007. Effects of acute tryptophan depletion on mood and facial emotion perception related brain activation and performance in healthy women with and without a family history of depression. *Neuropsychopharmacology* 32: 216–24.

Van Snellenberg, J. X., I. J. Torres, and A. E. Thornton. 2006. Functional neuroimaging of working memory in schizophrenia: Task performance as a moderating variable. *Neuropsychology* 20: 497–510.

Whalley, H. C., V. E. Gountouna, J. Hall, A. M. McIntosh, E. Simonotto, D. E. Job, D. G. Owens, E. C. Johnstone, and S. M. Lawrie. 2009. fMRI changes over time and reproducibility in unmedicated subjects at high genetic risk of schizophrenia. *Psychol Med*. 39: 1189–99.

Whalley, H. C., E. Simonotto, W. Moorhead, A. McIntosh, I. Marshall, K. P. Ebmeier, D. G. Owens, N. H. Goddard, E. C. Johnstone, and S. M. Lawrie. 2006. Functional imaging as a predictor of schizophrenia. *Biol Psychiatry* 60: 454–62.

Worsley, K. J., and K. J. Friston. 1995. Analysis of fMRI time-series revisited—again. *Neuroimage* 2: 173–81.

Yao, Z., L. Wang, Q. Lu, H. Liu, and G. Teng. 2009. Regional homogeneity in depression and its relationship with separate depressive symptom clusters: A resting-state fMRI study. *J Affect Disord*. 115: 430–38.

Yung, A. R., and P. D. McGorry. 1996. The initial prodrome in psychosis: Descriptive and qualitative aspects. *Australian and New Zealand Journal of Psychiatry* 30: 587–99.

Yung, A. R., P. D. McGorry, C. A. McFarlane, H. J. Jackson, G. C. Patton, and A. Rakkar. 1996. Monitoring and care of young people at incipient risk of psychosis. *Schizophrenia Bulletin* 22: 283–303.

Yung, A. R., L. J. Phillips, P. D. McGorry, C. A. McFarlane, S. Francey, S. Harrigan, G. C. Patton, and H. J. Jackson. 1998. Prediction of psychosis. A step toward indicated prevention of schizophrenia. *Br J Psychiatry Suppl* 172: 14–20.

Zang, Y. F., Y. He, C. Z. Zhu, Q. J. Cao, M. Q. Sui, M. Liang, L. X. Tian, T. Z. Jiang, and Y. F. Wang. 2007. Altered baseline brain activity in children with ADHD revealed by resting-state functional MRI. *Brain Dev*. 29: 83–91.

Zhou, Y., M. Liang, T. Jiang, L. Tian, Y. Liu, Z. Liu, H. Liu, and F. Kuang. 2007. Functional dysconnectivity of the dorsolateral prefrontal cortex in first-episode schizophrenia using resting-state fMRI. *Neurosci Lett*. 417: 297–302.

Zhu, C. Z., Y. F. Zang, Q. J. Cao, C. G. Yan, Y. He, T. Z. Jiang, M. Q. Sui, and Y. F. Wang. 2008. Fisher discriminative analysis of resting-state brain function for attention-deficit/hyperactivity disorder. *Neuroimage* 40: 110–20.

5

Stress and the Impact of Personalized Medicine

Charles F. Gillespie, MD, PhD, Elisabeth B. Binder, MD, PhD, Paul E. Holtzheimer, MD, and Charles B. Nemeroff, MD, PhD

Introduction

The developmental history of individual patients has long been recognized as a major variable contributing to disease vulnerability. Conservative estimates suggest that each year in the United States, more than one million children are exposed to sexual or physical abuse or neglect (Sedlack 1996). Exposure to trauma and neglect during early life has consistently been found to exert an adverse and graded influence (McCauley et al. 1997; Felitti et al. 1998) on adult risk for mood and anxiety disorders (McCauley et al. 1997; Felitti et al. 1998; Dube et al. 2001; Chapman et al. 2004; Gladstone et al. 2004). It also influences a variety of stress-related medical illnesses, including obesity (Lissau and Sorensen 1994; Felitti et al. 1998; Gunstad et al. 2006), cardiovascular disease (Batten et al. 2004; Dong et al. 2004; Goodwin and Stein 2004; Caspi et al. 2006), and diabetes mellitus (Goodwin and Stein 2004; Goodwin and Davidson 2005). A parallel—though much more circumscribed—study of individuals who emerge from such adverse environments without significant mood or anxiety disorder (Rutter 2006) has resulted in the identification of psychosocial and biological variables associated with psychological resilience (Feder, Nestler, and Charney 2009). Predisposing genetic factors, similar to environmental variables, have also been found to influence both vulnerability (Sullivan et al. 2000; Stein et al. 2002) and resilience (Rijsdijk et al. 2003) with respect to aggregate risk for mood and anxiety disorders, including depression and post-traumatic stress disorder (PTSD). In this chapter, we review psychobiological and genetic findings that have progressively refined our understanding of the role of stress, particularly developmental stressors, such as child abuse, neglect, or severe childhood illness, in the psychobiology of mood and anxiety disorders. These factors have a substantial impact on markers for shaping treatment programs in personalized medicine.

Neuroendocrinology of Stress and the Pathophysiology of Mood and Anxiety Disorders

The hypothalamic-pituitary-adrenal (HPA) axis is a collection of neural and endocrine structures that facilitate the adaptive response to stress. Parvocellular neurons of the paraventricular nucleus (PVN) of the hypothalamus project to the median eminence, where they release corticotropin-releasing factor (CRF) into the primary plexus of blood vessels that comprise the hypothalamo-hypophyseal portal system (Swanson et al. 1983). The secreted CRF is subsequently transported to the anterior pituitary gland, where it activates CRF_1 receptors on pituitary corticotrophs, resulting in increased secretion of adrenocorticotrophic hormone (ACTH). ACTH released from the anterior pituitary into the systemic circulation stimulates the production and release of cortisol from the adrenal cortex through actions on type 2 melanocortin receptors.

Within the brain, two glucocorticoid-responsive receptors, the mineralocorticoid receptor (MR) and the glucocorticoid receptor (GR), play multiple roles in regulating neuronal response to glucocorticoids and mineralocorticoids (Joels and de Kloet 1994). Mineralocorticoid receptors have a high affinity for glucocorticoids and are restricted within the brain to limbic structures, such as the septum, hippocampus, and amygdala, where they regulate basal activity of the HPA axis (Jacobson and Sapolsky 1991). In contrast, GRs have a lower affinity for glucocorticoids than MRs and are distributed throughout the brain, but are densely concentrated within the limbic system, parvocellular division of the PVN, pituitary gland, and ascending monoaminergic projection neurons. With elevated activity of the HPA axis, circulating levels of cortisol increase and the lower-affinity GRs become occupied, leading to suppression of HPA axis activity through effects on the hippocampus, PVN, and pituitary (de Kloet et al. 1991). Feedback inhibition—mediated in part by the action of cortisol on mineralocorticoid and glucocorticoid receptors at the hippocampus, PVN, and pituitary (de Kloet et al. 1991)—reduces stress-induced activation of the HPA axis and limits excess secretion of glucocorticoids, effectively dampening the stress response (Jacobson and Sapolsky 1991).

A large body of literature has implicated dysfunction of the HPA axis in the pathophysiology of mood and anxiety disorders. Excess secretion of cortisol (Gibbons and McHugh 1962; Carpenter and Bunney 1971) and its metabolites (Sachar et al. 1970) was first observed in depressed patients over forty years ago. The presence of depression and anxiety in patients with endocrinopathies affecting the HPA axis, such as Cushing's disease or syndrome (Dorn et al. 1995; 1997), in conjunction with the observation of increased secretion of cortisol in healthy patients exposed to stress, contributed in part to the development of the stress-diathesis model of depression. This model hypothesizes that individual predisposition to excess reactivity of the neural and endocrine systems that respond to stress plays a critical role in susceptibility to depression. The presence of acute or prolonged stress in vulnerable individuals is thus believed to play a significant role in both the onset and relapse of certain forms of depression and anxiety disorders, and may also contribute to the burden of medical illness in patients with mood and anxiety disorders, as well (Brown, Varghese, and McEwen 2004).

The initial finding that the experience of child abuse pathologically alters the response of the HPA axis and autonomic nervous system to stress (Heim et al. 2000) has fundamentally altered our understanding of the effects of childhood stress on the biology of mood and anxiety disorders in adults. It also has further refined our conceptualization of the stress-diathesis model of depression (Heim et al. 2008b). Given the powerful relationship between childhood trauma and risk of stress-related medical illness in adulthood, it now appears that the experience of severe childhood stress contributes to adult risk for chronic stress-related disease through pathological programming of the adult response to stress. Depressed patients who have a history of childhood adversity show elevated secretion of ACTH and cortisol and increased heart rate in response to a standardized laboratory stress test (Heim et al. 2000), as well as abnormal responses to neuroendocrine challenge tests, including the dexamethasone-CRF test (Heim et al. 2001; Heim et al. 2008a; Tyrka et al. 2009). In addition, healthy volunteers with variant forms of HPA axis-associated genes

(Ising et al. 2008) or healthy volunteers with a history of childhood trauma (Heim et al. 2000; Carpenter et al. 2007) also have been found to exhibit abnormal responses to laboratory social stress tests. Collectively, these data reveal the capacity of stressful experiences during childhood to persistently alter the endogenous stress response. Moreover, variant forms of HPA axis genes, independent of childhood trauma, mediate individual differences in social stress responsivity. In the sections below, we review the relationship between mood and anxiety disorders and particular elements of the HPA axis.

Corticotropin-releasing Factor and Mood and Anxiety Disorders

In addition to its central role in the regulation of the HPA axis, corticotropin-releasing factor (CRF) is widely distributed in areas of the brain outside of the hypothalamus, including the amygdala, septum, bed nucleus of the stria terminalis, and the cerebral cortex (Swanson et al. 1983), where it functions, along with the hypothalamic CRF system, as a neurotransmitter in coordinating the behavioral, autonomic, endocrine, and immune responses to stress (Arborelius et al. 1999). Two CRF receptor subtypes (CRF_1 and CRF_2), both coupled positively to the production of adenylate cyclase (Chalmers et al. 1996), appear to mediate divergent effects of CRF on behavior (Heinrichs et al. 1997). CRF_1 receptors are broadly distributed within the central nervous system (CNS) and have a high affinity for CRF. Decreased activity of CRF_1 receptors is associated with reduced anxiety in animal models. CRF_2 receptors are broadly distributed throughout the brain, overlapping topographically with CRF_1 receptors. CRF_2 receptors display a lower affinity for CRF than CRF_1 receptors and likely utilize urocortin as their endogenous ligand. However, in contrast to CRF_1 receptors, reduced activity of CRF_2 receptors results in increased anxiety in animal models.

Injection of CRF into the CNS of laboratory animals initiates changes in the activity of the autonomic nervous system that result in the elevation of peripheral catecholamines, reduced gastrointestinal activity, increased heart rate, and elevated blood pressure. Further, behavioral changes, such as disturbed sleep, diminished food intake, reduced grooming, decreased sexual behavior, and enhanced fear-conditioning—similar to those observed in patients with major depressive disorder (MDD)—also occur following direct CNS administration of CRF (Dunn and Berridge 1990; Owens and Nemeroff 1991). They are reduced by pretreatment with CRF receptor antagonists (Heinrichs et al. 1995; Gutman et al. 2003).

A large body of data using measures of CRFergic neural circuits—such as cerebrospinal fluid (CSF) CRF concentrations, postmortem brain CRF concentrations, CRF mRNA expression, and CRF receptor density—has been accumulated. These data suggest that the abnormal activity of the HPA axis often observed in depressed patients (Plotsky et al. 1998) is due, in part, to hypersecretion of CRF (Arborelius et al. 1999). Elevated CSF concentrations of CRF have repeatedly been reported in patients with depression (Nemeroff et al. 1984; Hartline et al. 1996), as well as in combat veterans suffering from PTSD (Bremner et al. 1997a; Baker et al. 1999). In addition, the transcription (Raadsheer et al. 1995), as well as the expression (Raadsheer et al. 1994), of CRF is increased in postmortem tissue in patients with depression. Further, postmortem studies of individuals who have committed suicide have revealed elevated concentrations of CSF CRF (Arato et al. 1989), decreased expression of CRF_1 receptor mRNA within the frontal cortex (Merali et al. 2004), and increased CRF concentrations and decreased density of CRF receptors within the frontal cortex in comparison to controls (Nemeroff et al. 1988; Merali et al. 2006). Successful treatment of depression, using either electroconvulsive therapy (Nemeroff et al. 1991) or fluoxetine (De Bellis et al. 1993), has been shown to reduce the high pretreatment concentrations of CSF CRF in patients with MDD. In contrast, persistent elevations of CSF CRF in symptomatically improved depressed patients are associated with early relapse of depression (Banki et al. 1992).

Several independent lines of investigation have identified single-nucleotide polymorphisms (SNPs) within the gene coding for the CRF_1 receptor, *CRHR1*, that are associated with the main effects for depression (Liu et al. 2006; Utge et al. 2009), antidepressant treatment response in depressed patients (Licinio et al. 2004; Liu et al. 2007), and depression and suicidality in male patients (Wasserman et al. 2008; Wasserman et al. 2009). Licinio and colleagues (2004) examined the association of *CRHR1* genotypes with the phenotype of antidepressant treatment response in depressed Mexican-Americans and identified a haplotype—consisting of SNPs rs1876828, rs242939 and rs242941—that was associated with a robust antidepressant treatment response. Subsequent investigation in a Chinese cohort identified a main effect of *CRHR1* haplotype status with major depression frequency (Liu et al. 2006) and partially replicated the effects of the previously identified haplotype (Licinio et al. 2004) on antidepressant treatment response (Liu et al. 2007). Using family trios with offspring that had attempted suicide, Wasserman and colleagues (2008; 2009) identified *CRHR1* polymorphisms that were associated with suicide attempt history in male patients as a function of depression.

In addition to the main effects of *CRHR1* on depression-related outcomes as described above, exposure to childhood maltreatment also appears to have a significant influence on adult depressive symptomatology (Bradley et al. 2008; Polanczyk et al. 2009) and HPA axis reactivity (Tyrka et al. 2009) that seems to be mediated by gene x environment interactions involving *CRHR1*. Bradley and colleagues (2008) performed an association study examining interaction between genetic polymorphisms at the *CRHR1* locus and measures of child abuse on adult depressive symptomatology in a sample of African-American, inner-city-based, primary care patients. Multiple individual *CRHR1* SNPs, as well as a common haplotype spanning intron 1 of the *CRHR1* locus, that modify adult risk of depressive symptomatology in the presence of childhood trauma exposure were identified. These initial findings were supported in the same study by similar results from a second independent sample (Bradley et al. 2008). Tyrka et al. (2009) which extended these findings by administration of the dexamethasone/corticotropin-releasing hormone (DEX/CRH) test, a measure of HPA axis reactivity, to healthy volunteers assessed for history of childhood trauma. They found that two of the *CRHR1* SNPs previously identified by Bradley and colleagues (2008) also interacted with a history of childhood maltreatment to predict abnormal responses to the DEX/CRH test. These gene x environment interactions on the DEX-CRH test were also reported by Heim and colleagues (2009), but only in males. Also, the interaction on depressive symptomatology seemed more pronounced in men in an expanded cohort of the one reported in Bradley and colleagues (2008). The gender-specific interactions in this cohort likely attributable to the interactions of *CRHR1* SNPs with physical abuse, but not other types of abuse, most robustly predicted adult depression. Physical abuse is the most common type of abuse in men but not in women in this sample, pointing to the importance of an in-depth characterization of environmental predictors. More recently, Polanczyk et al. (2009) replicated the findings of Bradley and colleagues in a sample of English women (E-Risk Study), but were unable to replicate the findings using subjects from the Dunedin Study.

A feature common to the assessment of childhood trauma in the three replicating samples (as well as the study by Tyrka and colleagues [2009]) was the use of the Childhood Trauma Questionnaire (CTQ) to assess childhood trauma. The CTQ was not used to assess childhood trauma in the Dunedin Study. It differs from the multisource evaluation of childhood maltreatment in the Dunedin Study because it assesses trauma during adolescence in addition to childhood, and appears to more directly evoke depression-relevant emotional memory. Such differences in the way the environment is assessed between studies may be critical in explaining inconsistent findings with respect to outcomes putatively related to gene x environment interactions. In addition to the complexity of the environment, additional moderating genetic factors also need to be taken into account when interpreting and analyzing gene x environment interactions. In fact, Ressler and colleagues (2009) reported a gene x gene x environment interaction of *CRHR1* SNPs; the serotonin transporter gene polymorphism, *5-HTTLPR*; and child abuse to predict adult depression. Individuals carrying both risk genotypes reported a higher severity of adult depressive symptoms in the presence of lower level of child abuse than carriers of the other genotype combinations. These findings highlight the importance of taking both environmental and genetic complexity into account in gene x environment interaction studies.

FKBP5 and Mood and Anxiety Disorders

FKBP5 is a co-chaperone component of the GR heterocomplex (Schiene-Fischer and Yu 2001) that plays a key role in the regulation of GR sensitivity and, hence, the expression of glucocorticoid-responsive genes by virtue of its participation in an ultra-short, intracellular negative feedback loop regulating GR activity (Vermeer et al. 2003). Several lines of data suggest a role for *FKBP5* variants in the pathophysiology and treatment of PTSD and depression (Binder 2009). Over-expression of *FKBP5* reduces the hormone binding affinity (Denny et al. 2000) and nuclear translocation of GR (Wochnik et al. 2005). New World monkeys with naturally occurring over-expression of *FKBP5* experience increased GR resistance and hypercortisolemia (Denny et al. 2000; Scammell et al. 2001). In addition, clinical research has identified *FKBP5* alleles associated with variation in GR resistance in depressed patients (Binder et al. 2004). These alleles are also associated with elevated peri-traumatic dissociation in medically injured children (Koenen et al. 2005), a psychological response to trauma that predicts PTSD risk in adults (Ozer et al. 2003), and delayed recovery from a laboratory-based social stress paradigm in healthy volunteers (Ising et al. 2008). Further, level of *FKBP5* expression in peripheral blood mononuclear cells at four months post-trauma exposure is predictive of PTSD diagnosis in trauma survivors (Segman et al. 2005). Binder et al. (2008) performed an association study examining interaction between genetic polymorphisms at the *FKBP5* locus and measures of child abuse on PTSD symptomatology in a sample of African-American, inner-city-based primary care patients. Multiple individual *FKBP5* SNPs were identified that modify adult risk of PTSD symptomatology in the presence of childhood trauma exposure. This genetic interaction was also paralleled by *FKBP5* genotype- and PTSD-dependent effects on glucocorticoid receptor sensitivity as measured by the dexamethasone suppression test. More recently, gene expression analysis from a study of PTSD subjects from the September 11, 2001 attack on the World Trade Center found that *FKBP5* showed reduced expression in PTSD, consistent with enhanced GR responsiveness (Yehuda et al. 2009). This has been observed in previous research demonstrating increased HPA axis reactivity (Yehuda et al. 1991; Yehuda 2001) and elevated GR sensitivity (Yehuda et al. 2004a; Yehuda et al. 2004b) in patients with PTSD.

In summary, polymorphisms in *FKBP5* have been shown to be associated with differential upregulation of *FKBP5* following GR activation and differences in GR sensitivity and stress hormone system regulation. Alleles associated with enhanced expression of *FKBP5* following GR activation lead to increased GR resistance and decreased efficiency of the negative feedback of the stress hormone axis in healthy controls. This results in a prolongation of stress hormone system activation following exposure to stress, which has been posited to be a risk factor for stress-related psychiatric disorders. In fact, the same *FKBP5* alleles associated with increased GR resistance are also over-represented in individuals with major depression (Lekman et al. 2008), bipolar disorder (Willour et al. 2009), and PTSD (Binder et al. 2008).

Brain Imaging Studies of Mood and Anxiety Disorders

Hippocampus

Owing to its important role in the regulation of both stress and learning, the hippocampus has been a major focus of neuroanatomical studies in patients with mood and anxiety disorders, including depression and PTSD. Functionally, the hippocampus plays a central role in adaptation to stress because it is central to the regulation of HPA axis activity, as well as the registration of explicit memory and learning context. Further, the hippocampus is a remarkably plastic structure and a major site of neurogenesis in the adult brain; it also possesses a well-demonstrated vulnerability to stress. Neurons of the CA3 region exhibit a loss of dendritic spines, reduced dendritic branching, and impaired neurogenesis in response to stress exposure (Fuchs and Gould 2000; Nestler et al. 2002; Duman and Monteggia 2006).

Hippocampal vulnerability to stress may, in part, be a consequence of its high concentration of type I and type II corticosteroid receptors, which enable the hippocampus to exert an inhibitory role on activity of the HPA axis (Jacobson and Sapolsky 1991). Prolonged exposure to high concentrations of corticosteroids during stress results in hippocampal damage in rodents (Magarinos and McEwen 1995a; Magarinos and McEwen 1995b; Magarinos, Verdugo, and McEwen 1997) and non-human primates (Uno et al. 1989; Sapolsky et al. 1990; Uno et al. 1994; Magarinos et al. 1996). In humans, bilateral hippocampal atrophy has been observed in patients with Cushing's disease, in whom the extent of hippocampal atrophy correlates with the magnitude of corticosteroid hypersecretion (Starkman et al. 1992).

Reduced hippocampal size has been documented in a wide variety of neuropsychiatric disorders (Geuze et al. 2005). With respect to mood disorders, reduced hippocampal volume has been found in some (Sheline 1996; Bremner et al. 2000; Frodl et al. 2002b; Sheline 2003), but not all, studies (Vakili et al. 2000; Rusch et al. 2001) of patients with unipolar depression (Campbell and Macqueen 2004). In patients with a history of depression who also have hippocampal atrophy, the extent of atrophy is greater in patients with a higher total lifetime duration of depression (Sheline et al. 1996; 1999), suggesting that antidepressant medications may have a protective effect against hippocampal atrophy (Sheline et al. 2003). The finding of reduced hippocampal volume in patients with remitted unipolar depression (Neumeister et al. 2005) has also raised the possibility that small hippocampal size may be a risk factor for depression.

Exposure to traumatic stress, either early in life or during adulthood, is associated with loss of hippocampal volume and has been documented in individuals with combat-induced (Bremner et al. 1995; Vythilingam et al. 2005), childhood-related (Bremner et al. 1997b; Bremner et al. 2003), and chronic (Kitayama et al. 2005) PTSD, as well as in dissociative identity disorder (Vermetten et al. 2006). Similarly, patients with a history of early life stress or trauma and MDD also have been found to exhibit decreased hippocampal volume (Stein et al. 1997; Driessen et al. 2000). Vythilingam et al. (2002) reported that depressed women with a history of childhood sexual abuse, but not depressed women without such a history, had a reduction in hippocampal volume compared with control subjects, suggesting that previous reports of reduced hippocampal size in patients with depression may, in fact, be secondary to childhood trauma rather than to depression per se, although other investigators (Lenze et al. 2008) have been unable to replicate this finding with women who experienced mild, as opposed to severe, childhood adversity. Finally, Gatt and colleagues (Gatt et al. 2009) recently identified an interaction between the BDNF *Val66Met* polymorphism and early life stress impacting brain structure. BDNF *Met* carriers exposed to early life stress exhibited smaller hippocampal and amygdala volumes, as well as reduced gray matter in the hippocampus and lateral prefrontal cortex. In contrast, BDNF *Val/Val* subjects exposed to early life stress exhibited increased gray matter in the amygdala and medial prefrontal cortex.

Subgenual Cingulate

The subgenual cingulate (Cg25) has received substantial attention in studies of patients with mood disorders. The importance of limbic-cortical pathways in the pathogenesis of major depression has been highlighted by several functional brain imaging studies (Mayberg 1994; Drevets 1999; Brody et al. 2001). Additional research (Mayberg et al. 1999; Seminowicz et al. 2004) has demonstrated the importance of Cg25 in the modulation of acute sadness and response to antidepressant treatment, indicating a role for Cg25 in the regulation of negative affect states. Cg25 possesses descending projections to a variety of brain regions (brainstem, hypothalamus, and insula) implicated in the regulation of neurovegetative and motivational states associated with depression (Jurgens and Muller-Preuss 1977; Ongur et al. 1998; Freedman et al. 2000; Barbas et al. 2003). Further, a number of pathways link Cg25 to the orbitofrontal, medial prefrontal, as well as anterior and posterior cingulate cortices, providing a means through which autonomic and homeostatic processes influence learning, reward, and motivation (Vogt and Pandya 1987; Carmichael and Price 1996; Barbas et al. 2003; Haber 2003). Genetic variation may significantly affect Cg25 function, as demonstrated by the recent report of Pezawas and colleagues (Pezawas et al. 2008) demonstrating epistatic interactions between the BDNF *Val66Met* polymorphism and the *5-HTTLPR* polymorphism on subgenual cingulate volume and connectivity with the amygdala in healthy subjects. Decreased activity of Cg25 measured prospectively has been associated with positive clinical response to a wide variety of antidepressant treatments (Mayberg et al. 2000; Nobler et al. 2001; Mottaghy et al. 2002; Dougherty et al. 2003; Goldapple et al. 2004). Finally, deep brain stimulation (DBS) of Cg25, originally developed for the treatment of Parkinson's disease (Benabid 2003), has been shown to produce significant clinical benefits for patients with severe treatment-resistant depression (Mayberg et al. 2005).

Amygdala

The amygdala plays a central role in the processing of affectively charged stimuli and emotional learning (Davidson et al. 2002). Within the brain, the amygdala appears to function as one component of a threat-responsive comparator system, which differentiates threatening from non-threatening environmental stimuli in real time on the basis of prototype matching to fear memories. It initiates adaptive behavior to deal with any perceived threat (Bishop 2008; Rodrigues et al. 2009). For example, Isenberg and colleagues (1999) found significantly greater bilateral amygdala activation during both the response to, and the perception of, emotionally charged stimuli when compared to non-emotionally charged stimuli. Several imaging studies of patients with mood disorders have focused on determining the association between mood alterations and abnormalities in the amygdala, as well as other structures (Phan et al. 2004; Leppanen 2006). An initial report found that increased resting metabolism within the right amygdala of depressed patients is predictive of negative affect (Abercrombie et al. 1998). Subsequent fMRI investigation using a masked face paradigm found relatively increased activity of the left amygdala among depressed patients that normalizes with antidepressant treatment (Sheline et al. 2001). Similarly, increased activation of the left amygdala in depressed adolescents using evocative face viewing has been associated with poor memory for faces relative to healthy and anxious controls (Roberson-Nay et al. 2006). In addition, during exposure to emotionally valenced visual stimuli, the amygdala also responds differentially to both positive and negative words (Hamann and Mao 2002). Using PET methodologies, Drevets et al. (2002) reported that left amygdala metabolism was increased and positively correlated with stress-induced plasma cortisol concentrations in subjects with unipolar and bipolar depression. Finally, genetic variation within *5-HTTLPR* has been shown to functionally impact amygdala activity (Hariri et al. 2002), amygdala-cingulate interactions (Pezawas et al. 2005), and amygdala-hippocampal interactions in patients exposed to stressful experiences during childhood (Canli et al. 2006).

Structural imaging studies have identified bilateral increases in amygdala volume in patients during the first episode of major depression (Frodl et al. 2002a; Lange and Irle 2004), though not in patients with recurrent depression (Frodl et al. 2003). Increased right amygdala volume has been reported in women with depression and a history of suicidality (Monkul et al. 2007). Decreased amygdala volume (Rosso et al. 2005) and blunted amygdala activation have been reported in children with depression (Thomas et al. 2001). Sustained activation of the amygdala observed in unipolar depressed patients, relative to non-depressed controls, in response to negative information correlates with self-reported rumination (Siegle et al. 2002), as well as personal relevance rating of words (Siegle et al. 2007). Anticipation of aversive stimuli increases activation of the extended amygdala in women with major depression, compared to healthy controls (Abler et al. 2007).

Inflammation, Mood and Anxiety Disorders, and Stress

A growing body of evidence suggests that systemic inflammation may contribute to the pathophysiology of mood and anxiety disorders (Miller et al. 2009). Inflammation is a key component of the adaptive response to stress (Chrousos 1995). Acute stress results in the secretion of proinflammatory cytokines (McEwen 1998; Elenkov et al. 2005; Glaser and Kiecolt-Glaser 2005); peripheral secretion of acute phase proteins, such as fibrinogen and C-reactive protein (CRP) that promote resistance to infection and tissue repair; and transcription of nuclear factor (NF) kappa-B, a candidate mediator for some of the CNS effects related to the inflammatory response (Bierhaus et al. 2003). The latter is also a potentially important mediator in synaptic development within the central and peripheral nervous systems (Gutierrez et al. 2005). Another component of the adaptive response to acute stress is increased HPA axis activity and its end product, the increased secretion of glucocorticoids that act upon intracellular signaling pathways. This helps to terminate the inflammatory response after cessation of acute stress (Chrousos 1995). This relationship between the HPA axis and systemic inflammation may be especially relevant to mood and anxiety disorders in view of the findings that patients with a history of childhood stress have an impaired capacity to regulate the HPA axis in the presence of psychosocial stress (Heim et al. 2000), as well as an increase in inflammatory cytokines, suggesting that control of the inflammatory response may be impaired, as well.

A diverse and growing number of investigations indicate that systemic inflammation may play a key role in the etiology of a number of common medical diseases, including diabetes (Wellen and Hotamisligil 2005), cardiovascular disease (Willerson and Ridker 2004), cancer (Li, Withoff, and Verma 2005), and depression (Maes 1999; Evans et al. 2005; Raison, Capuron, and Miller 2006). Administration of proinflammatory cytokines to laboratory animals results in a pattern of behavioral and neurovegetative symptoms, known as "sickness behavior," that is similar to the depression observed in human patients and includes psychomotor slowing, fatigue, elevated pain sensitivity, disrupted sleep, and anxiety (Kent et al. 1992). Clinical research in humans has identified an array of proinflammatory cytokines, including interleukin(IL)-6, IL-1, and tumor necrosis factor-alpha (TNF-α) (Musselman et al. 2001b; Miller et al. 2002; Penninx et al. 2003; Alesci et al. 2005; Kahl et al. 2005), and inflammatory markers such as CRP (Miller et al. 2002; Danner et al. 2003; Penninx et al. 2003; Ford and Erlinger 2004; Panagiotakos et al. 2004; Kahl et al. 2005) that are elevated in patients with depression. Conversely, dysfunction of the anti-inflammatory actions of T lymphocytes (T cells) may worsen the effects of proinflammatory cytokines and downstream pathological effects (Miller 2009). Humans treated with interferon-α, which promotes the release of IL-6, IL-1b and TNF-α, commonly develop symptoms of depression; a significant subset of these individuals fulfill diagnostic criteria for an episode of major depression (Musselman et al. 2001a; Raison et al. 2005), which may be effectively treated with SSRIs (Musselman et al. 2001a).

Inflammation may be particularly relevant to three populations of depressed patients: those with treatment-resistant depression (Sluzewska, Sobieska, and Rybakowski 1997; Maes 1999), chronic medical illness comorbid with depression (Evans et al. 2005), and depression associated with early-life stress (Pace et al. 2006; Danese et al. 2007). In some patients, exposure to trauma or stressful childhood experiences may be a variable connecting risk for depression, chronic medical illness, and inflammation. Thus, patients with childhood trauma have an elevated risk not only for depression (Kendler, Kuhn, and Prescott 2004), which is often chronic in course (Kaplan and Klinetob 2000; Lara, Klein, and Kasch 2000; Hayden and Klein 2001), but also for a variety of common, chronic, and progressive medical diseases, including coronary artery disease (Goodwin and Stein 2004). Elevated levels of inflammatory markers, including CRP, fibrinogen, and IL-6, are associated with psychosocial risk factors for cardiovascular disease (Ranjit et al. 2007), and a graded relationship exists between childhood maltreatment and elevations in plasma CRP concentrations in adulthood (Danese et al. 2007; 2008). Chronic low-grade elevation of CRP has also been observed in patients with PTSD, as has the presence of elevated levels of multiple proinflammatory cytokines in patients with PTSD and panic disorder (Hoge et al. 2009), conditions commonly seen in patients with depression and a history of child abuse. Finally, depressed men with a history of childhood stress demonstrate an increased inflammatory response to acute psychosocial stress (Pace et al. 2006).

Conclusions

Much of the standardized component of medical practice has historically relied to a great extent on data gathered from cohorts of patients recruited for participation in epidemiologic studies and clinical trials. Data derived from these investigations have been used to identify patients at risk for disease, formulate mechanisms of disease prevention, and generate evidence-based, disease-specific treatment guidelines. Although such population-scale methods have contributed significantly to our present understanding of the architecture of disease risk and the proper implementation of preventative measures and treatments in groups of patients, they have not always been universally effective in preventing and treating disease at the level of the individual patient. The combination of genomics, brain imaging, and other body biomarkers will likely result in the rapid advancement of a more personalized medicine in psychiatry—the right treatment, for the right patient, at the right time.

References

Abercrombie, H. C., S. M. Schaefer, C. L. Larson, T. R. Oakes, K. A. Lindgren, J. E. Holden, S. B. Perlman, P. A. Turski, D. D. Krahn, R. M. Benca, and R.J. Davidson. 1998. Metabolic rate in the right amygdala predicts negative affect in depressed patients. *Neuroreport.* 9: 3301–07.

Abler, B., S. Erk, U. Herwig, and H. Walter. 2007. Anticipation of aversive stimuli activates extended amygdala in unipolar depression. *Journal of Psychiatric Research* 41: 511–522.

Alesci, S., P. E. Martinez, S. Kelkar, I. Ilias, D. S. Ronsaville, S. J. Listwak, A. R. Ayala, J. Licinio, H. K. Gold, M. A. Kling, G. P. Chrousos, and P. W. Gold. 2005. Major depression is associated with significant diurnal elevations in plasma interleukin-6 levels, a shift of its circadian rhythm, and loss of physiological complexity in its secretion: Clinical implications. *The Journal of Clinical Endocrinology and Metabolism* 90: 2522–30.

Arato, M., C. M. Banki, G. Bissette, and C. B. Nemeroff. 1989. Elevated CSF CRF in suicide victims. *Biological Psychiatry* 25: 355–59.

Arborelius, L., M. J. Owens, P. M. Plotsky, and C. B. Nemeroff. 1999. The role of corticotropin-releasing factor in depression and anxiety disorders. *J Endocrinol.* 160: 1–12.

Baker, D. G., S. A. West, W. E. Nicholson, N. N. Ekhator, J. W. Kasckow, K. K. Hill, A. B. Bruce, D. N. Orth, and T. D. Geracioti, Jr. 1999. Serial CSF corticotropin-releasing hormone levels and adrenocortical activity in combat veterans with posttraumatic stress disorder. *The American Journal of Psychiatry* 156: 585–88.

Banki, C. M., L. Karmacsi, G. Bissette, and C. B. Nemeroff. 1992. CSF corticotropin-releasing hormone and somatostatin in major depression: Response to antidepressant treatment and relapse. *Eur Neuropsychopharmacol.* 2: 107–13.

Barbas, H., S. Saha, N. Rempel-Clower, and T. Ghashghaei. 2003. Serial pathways from primate prefrontal cortex to autonomic areas may influence emotional expression. *BMC Neuroscience* 4: 25.

Batten, S. V., M. Aslan, P. K. Maciejewski, and C. M. Mazure. 2004. Childhood maltreatment as a risk factor for adult cardiovascular disease and depression. *The Journal of Clinical Psychiatry* 65: 249–54.

Benabid, A. L. 2003. Deep brain stimulation for Parkinson's disease. *Current Opinion in Neurobiology* 13: 696–706.

Bierhaus, A., J. Wolf, M. Andrassy, N. Rohleder, P. M. Humpert, D. Petrov, et al. 2003. A mechanism converting psychosocial stress into mononuclear cell activation. *Proceedings of the National Academy of Sciences of the United States of America* 100: 1920–25.

Binder, E. B. 2009. The role of FKBP5, a co-chaperone of the glucocorticoid receptor in the pathogenesis and therapy of affective and anxiety disorders. *Psychoneuroendocrinology* 34: S186–95.

Binder, E. B., R. G. Bradley, W. Liu, M. P. Epstein, T. C. Deveau, K. B. Mercer, et al. 2008. Association of FKBP5 polymorphisms and childhood abuse with risk of posttraumatic stress disorder symptoms in adults. *JAMA.* 299: 1291–1305.

Binder, E. B., D. Salyakina, P. Lichtner, G. M. Wochnik, M. Ising, B. Putz, et al. 2004. Polymorphisms in FKBP5 are associated with increased recurrence of depressive episodes and rapid response to antidepressant treatment. *Nature Genetics* 36: 1319–25.

Bishop, S. J. 2008. Neural mechanisms underlying selective attention to threat. *Annals of the New York Academy of Sciences* 1129: 141–152.

Bradley, R. G., E. B. Binder, M. P. Epstein, Y. Tang, H. P. Nair, W. Liu, et al. 2008. Influence of child abuse on adult depression: Moderation by the corticotropin-releasing hormone receptor gene. *Archives of General Psychiatry* 65: 190–200.

Bremner, J., M. Narayan, E. Andersen, H. Miller, and D. Charney. 2000. Hippocampal volume reduction in major depression. *The American Journal of Psychiatry* 157: 115–18.

Bremner, J. D., J. Licinio, A. Darnell, J. H. Krystal, M. J. Owens, S. M. Southwick, et al. 1997a. Elevated CSF corticotropin-releasing factor concentrations in posttraumatic stress disorder. *The American Journal of Psychiatry* 154: 624–29.

Bremner, J. D., P. Randall, T. M. Scott, R. A. Bronen, J. P. Seibyl, S. M. Southwick, et al. 1995. MRI-based measurement of hippocampal volume in patients with combat-related posttraumatic stress disorder. *The American Journal of Psychiatry* 152: 973–81.

Bremner, J. D., P. Randall, E. Vermetten, L. Staib, R. A. Bronen, C. Mazure, et al. 1997b. Magnetic resonance imaging-based measurement of hippocampal volume in posttraumatic stress disorder related to childhood physical and sexual abuse a preliminary report. *Biological Psychiatry* 41: 23–32.

Bremner, J. D., M. Vythilingam, E. Vermetten, S. M. Southwick, T. McGlashan, A. Nazeer, et al. 2003. MRI and PET study of deficits in hippocampal structure and function in women with childhood sexual abuse and posttraumatic stress disorder. *The American Journal of Psychiatry* 160: 924–32.

Brody, A. L., M. W. Barsom, R. G. Bota, and S. Saxena. 2001. Prefrontal-subcortical and limbic circuit mediation of major depressive disorder. *Seminars in Clinical Neuropsychiatry* 6: 102–12.

Brown, E. S., F. P. Varghese, and B. S. McEwen. 2004. Association of depression with medical illness: Does cortisol play a role? *Biological Psychiatry* 55: 1–9.

Campbell, S., and G. Macqueen. 2004. The role of the hippocampus in the pathophysiology of major depression. *J Psychiatry Neurosci.* 29: 417–26.

Canli, T., M. Qiu, K. Omura, E. Congdon, B. W. Haas, Z. Amin, et al. 2006. Neural correlates of epigenesis. *Proceedings of the National Academy of Sciences of the United States of America* 103: 16033–38.

Carmichael, S. T., and J. L. Price. 1996. Connectional networks within the orbital and medial prefrontal cortex of macaque monkeys. *The Journal of Comparative Neurology* 371: 179–207.

Carpenter, L. L., J. P. Carvalho, A. R. Tyrka, L. M. Wier, A. F. Mello, M. F. Mello, et al. 2007. Decreased adrenocorticotropic hormone and cortisol responses to stress in healthy adults reporting significant childhood maltreatment. *Biological Psychiatry* 62: 1080–87.

Carpenter, W. T., Jr., and W. E. Bunney, Jr. 1971. Adrenal cortical activity in depressive illness. *The American Journal of Psychiatry* 128: 31–40.

Caspi, A., H. Harrington, T. E. Moffitt, B. J. Milne, and R. Poulton. 2006. Socially isolated children 20 years later: Risk of cardiovascular disease. *Arch Pediatr Adolesc Med.* 160: 805–11.

Chalmers, D. T., T. W. Lovenberg, D. E. Grigoriadis, D. P. Behan, and E. B. De Souza. 1996. Corticotrophin-releasing factor receptors: From molecular biology to drug design. *Trends Pharmacol Sci.* 17: 166–72.

Chapman, D. P., C. L. Whitfield, V. J. Felitti, S. R. Dube, V. J. Edwards, and R. F. Anda. 2004. Adverse childhood experiences and the risk of depressive disorders in adulthood. *Journal of Affective Disorders* 82: 217–25.

Chrousos, G. P. 1995. The hypothalamic-pituitary-adrenal axis and immune-mediated inflammation. *The New England Journal of Medicine* 332: 1351–62.

Danese, A., T. E. Moffitt, C. M. Pariante, A. Ambler, R. Poulton, and A. Caspi. 2008. Elevated inflammation levels in depressed adults with a history of childhood maltreatment. *Archives of General Psychiatry* 65: 409–15.

Danese, A., C. M. Pariante, A. Caspi, A. Taylor, and R. Poulton. 2007. Childhood maltreatment predicts adult inflammation in a life-course study. *Proceedings of the National Academy of Sciences of the United States of America* 104: 1319–24.

Danner, M., S. V. Kasl, J. L. Abramson, and V. Vaccarino. 2003. Association between depression and elevated C-reactive protein. *Psychosomatic Medicine* 65: 347–56.

Davidson, R. J., D. Pizzagalli, J. B. Nitschke, and K. Putnam. 2002. Depression: Perspectives from affective neuroscience. *Annual Review of Psychology* 53: 545–74.

De Bellis, M. D., P. W. Gold, T. D. Geracioti, Jr., S. J. Listwak, and M. A. Kling. 1993. Association of fluoxetine treatment with reductions in CSF concentrations of corticotropin-releasing hormone and arginine vasopressin in patients with major depression. *The American Journal of Psychiatry* 150: 656–57.

De Kloet, E. R., M. Joels, M. Oitzl, and W. Sutanto. 1991. Implication of brain corticosteroid receptor diversity for the adaptation syndrome concept. *Methods Achiev Exp Pathol.* 14: 104–32.

Denny, W. B., D. L. Valentine, P. D. Reynolds, D. F. Smith, and J. G. Scammell. 2000. Squirrel monkey immunophilin FKBP51 is a potent inhibitor of glucocorticoid receptor binding. *Endocrinology* 141: 4107–13.

Dong, M., W. H. Giles, V. J. Felitti, S. R. Dube, J. E. Williams, D. P. Chapman, and R. F. Anda. 2004. Insights into causal pathways for ischemic heart disease: Adverse childhood experiences study. *Circulation* 110: 1761–66.

Dorn, L. D., E. S. Burgess, B. Dubbert, S. E. Simpson, T. Friedman, M. Kling, P. W. Gold, and G. P. Chrousos. 1995. Psychopathology in patients with endogenous Cushing's syndrome: "Atypical" or melancholic features. *Clinical Endocrinology* 43: 433–42.

Dorn, L. D., E. S. Burgess, T. C. Friedman, B. Dubbert, P. W. Gold, and G. P. Chrousos. 1997. The longitudinal course of psychopathology in Cushing's syndrome after correction of hypercortisolism. *J Clin Endocrinol Metab.* 82: 912–19.

Dougherty, D. D., A. P. Weiss, G. R. Cosgrove, N. M. Alpert, E. H. Cassem, A. A. Nierenberg, B. H. Price, H. S. Mayberg, A. J. Fischman, and S. L. Rauch. 2003. Cerebral metabolic correlates as potential predictors of response to anterior cingulotomy for treatment of major depression. *Journal of Neurosurgery* 99: 1010–17.

Drevets, W. C. 1999. Prefrontal cortical-amygdalar metabolism in major depression. *Annals of the New York Academy of Sciences* 877: 614–37.

Drevets, W. C., J. L. Price, M. E. Bardgett, T. Reich, R. D. Todd, and M. E. Raichle. 2002. Glucose metabolism in the amygdala in depression: Relationship to diagnostic subtype and plasma cortisol levels. *Pharmacology Biochemistry and Behavior* 71: 431–47.

Driessen, M., J. Herrmann, K. Stahl, M. Zwaan, S. Meier, A. Hill, M. Osterheider, and D. Petersen. 2000. Magnetic resonance imaging volumes of the hippocampus and the amygdala in women with borderline personality disorder and early traumatization. *Archives of General Psychiatry* 57: 1115–22.

Dube, S. R., R. F. Anda, V. J. Felitti, D. P. Chapman, D. F. Williamson, and W. H. Giles. 2001. Childhood abuse, household dysfunction, and the risk of attempted suicide throughout the life span: Findings from the adverse childhood experiences study. *JAMA* 286: 3089–96.

Duman, R. S., and L. M. Monteggia. 2006. A neurotrophic model for stress-related mood disorders. *Biological Psychiatry* 59: 1116–27.

Dunn, A. J., and C. W. Berridge. 1990. Physiological and behavioral responses to corticotropin-releasing factor administration: Is CRF a mediator of anxiety or stress responses? *Brain Res Brain Res Rev.* 15: 71–100.

Elenkov, I. J., D. G. Iezzoni, A. Daly, A. G. Harris, and G. P. Chrousos. 2005. Cytokine dysregulation, inflammation and well-being. *Neuroimmunomodulation* 12: 255–69.

Evans, D. L., D. S. Charney, L. Lewis, R. N. Golden, J. M. Gorman, K. R. Krishnan, C. B. Nemeroff, J. D. Bremner, R. M. Carney, J. C. Coyne, M. R. Delong, N. Frasure-Smith, A. H. Glassman, P. W. Gold, I. Grant, L. Gwyther, G. Ironson, R. L. Johnson, A. M. Kanner, W. J. Katon, P. G. Kaufmann, F. J. Keefe, T. Ketter, T. P. Laughren, J. Leserman, C. G. Lyketsos, W. M. McDonald, B. S. McEwen, A. H. Miller, D. Musselman, C. O'Connor, J. M. Petitto, B. G. Pollock, R. G. Robinson, S. P. Roose, J. Rowland, Y. Sheline, D. S. Sheps, G. Simon, D. Spiegel, A. Stunkard, T. Sunderland, P. Tibbits, Jr., and W. J. Valvo. 2005. Mood disorders in the medically ill: Scientific review and recommendations. *Biological Psychiatry* 58: 175–89.

Feder, A., E. J. Nestler, and D. S. Charney. 2009. Psychobiology and molecular genetics of resilience. *Nat Rev Neurosci.* 10: 446–57.

Felitti, V. J., R. F. Anda, D. Nordenberg, D. F. Williamson, A. M. Spitz, V. Edwards, M. P. Koss, and J. S. Marks. 1998. Relationship of childhood abuse and household dysfunction to many of the leading causes of death in adults. The Adverse Childhood Experiences ACE) Study. *American Journal of Preventive Medicine* 14: 245–58.

Ford, D. E., and T. P. Erlinger. 2004. Depression and C-reactive protein in US adults: Data from the Third National Health and Nutrition Examination Survey. *Archives of Internal Medicine* 164: 1010–14.

Freedman, L. J., T. R. Insel, and Y. Smith. 2000. Subcortical projections of area 25 subgenual cortex) of the macaque monkey. *The Journal of Comparative Neurology* 421: 172–88.

Frodl, T., E. Meisenzahl, T. Zetzsche, R. Bottlender, C. Born, C. Groll, C. Born, C. Groll, M. Jager, G. Leinsinger, R. Bottlender, K. Hahn, and H. J. Moller. 2002a. Enlargement of the amygdala in patients with a first episode of major depression. *Biological Psychiatry* 51: 708–14.

Frodl, T., E. M. Meisenzahl, T. Zetzsche, C. Born, C. Groll, M. Jager, C. Groll, R. Bottlender, G. Leinsinger, and H. J. Moller. 2002b. Hippocampal changes in patients with a first episode of major depression. *The American Journal of Psychiatry* 159: 1112–18.

Frodl, T., E. M. Meisenzahl, T. Zetzsche, C. Born, M. Jager, C. Groll C, M. Jager, G. Leinsinger, K. Hahn, and H. J. Moller. 2003. Larger amygdala volumes in first depressive episode as compared to recurrent major depression and healthy control subjects. *Biological Psychiatry* 53: 338–44.

Fuchs, E., and E. Gould. 2000. Mini-review: In vivo neurogenesis in the adult brain: Regulation and functional implications. *The European Journal of Neuroscience* 12: 2211–14.

Gatt, J. M., C. B. Nemeroff, C. Dobson-Stone, R. H. Paul, R. A. Bryant, P. R. Schofield, E. Gordon, A. H. Kemp, and L. M. Williams. 2009. Interactions between BDNF Val66Met polymorphism and early life stress predict brain and arousal pathways to syndromal depression and anxiety. *Molecular Psychiatry* 14: 681–95.

Geuze, E., E. Vermetten, and J. D. Bremner. 2005. MR-based in vivo hippocampal volumetrics: 2. Findings in neuropsychiatric disorders. *Molecular Psychiatry* 10: 160–84.

Gibbons, J. L., and P. R. McHugh. 1962. Plasma cortisol in depressive illness. *Journal of Psychiatric Research* 1: 162–71.

Gladstone, G. L., G. B. Parker, P. B. Mitchell, G. S. Malhi, K. Wilhelm, and M. P. Austin. 2004. Implications of childhood trauma for depressed women: An analysis of pathways from childhood sexual abuse to deliberate self-harm and revictimization. *The American Journal of Psychiatry* 161: 1417–25.

Glaser, R., and J. K. Kiecolt-Glaser. 2005. Stress-induced immune dysfunction: Implications for health. *Nat Rev Immunol.* 5: 243–51.

Goldapple, K., Z. Segal, C. Garson, M. Lau, P. Bieling, S. Kennedy, and H. Mayberg. 2004. Modulation of cortical-limbic pathways in major depression: treatment-specific effects of cognitive behavior therapy. *Archives of General Psychiatry* 61: 34–41.

Goodwin, R. D., and J. R. Davidson. 2005. Self-reported diabetes and posttraumatic stress disorder among adults in the community. *Prev Med.* 40: 570–74.

Goodwin, R. D., and M. B. Stein. 2004. Association between childhood trauma and physical disorders among adults in the United States. *Psychological Medicine* 34: 509–20.

Gunstad, J., R. H. Paul, M. B. Spitznagel, R. A. Cohen, L. M. Williams, M. Kohn, and E. Gordon. 2006. Exposure to early life trauma is associated with adult obesity. *Psychiatry Research* 142: 31–37.

Gutierrez, H., V. A. Hale, X. Dolcet, and A. Davies. 2005. NF-kappaB signalling regulates the growth of neural processes in the developing PNS and CNS. *Development* 132: 1713–26.

Gutman, D. A., M. J. Owens, K. H. Skelton, K. V. Thrivikraman, and C. B. Nemeroff. 2003. The corticotropin-releasing factor1 receptor antagonist R121919 attenuates the behavioral and endocrine responses to stress. *J Pharmacol Exp Ther.* 304: 874–80.

Haber, S. N. 2003. The primate basal ganglia: Parallel and integrative networks. *Journal of Chemical Neuroanatomy* 26: 317–30.

Hamann, S., and H. Mao. 2002. Positive and negative emotional verbal stimuli elicit activity in the left amygdala. *Neuroreport* 13: 15–19.

Hariri, A. R., V. S. Mattay, A. Tessitore, B. Kolachana, F. Fera, D. Goldman, et al. 2002. Serotonin transporter genetic variation and the response of the human amygdala. *Science New York, NY* 297: 400–03.

Hartline, K. M., M. J. Owens, and C. B. Nemeroff. 1996. Postmortem and cerebrospinal fluid studies of corticotropin-releasing factor in humans. *Annals of the New York Academy of Sciences* 780: 96–105.

Hayden, E. P., and D. N. Klein. 2001. Outcome of dysthymic disorder at 5-year follow-up: The effect of familial psychopathology, early adversity, personality, comorbidity, and chronic stress. *The American Journal of Psychiatry* 158: 1864–70.

Heim, C., T. Mletzko, D. Purselle, D. L. Musselman, and C. B. Nemeroff. 2008a. The dexamethasone/corticotropin-releasing factor test in men with major depression: Role of childhood trauma. *Biological Psychiatry* 63: 398–405.

Heim, C., D. J. Newport, R. Bonsall, A. H. Miller, and C. B. Nemeroff. 2001. Altered pituitary-adrenal axis responses to provocative challenge tests in adult survivors of childhood abuse. *The American Journal of Psychiatry* 158: 575–81.

Heim, C., D. J. Newport, S. Heit, Y. P. Graham, M. Wilcox, R. Bonsall, A. H. Miller, and C. B. Nemeroff. 2000. Pituitary-adrenal and autonomic responses to stress in women after sexual and physical abuse in childhood. *JAMA* 284: 592–97.

Heim, C., D. J. Newport, T. Mletzko, A. H. Miller, and C. B. Nemeroff. 2008b. The link between childhood trauma and depression: Insights from HPA axis studies in humans. *Psychoneuroendocrinology* 33: 693–710.

Heim, C. M., B. Bradley, T. Mletzko, T. C. Deveau, D. L. Musselman, C. B. Nemeroff, K.J. Ressler, and E.B. Binder. 2009. Effect of childhood trauma on adult depression and neuroendocrine function: Sex-specific moderation by CRH receptor 1 gene. *Front Behav Neurosci.* 3: 41.

Heinrichs, S. C., J. Lapsansky, T. W. Lovenberg, E. B. De Souza, and D. T. Chalmers. 1997. Corticotropin-releasing factor CRF1, but not CRF2, receptors mediate anxiogenic-like behavior. *Regul Pept.* 71: 15–21.

Heinrichs, S. C., F. Menzaghi, E. Merlo Pich, K. T. Britton, and G. F. Koob. 1995. The role of CRF in behavioral aspects of stress. *Annals of the New York Academy of Sciences* 771: 92–104.

Hoge, E. A., K. Brandstetter, S. Moshier, M. H. Pollack, K. K. Wong, and N. M. Simon. 2009. Broad spectrum of cytoK. kine abnormalities in panic disorder and posttraumatic stress disorder. *Depression and anxiety* 26: 447–455.

Isenberg, N., D. Silbersweig, A. Engelien, S. Emmerich, K. Malavade, B. Beattie, A. C. Leon, and E. Stern. 1999. Linguistic threat activates the human amygdala. *Proceedings of the National Academy of Sciences of the United States of America* 96: 10456–59.

Ising, M., A. M. Depping, A. Siebertz, S. Lucae, P. G. Unschuld, S. Kloiber, S. Horstmann, M. Uhr, B. Muller-Myhsok, and F. Holsboer. 2008. Polymorphisms in the FKBP5 gene region modulate recovery from psychosocial stress in healthy controls. *The European Journal of Neuroscience* 28: 389–98.

Jacobson, L., and R. Sapolsky. 1991. The role of the hippocampus in feedback regulation of the hypothalamic-pituitary-adrenocortical axis. *Endocr Rev.* 12: 118–34.

Joels, M., and E. R. de Kloet. 1994. Mineralocorticoid and glucocorticoid receptors in the brain. Implications for ion permeability and transmitter systems. *Prog Neurobiol.* 43: 1–36.

Jurgens, U., and P. Muller-Preuss. 1977. Convergent projections of different limbic vocalization areas in the squirrel monkey. *Exp* Brain Res. 29: 75–83.

Kahl, K. G., S. Rudolf, B. M. Stoeckelhuber, L. Dibbelt, H. B. Gehl, K. Markhof, F. Hohagen, and U. Schweiger. 2005. Bone mineral density, markers of bone turnover, and cytokines in young women with borderline personality disorder with and without comorbid major depressive disorder. *The American Journal of Psychiatry* 162: 168–74.

Kaplan, M. J., and N. A. Klinetob. 2000. Childhood emotional trauma and chronic posttraumatic stress disorder in adult outpatients with treatment-resistant depression. *The Journal of Nervous and Mental Disease* 188: 596–601.

Kendler, K. S., J. W. Kuhn, and C. A. Prescott. 2004. Childhood sexual abuse, stressful life events and risk for major depression in women. *Psychological Medicine* 34: 1475–82.

Kent, S., R. M. Bluthe, K. W. Kelley, and R. Dantzer. 1992. Sickness behavior as a new target for drug development. *Trends Pharmacol Sci* .13: 24–28.

Kitayama, N., V. Vaccarino, M. Kutner, P. Weiss, and J. D. Bremner. 2005. Magnetic resonance imaging MRI) measurement of hippocampal volume in posttraumatic stress disorder: A meta-analysis. *Journal of Affective Disorders* 88: 79–86.

Koenen, K. C., G. Saxe, S. Purcell, J. W. Smoller, D. Bartholomew, A. Miller, E. Hall, J. Kaplow, M. Bosquet, S. Moulton, and C. Baldwin. 2005. Polymorphisms in FKBP5 are associated with peritraumatic dissociation in medically injured children. *Molecular Psychiatry* 10: 1058–59.

Lange, C., and E. Irle. 2004. Enlarged amygdala volume and reduced hippocampal volume in young women with major depression. *Psychological Medicine* 34: 1059–64.

Lara, M. E., D. N. Klein, and K. L. Kasch. 2000. Psychosocial predictors of the short-term course and outcome of major depression: A longitudinal study of a nonclinical sample with recent-onset episodes. *Journal of Abnormal Psychology* 109: 644–50.

Lekman, M., G. Laje, D. Charney, A. J. Rush, A. F. Wilson, A. J. Sorant, R. Lipsky, S. R. Wisniewski, H. Manji, F. J. McMahon, and S. Paddock. 2008. The FKBP5-gene in depression and treatment response an association study in the Sequenced Treatment Alternatives to Relieve Depression (STAR*D) Cohort. *Biological Psychiatry* 63: 1103–10.

Lenze, S. N., C. Xiong, and Y. I. Sheline. 2008. Childhood adversity predicts earlier onset of major depression but not reduced hippocampal volume. *Psychiatry Research* 162: 39–49.

Leppanen, J. M. 2006. Emotional information processing in mood disorders: A review of behavioral and neuroimaging findings. *Current Opinion in Psychiatry* 19: 34–39.

Li, Q., S. Withoff, and I. M. Verma. 2005. Inflammation-associated cancer: NF-kappaB is the lynchpin. *Trends in Immunology* 26: 318–25.

Licinio, J., F. O'Kirwan, K. Irizarry, B. Merriman, S. Thakur, R. Jepson, S. Lake, K. G. Tantisira, S. T. Weiss, and M. L. Wong. 2004. Association of a corticotropin-releasing hormone receptor 1 haplotype and antidepressant treatment response in Mexican-Americans. *Molecular Psychiatry* 9: 1075–82.

Lissau, I., and T. I. Sorensen. 1994. Parental neglect during childhood and increased risk of obesity in young adulthood. *Lancet* 343: 324–27.

Liu, Z., F. Zhu, G. Wang, Z. Xiao, J. Tang, W. Liu, H. Wang, H. Liu, X. Wang, Y. Wu, Z. Cao, and W. Li. 2007. Association study of corticotropin-releasing hormone receptor1 gene polymorphisms and antidepressant response in major depressive disorders. *Neuroscience Letters* 414: 155–58.

Liu, Z., F. Zhu, G. Wang, Z. Xiao, H. Wang, J. Tang J, X. Wang, D. Qiu, W. Liu, Z. Cac, and W. Li. 2006. Association of corticotropin-releasing hormone receptor1 gene SNP and haplotype with major depression. *Neuroscience Letters* 404: 358–62.

Maes, M. 1999. Major depression and activation of the inflammatory response system. *Advances in Experimental Medicine and Biology* 461: 25–46.

Magarinos, A. M., and B. S. McEwen. 1995a. Stress-induced atrophy of apical dendrites of hippocampal CA3c neurons: Comparison of stressors. *Neuroscience* 69: 83–88.

Magarinos, A. M., and B. S. McEwen. 1995b. Stress-induced atrophy of apical dendrites of hippocampal CA3c neurons: Involvement of glucocorticoid secretion and excitatory amino acid receptors. *Neuroscience* 69: 89–98.

Magarinos, A. M., B. S. McEwen, G.Flugge, and E. Fuchs. 1996. Chronic psychosocial stress causes apical dendritic atrophy of hippocampal CA3 pyramidal neurons in subordinate tree shrews. *J Neurosci.* 16: 3534–40.

Magarinos, A. M., J. M. Verdugo, and B. S. McEwen. 1997. Chronic stress alters synaptic terminal structure in hippocampus. *Proceedings of the National Academy of Sciences of the United States of America* 94: 14002–08.

Mayberg, H. S. 1994. Frontal lobe dysfunction in secondary depression. *The Journal of Neuropsychiatry and Clinical Neurosciences* 6: 428–42.

Mayberg, H. S., S. K. Brannan, J. L. Tekell, J. A. Silva, R. K. Mahurin, S. McGinnis, and P. A. Jerabek. 2000. Regional metabolic effects of fluoxetine in major depression: serial changes and relationship to clinical response. *Biological Psychiatry* 48: 830–43.

Mayberg, H. S., M. Liotti, S. K. Brannan, S. McGinnis, R. K. Mahurin, P. A. Jerabek, J. A. Silva, J. L. Tekell, C. C. Martin, J. L. Lancaster, and P. T. Fox. 1999. Reciprocal limbic-cortical function and negative mood: Converging PET findings in depression and normal sadness. *The American Journal of Psychiatry* 156: 675–82.

Mayberg, H. S., A. M. Lozano, V. Voon, H. E. McNeely, D. Seminowicz, C. Hamani, J. M. Schwalb, and S. H. Kennedy. 2005. Deep brain stimulation for treatment-resistant depression. *Neuron* 45: 651–60.

McCauley, J., D. E. Kern, K. Kolodner, L. Dill, A. F. Schroeder, H. K. DeChant, J. Ryden, L. R. Derogatis, and E. B. Bass. 1997. Clinical characteristics of women with a history of childhood abuse: Unhealed wounds. *JAMA* 277: 1362–68.

McEwen, B. S. 1998. Protective and damaging effects of stress mediators. *The New England Journal of Medicine* 338: 171–79.

Merali, Z., L. Du, P. Hrdina, M. Palkovits, G. Faludi, M. O. Poulter, and H. Anisman. 2004. Dysregulation in the suicide brain: mRNA expression of corticotropin-releasing hormone receptors and GABAA) receptor subunits in frontal cortical brain region. *J Neurosci.* 24: 1478–85.

Merali, Z., P. Kent, L. Du, P. Hrdina, M. Palkovits, G. Faludi, M. O. Poulter, T. Bedard, and H. Anisman. 2006. Corticotropin-releasing hormone, arginine vasopressin, gastrin-releasing peptide, and neuromedin B alterations in stress-relevant brain regions of suicides and control subjects. *Biological Psychiatry* 59: 594–602.

Miller, A. H. 2009. Depression and immunity: A role for T cells? *Brain, behavior, and immunity* 24: 1–8.

Miller, A. H., V. Maletic, and C. L. Raison. 2009. Inflammation and its discontents: The role of cytokines in the pathophysiology of major depression. *Biological Psychiatry* 65: 732–41.

Miller, G. E., C. A. Stetler, R. M. Carney, K. E. Freedland, and W. A. Banks. 2002. Clinical depression and inflammatory risk markers for coronary heart disease. *Am J Cardiol.* 90: 1279–83.

Monkul, E. S., J. P. Hatch, M. A. Nicoletti, S. Spence, P. Brambilla, A. L. Lacerda, R. B. Sassi, A. G. Mallinger, M. S. Keshavan, and J. C. Soares. 2007. Fronto-limbic brain structures in suicidal and non-suicidal female patients with major depressive disorder. *Molecular Psychiatry* 12: 360–66.

Mottaghy, F. M., C. E. Keller, M. Gangitano, J. Ly, M. Thall, J. A. Parker, and A. Pascual-Leone. 2002. Correlation of cerebral blood flow and treatment effects of repetitive transcranial magnetic stimulation in depressed patients. *Psychiatry Research* 115: 1–14.

Musselman, D. L., D. H. Lawson, J. F. Gumnick, A. K. Manatunga, S. Penna, R. S. Goodkin, K. Greiner, C. B. Nemeroff, and A. H. Miller. 2001a. Paroxetine for the prevention of depression induced by high-dose interferon alfa. *The New England Journal of Medicine* 344: 961–66.

Musselman, D. L., A. H. Miller, M. R. Porter, A. Manatunga, F. Gao, S. Penna, B. D. Pearce, J. Landry, S. Glover, J. S. McDaniel, and C. B. Nemeroff. 2001b. Higher than normal plasma interleukin-6 concentrations in cancer patients with depression: preliminary findings. *The American Journal of Psychiatry* 158: 1252–57.

Nemeroff, C. B., G. Bissette, H. Akil, and M. Fink. 1991. Neuropeptide concentrations in the cerebrospinal fluid of depressed patients treated with electroconvulsive therapy. Corticotrophin-releasing factor, beta-endorphin and somatostatin. *Br J Psychiatry* 158: 59–63.

Nemeroff, C. B., M. J. Owens, G. Bissette, A. C. Andorn, and M. Stanley. 1988. Reduced corticotropin releasing factor binding sites in the frontal cortex of suicide victims. *Archives of General Psychiatry* 45: 577–79.

Nemeroff, C. B., E. Widerlov, G. Bissette, H. Walleus, I. Karlsson, K. Eklund, C. D. Kilts, P. T. Loosen, and W. Vale. 1984. Elevated concentrations of CSF corticotropin-releasing factor-like immunoreactivity in depressed patients. *Science New York, NY* 226: 1342–44.

Nestler, E. J., M. Barrot, R. J. DiLeone, A. J. Eisch, S. J. Gold, and L. M. Monteggia. 2002. Neurobiology of depression. *Neuron* 34: 13–25.

Neumeister, A., S. Wood, O. Bonne, A. C. Nugent, D. A. Luckenbaugh, T. Young, E. E. Bain, D. S. Charney, and W. C. Drevets. 2005. Reduced hippocampal volume in unmedicated, remitted patients with major depression versus control subjects. *Biological Psychiatry* 57: 935–37.

Nobler, M. S., M. A. Oquendo, L. S. Kegeles, K. M. Malone, C. C. Campbell, H. A. Sackeim, and J. J. Mann. 2001. Decreased regional brain metabolism after ECT. *The American Journal of Psychiatry* 158: 305–08.

Ongur, D., X. An, and J. L. Price. 1998. Prefrontal cortical projections to the hypothalamus in macaque monkeys. *The Journal of Comparative Neurology* 401: 480–505.

Owens, M. J., and C. B. Nemeroff. 1991. Physiology and pharmacology of corticotropin-releasing factor. *Pharmacol Rev.* 43: 425–73.

Ozer, E. J., S. R. Best, T. L. Lipsey, and D. S. Weiss. 2003. Predictors of posttraumatic stress disorder and symptoms in adults: A meta-analysis. *Psychological Bulletin* 129: 52–73.

Pace, T. W., T. C. Mletzko, O. Alagbe, D. L. Musselman, C. B. Nemeroff, A. H. Miller, et al. 2006. Increased stress-induced inflammatory responses in male patients with major depression and increased early life stress. *The American Journal of Psychiatry* 163: 1630–33.

Panagiotakos, D. B., C. Pitsavos, C. Chrysohoou, E. Tsetsekou, C. Papageorgiou, G. Christodoulou, and C. Stefanadis. 2004. Inflammation, coagulation, and depressive symptomatology in cardiovascular disease-free people: The ATTICA study. *Eur Heart J.* 25: 492–99.

Penninx, B. W., S. B. Kritchevsky, K. Yaffe, A. B. Newman, E. M. Simonsick, S. Rubin, L. Ferrucci, T. Harris, and M. Pahor. 2003. Inflammatory markers and depressed mood in older persons: Results from the Health, Aging and Body Composition study. *Biological Psychiatry* 54: 566–72.

Pezawas, L., A. Meyer-Lindenberg, E. M. Drabant, B. A. Verchinski, K. E. Munoz, B. S. Kolachana, M. F. Egan, V. S. Mattay, A. R. Hariri, and D. R. Weinberger. 2005. 5-HTTLPR polymorphism impacts human cingulate-amygdala interactions: A genetic susceptibility mechanism for depression. *Nature Neuroscience* 8: 828–34.

Pezawas, L., A. Meyer-Lindenberg, A. L. Goldman, B. A. Verchinski, G. Chen, B. S. Kolachana, M. F. Egan, V. S. Mattay, A. R. Hariri, and D. R. Weinberger. 2008. Evidence of biologic epistasis between BDNF and SLC6A4 and implications for depression. *Molecular Psychiatry* 13: 709–16.

Phan, K. L., T. D. Wager, S. F. Taylor, and I. Liberzon. 2004. Functional neuroimaging studies of human emotions. *CNS Spectrums* 9: 258–66.

Plotsky, P. M., M. J. Owens, and C. B. Nemeroff. 1998. Psychoneuroendocrinology of depression. Hypothalamic-pituitary-adrenal axis. *The Psychiatric Clinics of North America* 21: 293–307.

Polanczyk, G., A. Caspi, B. Williams, T. S. Price, A. Danese, K. Sugden, R. Uher, R. Poulton, and T. E. Moffitt. 2009. Protective effect of CRHR1 gene variants on the development of adult depression following childhood maltreatment: Replication and extension. *Archives of General Psychiatry* 66: 978–85.

Raadsheer, F. C., W. J. Hoogendijk, F. C. Stam, F. J. Tilders, and D. F. Swaab. 1994. Increased numbers of corticotropin-releasing hormone expressing neurons in the hypothalamic paraventricular nucleus of depressed patients. *Neuroendocrinology* 60: 436–44.

Raadsheer, F. C., J. J. van Heerikhuize, P. J. Lucassen, W. J. Hoogendijk, F. J. Tilders, and D. F. Swaab. 1995. Corticotropin-releasing hormone mRNA levels in the paraventricular nucleus of patients with Alzheimer's disease and depression. *The American Journal of Psychiatry* 152: 1372–76.

Raison, C. L., A. S. Borisov, S. D. Broadwell, L. Capuron, B. J. Woolwine, I. M. Jacobson, C. B. Nemeroff, and A. H. Miller. 2005. Depression during pegylated interferon-alpha plus ribavirin therapy: Prevalence and prediction. *The Journal of Clinical Psychiatry* 66: 41–48.

Raison, C. L., L. Capuron, and A. H. Miller. 2006. Cytokines sing the blues: Inflammation and the pathogenesis of depression. *Trends in Immunology* 27: 24–31.

Ranjit, N., A. V. Diez-Roux, S. Shea, M. Cushman, T. Seeman, S. A. Jackson, and H. Ni. 2007. Psychosocial factors and inflammation in the multi-ethnic study of atherosclerosis. *Archives of Internal Medicine* 167: 174–81.

Ressler, K. J., B. Bradley, K. B. Mercer, T. C. Deveau, A. K. Smith, C. F. Gillespie, C. B. Nemeroff, J. F. Cubells, and E. B. Binder. 2009. Polymorphisms in *CRHR1* and the serotonin transporter loci: Gene x gene x environment interactions on depressive symptoms. *American Journal of Medical Genetics Neuropsychiatric Genetics.* 153B: 112–24.

Rijsdijk, F. V., H. Snieder, J. Ormel, P. Sham, D. P. Goldberg, and T. D. Spector. 2003. Genetic and environmental influences on psychological distress in the population: General Health Questionnaire analyses in UK twins. *Psychological Medicine* 33: 793–801.

Roberson-Nay, R., E. B. McClure, C. S. Monk, E. E. Nelson, A. E. Guyer, S. J. Fromm, D. S. Charney, E. Leibenluft, J. Blair, M. Ernst, and D. S. Pine. 2006. Increased amygdala activity during successful memory encoding in adolescent major depressive disorder: An FMRI study. *Biological Psychiatry* 60: 966–73.

Rodrigues, S. M., J. E. LeDoux, and R. M. Sapolsky. 2009. The influence of stress hormones on fear circuitry. *Annual Review of Neuroscience* 32: 289–313.

Rosso, I. M., C. M. Cintron, R. J. Steingard, P. F. Renshaw, A. D. Young, and D. A. Yurgelun-Todd. 2005. Amygdala and hippocampus volumes in pediatric major depression. *Biological Psychiatry* 57: 21–26.

Rusch, B. D., H. C. Abercrombie, T. R. Oakes, S. M. Schaefer, and R. J. Davidson. 2001. Hippocampal morphometry in depressed patients and control subjects: Relations to anxiety symptoms. *Biological Psychiatry* 50: 960–64.

Rutter, M. 2006. Implications of resilience concepts for scientific understanding. *Annals of the New York Academy of Sciences* 1094: 1–12.

Sachar, E. J., L. Hellman, D. K. Fukushima, and T. F. Gallagher. 1970. Cortisol production in depressive illness. A clinical and biochemical clarification. *Archives of General Psychiatry* 23: 289–98.

Sapolsky, R. M., H. Uno, C. S. Rebert, and C. E. Finch. 1990. Hippocampal damage associated with prolonged glucocorticoid exposure in primates. *J Neurosci.* 10: 2897–2902.

Scammell, J. G., W. B. Denny, D. L. Valentine, and D. F. Smith. 2001. Overexpression of the FK506-binding immunophilin FKBP51 is the common cause of glucocorticoid resistance in three New World primates. *General and Comparative Endocrinology* 124: 152–165.

Schiene-Fischer, C., and C. Yu. 2001. Receptor accessory folding helper enzymes: The functional role of peptidyl prolyl cis/trans isomerases. *FEBS Lett.* 495: 1–6.

Sedlack, A. J., and D. D. Broadhurst 1996. *Third national incidence study of child abuse and neglect*. Washington, D. C.: U. S. Department of Health and Human Services.

Segman, R. H., N. Shefi, T. Goltser-Dubner, N. Friedman, N. Kaminski, and A. Y. Shalev. 2005. Peripheral blood mononuclear cell gene expression profiles identify emergent post-traumatic stress disorder among trauma survivors. *Molecular Psychiatry* 10: 500–13.

Seminowicz, D. A., H. S. Mayberg, A. R. McIntosh, K. Goldapple, S. Kennedy, Z. Segal, and S. Rafi-Tari. 2004. Limbic-frontal circuitry in major depression: A path modeling metanalysis. *NeuroImage* 22: 409–18.

Sheline, Y. I. 1996. Hippocampal atrophy in major depression: A result of depression-induced neurotoxicity? *Molecular Psychiatry* 1: 298–99.

Sheline, Y. I. 2003. Neuroimaging studies of mood disorder effects on the brain. *Biological Psychiatry* 54: 338–52.

Sheline, Y. I., D. M. Barch, J. M. Donnelly, J. M. Ollinger, A. Z. Snyder, and M. A. Mintun. 2001. Increased amygdala response to masked emotional faces in depressed subjects resolves with antidepressant treatment: An fMRI study. *Biological Psychiatry* 50: 651–58.

Sheline, Y. I., M. H. Gado, and H. C. Kraemer. 2003. Untreated depression and hippocampal volume loss. *The American Journal of Psychiatry* 160: 1516–18.

Sheline, Y. I., M. Sanghavi, M. A. Mintun, and M. H. Gado. 1999. Depression duration but not age predicts hippocampal volume loss in medically healthy women with recurrent major depression. *J Neurosci.* 19: 5034–43.

Sheline, Y. I., P. W.Wang, M. H. Gado, J. G. Csernansky, and M. W. Vannier. 1996. Hippocampal atrophy in recurrent major depression. *Proceedings of the National Academy of Sciences of the United States of America* 93: 3908–13.

Siegle, G. J., S. R. Steinhauer, M. E. Thase, V. A. Stenger, and C. S. Carter. 2002. Can't shake that feeling: Event-related fMRI assessment of sustained amygdala activity in response to emotional information in depressed individuals. *Biological Psychiatry* 51: 693–707.

Siegle, G. J., W. Thompson, C. S. Carter, S. R. Steinhauer, and M. E. Thase. 2007. Increased amygdala and decreased dorsolateral prefrontal BOLD responses in unipolar depression: Related and independent features. *Biological Psychiatry* 61: 198–209.

Sluzewska, A., M. Sobieska, and J. K. Rybakowski. 1997. Changes in acute-phase proteins during lithium potentiation of antidepressants in refractory depression. *Neuropsychobiology* 35: 123–27.

Starkman, M. N., S. S. Gebarski, S. Berent, and D. E. Schteingart. 1992. Hippocampal formation volume, memory dysfunction, and cortisol levels in patients with Cushing's syndrome. *Biological Psychiatry* 32: 756–65.

Stein, M. B., K. L. Jang, S. Taylor, P. A. Vernon, and W. J. Livesley. 2002. Genetic and environmental influences on trauma exposure and posttraumatic stress disorder symptoms: A twin study. *The American Journal of Psychiatry* 159: 1675–81.

Stein, M. B., C. Koverola, C. Hanna, M. G. Torchia, and B. McClarty. 1997. Hippocampal volume in women victimized by childhood sexual abuse. *Psychological Medicine* 27: 951–59.

Sullivan, P. F., M. C. Neale, and K. S. Kendler. 2000. Genetic epidemiology of major depression: Review and meta-analysis. *The American Journal of Psychiatry* 157: 1552–62.

Swanson, L. W., P. E. Sawchenko, J. Rivier, and W. W. Vale. 1983. Organization of ovine corticotropin-releasing factor immunoreactive cells and fibers in the rat brain: An immunohistochemical study. *Neuroendocrinology* 36: 165–86.

Thomas, K. M., W. C. Drevets, R. E. Dahl, N. D. Ryan, B. Birmaher, C. H. Eccard, D. Axelson, P. J. Whalen, and B. J. Casey. 2001. Amygdala response to fearful faces in anxious and depressed children. *Archives of General Psychiatry* 58: 1057–63.

Tyrka, A. R., L. H. Price, J. Gelernter, C. Schepker, G. M. Anderson, and L. L. Carpenter. 2009. Interaction of childhood maltreatment with the corticotropin-releasing hormone receptor gene: Effects on hypothalamic-pituitary-adrenal axis reactivity. *Biological Psychiatry* 66: 681–85.

Uno, H., S. Eisele, A. Sakai, S. Shelton, E. Baker, Q. DeJesus, and J. Holden. 1994. Neurotoxicity of glucocorticoids in the primate brain. *Hormones and Behavior* 28: 336–48.

Uno, H., R. Tarara, J. G. Else, M. A. Suleman, and R. M. Sapolsky. 1989. Hippocampal damage associated with prolonged and fatal stress in primates. *J Neurosci.* 9: 1705–11.

Utge, S., P. Soronen, T. Partonen, A. Loukola, E. Kronholm, S. Pirkola, E. Nyman, T. Porkka-Heiskanen, and T. Paunio. 2009. A population-based association study of candidate genes for depression and sleep disturbance. *Am J Med Genet B Neuropsychiatr Genet.* 153B: 468–76.

Vakili, K., S. S. Pillay, B. Lafer, M. Fava, P. F. Renshaw, C. M. Bonello-Cintron, and D. A. Yurgelun-Todd. 2000. Hippocampal volume in primary unipolar major depression: A magnetic resonance imaging study. *Biological Psychiatry* 47: 1087–90.

Vermeer, H., B. I. Hendriks-Stegeman, B. van der Burg, S. C. van Buul-Offers, and M. Jansen. 2003. Glucocorticoid-induced increase in lymphocytic FKBP51 messenger ribonucleic acid expression: A potential marker for glucocorticoid sensitivity, potency, and bioavailability. *The Journal of Clinical Endocrinology and Metabolism* 88: 277–84.

Vermetten, E., C. Schmahl, S. Lindner, R. J. Loewenstein, and J. D. Bremner. 2006. Hippocampal and amygdalar volumes in dissociative identity disorder. *The American Journal of Psychiatry* 163: 630–36.

Vogt, B. A., and D. N. Pandya. 1987. Cingulate cortex of the rhesus monkey: II. Cortical afferents. *The Journal of Comparative Neurology* 262: 271–89.

Vythilingam, M., C. Heim, J. Newport, A. H. Miller, E. Anderson, R. Bronen, M. Brummer, L. Staib, E. Vermetten, D. S. Charney, C. B. Nemeroff, and J. D. Bremner. 2002. Childhood trauma associated with smaller hippocampal volume in women with major depression. *The American Journal of Psychiatry* 159: 2072–80.

Vythilingam, M., D. A. Luckenbaugh, T. Lam, C. A. Morgan, III, D. Lipschitz, D. S. Charney, J. D. Bremner, and S. M. Southwick. 2005. Smaller head of the hippocampus in Gulf War-related posttraumatic stress disorder. *Psychiatry Research* 139: 89–99.

Wasserman, D., M. Sokolowski, V. Rozanov, and J. Wasserman. 2008. The CRHR1 gene: A marker for suicidality in depressed males exposed to low stress. *Genes, Brain, and Behavior* 7: 14–19.

Wasserman, D., J. Wasserman, V. Rozanov, and M. Sokolowski. 2009. Depression in suicidal males: Genetic risk variants in the CRHR1 gene. *Genes, Brain, and Behavior* 8: 72–79.

Wellen, K. E., and G. S. Hotamisligil. 2005. Inflammation, stress, and diabetes. *Journal of Clinical Investigation* 115: 1111–19.

Willerson, J. T., and P. M. Ridker. 2004. Inflammation as a cardiovascular risk factor. *Circulation* 109: II2–10.

Willour, V. L., H. Chen, J. Toolan, P. Belmonte, D. J. Cutler, F. S. Goes, P. P. Zandi, R. S. Lee, D. F. MacKinnon, F. M. Mondimore, B. Schweizer, J. R. DePaulo, Jr., E. S. Gershon, F. J. McMahon, and J. B. Potash. 2009. Family-based association of FKBP5 in bipolar disorder. *Molecular Psychiatry* 14: 261–68.

Wochnik, G. M., J. Ruegg, G. A. Abel, U. Schmidt, F. Holsboer, and T. Rein. 2005. FK506-binding proteins 51 and 52 differentially regulate dynein interaction and nuclear translocation of the glucocorticoid receptor in mammalian cells. *J Biol Chem.* 280: 4609–16.

Yehuda, R. 2001. Biology of posttraumatic stress disorder. *The Journal of Clinical Psychiatry* 62: 41–46.

Yehuda, R., G. Cai, J. A. Golier, C. Sarapas, S. Galea, M. Ising, T. Rein, J. Schmeidler, B. Muller-Myhsok, F. Holsboer, and J. D. Buxbaum. 2009. Gene expression patterns associated with posttraumatic stress disorder following exposure to the World Trade Center attacks. *Biological Psychiatry* 66: 708–11.

Yehuda, R., E. L. Giller, S. M. Southwick, M. T. Lowy, and J. W. Mason. 1991. Hypothalamic-pituitary-adrenal dysfunction in posttraumatic stress disorder. *Biological Psychiatry* 30: 1031–48.

Yehuda, R., J. A. Golier, S. L. Halligan, M. J. Meaney, and L. M. Bierer. 2004a. The ACTH response to dexamethasone in PTSD. *The American Journal of Psychiatry* 161: 1397–1403.

Yehuda, R., J. A. Golier, R. K. Yang, and L. Tischler. 2004b. Enhanced sensitivity to glucocorticoids in peripheral mononuclear leukocytes in posttraumatic stress disorder. *Biological Psychiatry* 55: 1110–16.

6 Personalized Medicine for Schizophrenia

Jacob S. Ballon, MD, Ragy R. Girgis, MD, and
Jeffrey A. Lieberman, MD

Introduction

Personalized medicine has become the ideal model for the future of modern health care. We foresee enormous benefits from the advent of personalized medicine techniques in psychiatry and the treatment of mental illnesses. Historically, patients have been treated according to their diagnosis, even though there is tremendous heterogeneity within a given diagnosis, and significant individual variability within a given population. Nowhere is this more true than with mental illnesses, including schizophrenia, given the limited validity of these diagnoses in view of the underlying etiology and pathophysiology of each disorder. Because the diagnosis of mental illness and the prediction of response to treatment is so complex and difficult when based on signs and symptoms, the use of techniques involving genetics, cognitive assessment, and neuroimaging is anxiously anticipated, in the hope of realizing the promise of personalized medicine.

Schizophrenia

Schizophrenia is a devastating mental illness. It is associated with symptoms that disrupt mental functions, cause significant functional disability, shorten lifespan, and result in significant costs (Rossler et al. 2005). Schizophrenia occurs both spontaneously and as a heritable disease that presents in adolescence and early adulthood (Kessler et al. 2007). The characteristic pathology includes positive symptoms (delusions, hallucinations, and thought-form disorder), negative symptoms (amotivation, social withdrawal, and poor social relatedness), and cognitive deficits (poor working memory, verbal memory, impaired IQ, and executive function; American Psychiatric Association 2000). Although hallucinations and delusions are considered to be the most common symptoms of schizophrenia, the negative symptoms and cognitive changes are the greatest predictors of overall function (Green 1996).

Schizophrenia is increasingly understood as a neurodevelopmental disease, with a clear genetic risk and subtle neuropathology (Harrison and Weinberger 2005). In the majority of cases, a period of decreased functioning and mild mental and behavioral changes precedes the full schizophrenic syndrome, leading to questions as to when the disorder truly begins, and exactly how genetic and developmental factors play roles prior to its full presentation (Miller et al. 1999). Treatments remain poorly tolerated (Lieberman et al. 2005) and palliative in nature. Furthermore, there are no precise diagnostic tests available for schizophrenia. All of these problems persist in spite of recognized biologic features in patients, including brain ventricular enlargement, reduced medial temporal lobe volume, and increased striatal dopamine storage and release (Braff et al. 2007). Research using endophenotypes—quantitative, heritable, trait-related deficits typically assessed by laboratory-based methods—shows promise for clinical utility (Greenwood et al. 2007).

The search for personalized medicine measures for patients with schizophrenia is likely to focus on studies from multiple modalities including genetics, functional and structural imaging, and of other markers including cognitive testing and electrophysiological measures.

Genetic Factors in Schizophrenia Psychopathology

Attempts to personalize schizophrenia treatment have evolved over the last three decades. Much has been learned about the genetics of schizophrenia, and this has allowed researchers to begin to home in on the specific variations, and ultimately, what might differentiate one person's illness from another's. With this information, ultimately, practitioners could tailor treatment to the individual based on specific markers.

Initial genetic findings were derived from studies of twins. Beginning with work by Gottesman and Shields (1972), the rate of concordance for schizophrenia among monozygotic twins with schizophrenia was first reported to be approximately 50 percent. Replication studies reported similar rates of 52 percent (McGuffin et al. 1984) and 41 percent. Similar numbers were seen for schizoaffective disorder (Cardno et al. 1999). As these numbers do not show 100 percent concordance to demonstrate an abolute genetic basis, they are large enough to demonstrate that there are both genetic and environmental factors involved in the development of the illness.

While twin concordance studies gave a reason to suspect that genetics provided at least a partial explanation for why certain individuals developed schizophrenia, they did not provide the specific genetic information needed to identify particular loci of interest for determining individual differences. While this chapter is not intended to be a complete review of schizophrenia genetics, many studies have begun to identify targets that are worth mentioning for future investigation. Understanding the various genes that may have bearing on the development of schizophrenia would make it easier to ascertain an individual's level of susceptibility to the disease, even before the onset of symptoms (Heckers 2009). As the clinical symptoms of schizophrenia are often not seen until adolescence or early adulthood, it is difficult to pinpoint the precise onset of the disorder. Additionally, schizophrenia appears to be the result of multiple gene interactions - it does not appear that it is caused by any single or specific genetic mutation or variant. Although schizophrenia is not a rare disorder, with a prevalence in the range of 1.4 to 4.6 per 1000, and population and incidence rates in the range of 0.16 to 0.42 per 1000 people per year worldwide (Jablensky 2000), large samples are required to spot the possibly numerous genetic differences in affected people.

There are several chromosomal abnormalities that have been noted in schizophrenia samples (Bassett et al. 2000). One of the most established involves microdeletions at chromosome 22q11.2. These microdeletions have been seen in approximately 2 percent of schizophrenia patients (Karayiorgou et al. 1995); patients with schizophrenia have been shown to have a nearly eighty-fold increased risk for having a 22q11 microdeletion compared to the general population (Tezenas Du Montcel et al. 1996). In a recent study, three additional loci were identified as being related to schizophrenia: 1q21.1, 15q11.2 and

15q13.3 (Stefansson et al. 2008). Despite the association with schizophrenia, all of these genetic abnormalities are also associated with other psychopathologies, including autism and mental retardation.

Another potential source of genetic information related to schizophrenia has been identification of copy number variants (CNVs), which represent the loss or repetition of pieces of DNA ranging from kilobases to megabases in length, and can include deletions, insertions, overlapping regions, duplications, or replications in adjacent or nonadjacent regions. Four loci on chromosome 6, associated with genes for the major histocompatibility complex (MHC) proteins, have been identified as occurring with greater frequency in European schizophrenia patients. This leads to some speculation regarding previous hypotheses about infectious disease and schizophrenia, and whether these MHC changes may be associated with those findings (Stefansson et al. 2009). Although several CNVs have been identified, there is not yet a definitive mechanism to explain these findings and their role in the development of schizophrenia.

Pharmacogenetics of Treatment Response

While the genetics of diagnosis remain unresolved, there is some evidence of differential response to treatment based upon genetic factors (Arranz and de Leon 2007; Foster et al. 2007; Nnadi and Malhotra 2007; Reynolds 2007). However, this too has been limited by heterogeneity of diagnosis, multiple complex genetic interactions (Abdolmaleky et al. 2005), and technological limitations (Kim et al. 2007; Reynolds 2007; Blanc et al. 2009). The ability to identify patients as at risk or in the early stages of schizophrenia and to predict treatment response based on genetics could significantly reduce morbidity and shorten the duration of untreated psychosis, which factors into overall prognosis.

Genome-wide association (GWA) studies have proven difficult to carry out, due to the number of potential associations. The NIMH CATIE study (Clinical Antipsychotic Trials of Intervention Effectiveness; Lieberman et al. 2005), which compared four second-generation antipsychotics (olanzapine, risperidone, quetiapine, and ziprasidone) with a first-generation antipsychotic (perphenazine), presented an opportunity to look prospectively at factors that could predict differential response. Although the study was not specifically designed to address pharmacogenetic responses, the researchers devised a priori hypotheses to look at particular single-nucleotide polymorphisms (SNPs, or changes in one nucleotide in the genetic sequence) previously identified as potentially influencing treatment response. As CATIE compared multiple medications, it provided an opportunity to make inferences about why particular medications fared better in some individuals than in others (Stroup 2007). In a study looking at three thousand potential SNPs with twenty-one clinical and neurocognitive phenotypes, no SNPs demonstrated sufficient correlation with any particular phenotypes or medication responses after statistical correction (Need et al. 2009). In a GWA study, also from the CATIE data, McClay et al. examined 738 subjects for associations between SNPs and antipsychotic response. One region, rs17390445 on chromosome 4, was statistically significant and associated with decreased positive symptoms in patients taking ziprasidone. Despite the strong statistical significance of this locus, it is over 1 Mb away from the nearest coding region, and thus it is unclear mechanistically how this finding could demonstrate a biologically plausible explanation for improving symptoms, and why the benefit was specifically related to ziprasidone over other medications (McClay et al. 2009).

Looking at candidate genes, particularly those that most relate to the pathophysiology of schizophrenia, has yielded varying successes. An initial target in identifying the differential response to antipsychotic medication has been the catechol-O-methyltransferase (COMT) enzyme. This enzyme is responsible for breaking down catecholamines, such as dopamine. COMT has a common polymorphism at codon 108/158, in which a valine is substituted by a methionine, leading to a four-fold decrease in catecholamine metabolism. Heterozygotes demonstrate intermediate enzymatic activity (Egan et al. 2001).

The dopamine hypothesis in schizophrenia remains a leading model for the pathogenesis of the syndrome. Increased dopaminergic activity in the mesolimbic pathway, and hypodopaminergic activity in the mesocortical pathway in the brain, have been hypothesized to correlate with both the positive and negative symptoms of schizophrenia. Understanding the metabolism of dopamine in the brain could provide clues to the overall pathogenesis of schizophrenia. With regards to personalized medicine, understanding how people differ in their ability to break down these neurotransmitters may explain why some people receive more benefits from some medications than others do (Lachman et al. 1996).

Differential benefits of individual medications based on genotypes have been difficult to measure. In an eight-week trial with risperidone, Japanese researchers looked at seventy-three subjects who were either naive to antipsychotics or switched to risperidone. The subjects did not demonstrate any differential response to treatment based on whether they had a val/val, val/met, or met/met polymorphism for COMT, although small effects were noticed for other genetic variations. However, the study was underpowered to make definitive statements about the results (Yamanouchi et al. 2003). In another small study looking at negative symptoms, olanzapine was shown to be more likely to have an effect on negative symptoms, defined as 30 percent improvement in PANSS (Positive and Negative Syndrome Scale) score, in met/met or val/met subjects compared with val/val subjects (Bertolino et al. 2007).

Patients show differing responses to the positive symptoms of schizophrenia when taking different antipsychotics; the search for a medication to treat negative symptoms, however, has been generally unsuccessful. With the advent of second-generation antipsychotics—that is, those that are antagonists at both dopamine and serotonin receptors—it was hoped that treatment of negative symptoms would improve. While these expectations have not come to pass clinically, several studies have looked at polymorphisms in serotonin 5-hydroxytryptamine (5-HT) receptors to differentiate medication effects on negative symptoms.

Treatment with risperidone has been evaluated for several genetic loci for differential efficacy for a variety of symptom clusters. The primary targets have been polymorphisms in serotonergic and dopaminergic receptor genes (Lane et al. 2005). SNPs for the dopamine receptor D2 (*DRD2*) and *AKT1* genes were shown to confer a different likelihood of response to risperidone in a study of 120 first-episode, and previously antipsychotic-naive, patients. In the eight-week study, these two loci were associated with a differential response to risperidone, while twenty-eight other loci were not associated with any difference in response (Ikeda et al. 2008).

In an initial study of the serotonergic system, Wang and colleagues (2008) looked at 130 atypical antipsychotic-naïve patients from Shanghai to evaluate if the 1019 C/G and the C825T polymorphisms of the 5-HT1A receptor would show different responses to risperidone for negative symptoms. While the C825T polymorphism did not show any differential treatment response, the authors found that the 1019 C/G polymorphism was associated with improved response in those with the CC genotype compared with heterozygotes or those with the GG genotype (Wang et al. 2008). These results were corroborated by further work out of Spain on the same target (Reynolds et al. 2006). In a study comparing haloperidol and risperidone, it was seen that the CC genotype was also associated with improvements in negative symptoms in the risperidone group, but not in the haloperidol group. This gave rise to the hypothesis that the serotonergic effects of risperidone may interact with this genotypic variation and be a biologically plausible mechanism for the improvement

(Mossner et al. 2009). In a sub-analysis, Mossner and colleagues were able to attribute 7 to 12 percent of the variance in negative symptom response to risperidone, over the course of the four-week study, to the patients' genotype (Mossner et al. 2009). While these data point to response differences for risperidone at the 5-HT1 receptor, there was no difference seen for risperidone efficacy with polymorphisms at the 5-HT7 receptor (Wei et al. 2009).

Additional work on negative symptoms has also been conducted on the 5-HT2A receptor. Aripiprazole, which is a partial agonist for the D2 receptor, also has antagonist activity at the 5-HT2A receptor, as do almost all of the second-generation antipsychotic drugs. Chen et al. (2009) looked at polymorphisms of the 5-HT2A receptor to see if there might be a differential response to aripiprazole for negative symptoms. The GG/CC genotype group of *HTR2A* A-1438G/T102C showed a poorer response for negative symptoms than did other genotypes (Chen et al. 2009); *HTR2A* is the gene that encodes for the 5-HT2A receptor.

Iloperidone is a medication recently approved by the FDA for the treatment of schizophrenia. The mechanism for iloperidone is not novel; it is a mixed D2/5HT2 antagonist. It came to the market with data regarding particular efficacies in people with particular genotypes (Scott 2009). Studies conducted by researchers at Vanda Pharmaceuticals Inc. have identified SNPs in six genes (*NPAS3, XKR4, TNR, GRIA4, GFRA2,* and *NUDT9P1*) that are associated with a greater likelihood of benefit from iloperidone in people who have those particular genotypes (Lavedan et al. 2009; Volpi et al. 2009). Additionally, those who are homozygous for the G/G allele for the CNTR rs1800169 polymorphism show enhanced benefits from iloperidone (Lavedan et al. 2008). However, these results are based on relatively small patient samples and must be replicated and confirmed before being considered definitive. Nevertheless, this approach, highlighted by iloperidone, forges the path that is needed for the development of other drugs in the sub-group and a personalized medical future, i.e., conducting GWA studies prospectively for signature patterns of efficacy.

NOTCH4 is another gene that has been implicated as potentially associated with schizophrenia, although the association has only been weakly replicated. The gene is responsible for regulating signaling for the development of glia and neuronal cells from neural stem cells. *NOTCH4* polymorphisms have also been associated with earlier onset schizophrenia in people who carry this polymorphism. In a study that looked at a T->C polymorphism in the promoter region of *NOTCH4* along with the val->met polymorphism in COMT, it was shown that those with low-functioning COMT activity and the *NOTCH4* polymorphism were ten times less likely to respond to antipsychotics, and nonresponders were three times more likely to show the combined high-risk genotype (Anttila et al. 2004).

In a study looking at the rate of symptom improvement based on two polymorphisms in the *SLC6A2* gene that encodes the norepinephrine (NE) transporter, G1287A or T-182C, no difference in PANSS scores was found between genotypes after correcting for multiple tests (Meary et al. 2008).

CNR1 is the gene that encodes the CB1 cannabinoid receptor. Cannabis usage is known to be a risk factor for schizophrenia. Usage of cannabis has been shown to worsen both positive and negative symptoms and decrease cognitive performance. *CNR1* is located on chromosome 6.14-6.15, a putative region of association with schizophrenia, and several SNPs have been associated with increased risk of the hebephrenia (disorganized schizophrenia) subtype of schizophrenia, as well as increased likelihood for cannabis usage in those with schizophrenia. In a study of fifty-nine treatment-refractory patients, Hamdani et al. (2008) looked at the 1359 G/A polymorphism and its role in treatment refractoriness measured by changes in PANSS scores. Those with the G allele were more likely to be refractory to antipsychotic treatments (up to 76 percent for those with the GG genotype), while those with the AA genotype had a refractory rate of 5 percent (Hamdani et al. 2008).

Also looking at the risk of treatment response in a small (n=59) naturalistic study from Brazil, Kohlrausch et al. (2008a) found that there were no differences in refractoriness for polymorphisms in the *CYP2D6* or *DRD2* genes, but that the likelihood of a poor response to treatment was higher in those with the *CYP3A5* low-expresser genotype (*CYP3A5*3/CYP3A5*3*) and with the T/A/G/A/C haplotype of the *DRD3* gene. This same group also found that homozygosity for the T825 allele of the 825C>T polymorphism in the *GNB3* gene was associated with a poorer response to clozapine and an increased risk of convulsions (Kohlrausch et al. 2008b).

Pharmacogenetics of Antipsychotic Side Effects

In addition to their relationship to therapeutic responses, genotypic differences have been studied in relation to side effects from antipsychotic medications (Basile et al. 2002). The first-generation antipsychotic medications, while effective against the positive symptoms of schizophrenia, carry varying levels of risk for the development of extrapyramidal movement disorders, including tardive dyskinesia (TD) and sedation. The risk for neurologic effects and sedation was generally in proportion to the potency of these medications at the D2 receptor. Later, the second-generation, or atypical, antipsychotic medications also had varying potency for the D2 receptor, but included activity at the 5-HT2A receptor. This property has been hypothesized to provide a lower risk for movement disorders. Despite this advantage, many of the second-generation antipsychotics can produce neurologic side effects, but more significantly, they have a propensity for metabolic problems, including weight gain, diabetes, and dyslipidemia. The balance of side effects between these two classes of medication has led many practitioners to work with patients to determine the side effects to which they are susceptible and are more willing to risk incurring. The elucidation of genetic differences that confer greater risk for one side effect cluster or the other would enormously benefit practitioners and patients, helping them to make more rational and safer treatment selections.

Tardive dyskinesia (TD) is a neurological disorder characterized by abnormal movements; it arises in approximately 20 to 50 percent of patients after long-term exposure to antipsychotic medications (Ozdemir et al. 2006). Several polymorphisms have been evaluated for their role in the development of TD (Thelma et al. 2008). The search for genetic targets, associated with side effect liabilities, is as difficult as the search for syndrome determinant factors, as multiple genetic loci may be involved, and many hypothesized targets do not produce positive findings (Liou et al. 2006; Srivastava et al. 2006).

The dopamine-3 receptor has polymorphisms that have been implicated in genetic studies that differentiate risk for developing TD. Ozdemir and colleagues (2001) evaluated SNPs in the D3 receptor and the *CYP1A2* genes for risk of developing TD. The Ser9Gly SNP was shown to differentiate risk: those with the Gly/Gly variant were significantly more likely to develop TD in comparison with the Ser/Ser or the Ser/Gly variants. This was thought to be related to changes in the sensitivity to dopamine with the Gly/Gly polymorphism. This finding was in contrast to a study that did not show an association with D3 polymorphisms among seventy-eight people with TD and seventy-nine without TD (Rietschel et al. 2000). However, a recent study from Russia found a modest worsening of TD with the same polymorphism (Al Hadithy et al. 2009).

The cytochrome P450 genes, which code for drug metabolizing enzymes, have been examined for association with differences in the rates of side effects. It was thought that different rates of metabolism could yield higher serum drug levels and increase the propensity for side effects. It was seen that people with the C/C genotype of the *CYP1A2*1F* (C-> A) polymorphism have a greater than three-fold increase in severity of TD symptoms compared with those with the A/C or the A/A genotypes. Findings in the cytochrome system, however, have been mixed. For example, one group did not find differences in polymorphisms for the *CYP3A4* or *CYP2D6* genes (Tiwari et al. 2005).

As the newer antipsychotic agents have decreased the risk of TD, the risk for metabolic complications from these newer medications has become a greater concern for patients and clinicians. Although the medications vary in their propensity to cause these disturbances, understanding if there are individual variations in risk could allow psychiatrists to choose antipsychotics that are of less metabolic risk to individual patients (Reynolds 2007; Rege 2008). Many loci have been studied, including genes for the alpha 2A adrenergic receptor, serotonin 2C (5-HT2C) receptor, leptin, and guanine nucleotide-binding protein (*GNB3*) genes (Muller and Kennedy 2006).

One of the first wide-ranging genetic studies of weight gain and antipsychotics was conducted by Basile and colleagues (2001). They looked at 10 SNPs over nine candidate genes in patients taking clozapine. In following the patients' weight gain prospectively, they found a trend level significance for increased weight gain at four loci: *ADRB3, ADRA1A, TNF-α,* and *5-HTR2C* (Basile et al. 2001).

The adipocyte-derived hormone leptin has been associated with appetite, changes in body mass, and energy regulation. In a study of Hispanic patients taking olanzapine, Templeman et al. (2005) evaluated weight gain and leptin levels. Patients were typed for the *5-HT2C* receptor 759C/T polymorphism and the leptin 2548A/G polymorphism. Those with the T allele for the 5-HT2C receptor demonstrated decreased weight gain at all time points of the study, up to nine months. (The 5-HT2C receptor gene is located on the X chromosome, so males were hemizygous for T or C and females were seen to be either CC or CT, but no TT females were identified in the sample.) For the leptin gene, those with the GG genotype showed greater change in BMI than those with the AA or AG genotypes after nine months, but there was no difference between the groups in the short-term analyses.

In an initial study looking at the relationship of weight gain and GNB3, Bishop et al. (2006) looked at the C825T transitional polymorphism in exon 10 of the GNB3 gene. The patients in the study gained an average of 10 percent of their starting body weight in the study, but no difference was detected between those with the T/T, T/C, or C/C alleles. In a large meta-analysis, a trend level of significance was observed for decreased propensity for weight gain with the C/C genotype despite equivocal findings in other, smaller studies (Souza et al. 2008).

Looking at a Han Chinese sample taking clozapine, Zhang et al. (2007) found a significant difference in weight gain for those with the A/A genotype compared to patients with the G/A and G/G genotype of the G2548A polymorphism of the leptin gene. This finding, however, was only seen in males and not females, once gender was considered. In a similar study from North India, Srivastava et al. (2008) found an association with severe weight gain in those with the rs4731426C/G polymorphism, which they hypothesize to be related to changes in transcription due to configuration changes in the receptor.

As patients gain weight, they are at an increased risk for developing type II diabetes mellitus. While there are competing hypotheses for the etiology of antipsychotic-induced weight gain, it has been shown that the atypical antipsychotics—olanzapine and clozapine most notably—increase the risk for diabetes (Wirshing et al. 2002). Irvin et al. (2009) looked at three candidate SNPs in a sample of African American patients. The SNPs selected were based on the diabetes literature: transcription factor 7-like 2 (TCF7L2), calpain 10 (CAPN10), and ectoenzyme nucleotide pyrophosphatase phosphodiesterase 1 (ENNP1). They found an additive association for the T allele and increased risk of diabetes in the TCF7L2 SNP. The other two SNPs did not yield a significant association with diabetes risk.

Structural Brain Abnormalities

Diagnosis and treatment based on an individual's known genetic makeup represent the primary goals of personalized medicine; however, our understanding of the genetics of schizophrenia, let alone our failure to replicate many of the preliminary findings in people with schizophrenia, remains in relative infancy and inadequate for clinical use.

Early studies sought peripheral markers of disease, such as potential differences in phenolic or indolic amines in the urine of individuals with schizophrenia (Takesada et al. 1965). In the past several decades, thanks in large part to technological advances, researchers have been able to turn their focus to explicit markers and detailed investigation of the brain (Patel 2003). Advances in physics and instrumentation paved the way for computed tomography (CT), functional MRI (fMRI), PET, and sophisticated signal processing of electrical brain function (EEG and ERP), which have allowed these investigations to take place.

This section will review some of the more established endophenotypic characteristics and biomarkers associated with schizophrenia, and will discuss their potential relevance to personalized medicine. While their definitions vary depending on the source, for the purposes of this chapter, we take endophenotypes to refer to biologically identified characteristics that distinguish groups of persons and are putatively related to genetic factors (Gottesman and Gould 2003), while biomarkers are similar and may or may not have genetic bases.

Structural MRI Findings

There are several findings drawn from structural MRI that are associated with schizophrenia. The most consistently replicated is larger ventricular size in schizophrenia compared to control populations, which was initially identified via CT (Johnstone et al. 1976), and subsequently confirmed by MRI-based studies (Kelsoe et al. 1988; Shenton et al. 2001). Smaller whole-brain volume has also been reported (Wright et al. 2000; Shenton et al. 2001; Keshavan et al. 2008), as have relatively consistent volume decreases in temporal lobe structures, including the whole temporal lobe, the medial temporal lobe (i.e., amygdala, hippocampus, and parahippocampal gyrus), and the superior temporal gyrus (Barta et al. 1990; Breier et al. 1992; McCarley et al. 1993; Shenton et al. 2001). Antipsychotic treatment has been associated with increased caudate volumes (Chakos et al. 1994), while antipsychotic-naive patients and children of individuals with schizophrenia demonstrate decreased volumes of the caudate nuclei (Keshavan et al. 1998; Corson et al. 1999; Rajarethinam et al. 2007). Decreased thalamic volumes have also been noted (Konick and Friedman 2001), as has an association with cavum septi pallucidi (Kwon et al. 1998). Recent studies are also suggesting abnormalities in gyrification patterns in individuals with schizophrenia (Sallet et al. 2003; Harris et al. 2004), along with reduced cortical thickness in temporal and frontal/prefrontal cortices (Kuperberg et al. 2003; Goldman et al. 2009). Importantly, several studies suggest that some of the morphological changes in the brains of individuals with schizophrenia may be progressive for both gray and white matter (Farrow et al., 2005; Whitford et al., 2007).

In summary, schizophrenia is associated with both global morphometric abnormalities (e.g., increased lateral ventricular volume) and regional abnormalities (e.g., decreased volumes in the medial temporal lobe; Wright et al. 2000). While several gross, neuroimaging-based structural findings in schizophrenia are relatively robust (e.g., increased ventricular volume), many of these findings are not specific to schizophrenia and can be seen in a variety of other psychiatric illnesses, such as bipolar disorder, for example (Shenton et al. 2001; Keshavan et al. 2008).

Diffusion Tensor Imaging (DTI) Findings

Abnormalities have also been identified in the connections between brain regions identified in structural imaging studies—that is, in the white matter tracts that connect these regions. Diffusion tensor imaging (DTI) measures the degree of isotropy (i.e., random diffusion) of water in white matter tracts and evaluates their coherence (Kubicki et al. 2007). DTI has helped to identify abnormalities (i.e., frequently decreased fractional anisotropy) in numerous white matter areas of the brain, most consistently in the uncinate fasciculus (Kubicki et al. 2002; Kubicki et al. 2007), cingulum (Wang et al. 2004; Kubicki et al. 2007), and arcuate fasciculus (Burns et al. 2003; Kubicki et al. 2007), along with the corpus callosum, (Gasparotti et al. 2009), inferior longitudinal fasciculus (Ashtari et al. 2007), and in other white matter tracts, frequently in those radiating to or from frontal or temporal regions, such as the fornix (Kuroki et al. 2006). DTI has also identified microstructural alterations in regions of the hippocampus, thalamus, and nucleus accumbens (Spoletini et al. 2009), as well as in cerebellar peduncles (Okugawa et al. 2006), the superior temporal gyrus (Lee et al. 2009), and internal capsule and thalamic radiations (Sussmann et al. 2009).

Functional Abnormalities in Schizophrenia

fMRI, PET, and single-photon emission computed tomography (SPECT) have enabled investigators to study the functional characteristics of brains of individuals with schizophrenia. A primary focus of findings from the functional neuroimaging literature has been on activation of the frontal/prefrontal cortex. Although the results have not been fully consistent, several meta-analyses have reported decreased blood flow/activation (originally termed "hypofrontality") in both resting and activated (i.e., when performing a neuropsychological task of working memory) states (Davidson and Heinrichs 2003; Hill et al. 2004; Glahn et al. 2005; Ragland et al. 2009). However, it is beginning to appear that "hypofrontality" does not adequately convey the functional complexity of the prefrontal cortex in schizophrenia (Moghaddam and Homayoun 2008). In particular, overactivity of the orbitofrontal cortex, the ventral prefrontal cortex, and surrounding areas during cognitively engaging tasks may accompany deficits in the dorsolateral prefrontal cortex (Ragland et al. 2004; Moghaddam and Homayoun 2008).

Functional abnormalities in schizophrenia are not limited to the prefrontal cortex, although findings from studies of other brain regions have been less consistent (Keshavan et al. 2008). For example, there appear to be abnormalities of other frontal cortical areas in schizophrenia (Glahn et al. 2005), as well as of the temporal lobe, where several findings have been reported (Zakzanis et al. 2000; Davidson and Heinrichs 2003; Achim and Lepage 2005; Ragland et al. 2009). A very recent meta-analysis identified a network of regions involved in executive functioning both in individuals with schizophrenia and controls; the regions foremost include the dorsolateral prefrontal cortex, ventrolateral prefrontal cortex, anterior cingulate cortex, and mediodorsal thalamus (Minzenberg et al. 2009). Researchers also reported decreased activation of the left dorsolateral prefrontal cortex, rostral/dorsal anterior cingulate cortex, left thalamus, and other inferior/posterior cortical areas. They reported increased activation in areas including the ventrolateral prefrontal cortex, temporal/parietal cortical areas, the insula, and the amygdala (Minzenberg et al. 2009). Finally, a study by Schobel et al. (2009) used a high-resolution variant of fMRI to identify brain regions differentially targeted by schizophrenia. In their analysis, they found that baseline abnormalities of blood flow in the CA1 region of the hippocampus predicted progression from the prodromal state to psychosis.

As with structural imaging, the functional neuroimaging literature is complicated by different methodologies between and among studies, including, but not limited to, different scanning parameters and modalities, different neuropsychological paradigms, and different study subject demographics, medications used, and durations of illness (Davis et al. 2005). However, these studies have helped to delineate dysfunctional areas and circuits in schizophrenia, involving, in general, opposite dysregulation in dorsal versus ventral cortical networks.

Magnetic Resonance Spectroscopy Findings

Magnetic resonance spectroscopy (MRS) allows *in vivo* measurements of brain neurochemistry. N-acetylaspartate (NAA) is a measure of neuronal viability, and is one chemical that can be measured. The most consistent findings in the literature are decreased NAA or NAA to creatinine (NAA/Cr) ratios in both the gray and white matter of the frontal/prefrontal cortex and hippocampus (Steen et al. 2005). These decreases in NAA have prognostic value (Wood et al. 2006), may change over time (Ohrmann et al. 2005), and may correlate with cognitive impairments (Ohrmann et al. 2007). Abnormalities in metabolite levels have been less consistently found in other brain regions, such as the anterior cingulate (Wood et al. 2007), the thalamus (Jakary et al. 2005), and the cerebellum (Ende et al. 2005), among others (Steen et al. 2005).

Abnormalities in glutamate/glutamine concentrations of the dorsolateral prefrontal cortex have also been identfied. (Ohrmann et al. 2007; Lutkenhoff et al. 2008; Rusch et al. 2008). Elevated glutamine levels have been found in the thalamus and anterior cingulate in individuals who are antipsychotic-naive. Correlations of glutamate/glutamine levels with cognitive impairment in the dorsolateral prefrontal cortex (Ohrmann et al. 2007) and hippocampus(Rusch, Tebartz van Elst et al. 2008) have been described. Recent data also suggest abnormalities of gamma-aminobutyric acid (GABA)/Cr ratios in basal ganglia (Goto et al. 2009).

The integrity of neuronal cell membranes can be investigated using phosphorus-based MRS (Reddy and Keshavan 2003). Although not as much work has been done in this area compared to that done with structural and functional neuroimaging, membrane deficits have been noted in several areas, including the basal ganglia (Jayakumar et al. 2003), anterior cingulate (Jensen et al. 2004), and frontal/prefrontal cortex (Reddy and Keshavan 2003), among others. (Smesny et al. 2007)

Findings from Studies of Neurotransmitters

Dopamine

Early clues that suggested a link between the dopamine system and schizophrenia were the relationships between the potency of dopamine D2 antagonism and the effective doses of antipsychotics (Seeman and Lee 1975; Creese et al. 1976), along with the psychotogenic effects of stimulant medications (Lieberman et al. 1987). The relationship between dopamine and schizophrenia—the so-called "dopamine hypothesis"—as reformulated by Weinberger and Davis et al. (Weinberger 1987; Davis et al. 1991) and based on clinical cognitive and preclinical data, is an important part of the neurochemical signature of schizophrenia.

Individuals with schizophrenia demonstrate increased striatal dopamine synthesis (Hietala et al. 1995; Abi-Dargham and Laruelle 2005). Other studies demonstrated that amphetamine-stimulated dopamine release was greater in individuals with schizophrenia than in controls, further supporting the existence of dopamine dysregulation in the striatum (Laruelle et al. 1996; Abi-Dargham and Laruelle 2005). Finally, Abi-Dargham et al. (2000) showed increased baseline occupancy of striatal dopamine D2 receptors by dopamine in individuals with schizophrenia. Together, these data suggest a hyperstimulation of striatal dopamine receptors. Importantly, Abi-Dargham et al. also reported that greater striatal dopamine levels at baseline was associated with greater improvement in positive symptoms after antipsychotic treatment, suggesting that the dopamine-D2-blocking characteristics of antipsychotic medications are important to their mechanism of action (Abi-Dargham et al. 2000).

Hyperstimulation of dopaminergic receptors in the striatum likely does not adequately explain negative and cognitive symptoms of schizophrenia. These symptom domains may be more related to hypodopaminergic functioning in the prefrontal cortex (Abi-Dargham and Laruelle 2005). Results from a study of the dopamine D1 receptor in vivo in schizophrenia showed greater availability of this receptor in the prefrontal cortex in individuals with schizophrenia, which was associated with worse performance on tasks of working memory (Abi-Dargham et al. 2002). The authors suggested that this greater availability may be the result of upregulation of D1 receptors secondary to chronic dopamine depletion (Abi-Dargham et al. 2002). However, these findings are not consistent with other studies that used different radioligands (Okubo et al. 1997; Karlsson et al. 2002). Therefore, replication of these findings is needed.

In summary, a simplification of the association between dopamine and schizophrenia may be that positive symptoms are related to hyperdopaminergic activity in the striatum, while negative and cognitive symptoms relate to hypodopaminergic activity in the prefrontal cortex. However, as stated above, the findings in the striatum are more robust, while those in the prefrontal cortex require additional study and confirmation.

Other Neurochemical Findings in Schizophrenia

Abnormalities in several other neurochemical systems have been identified in individuals with schizophrenia, such as glutamate and GABA, major excitatory and inhibitory neurotransmitters in the human brain. Hypofunction of the N-methyl-D-aspartate (NMDA) glutamate system has been hypothesized to play a role in the pathophysiology of schizophrenia, and has been supported by numerous data (Olney et al. 1999). Postmortem investigation of the hippocampus has suggested decreased levels of NMDA and other glutamate receptors, such as α-amino-3-hydroxy-5-methyl-4-isoxazolepropionic acid receptor (AMPA) and kainite (Harrison et al. 2003). Postmortem studies have also suggested a decrease in the synthesis and concentration of GABA, an important inhibitory neurotransmitter, in the dorsolateral prefrontal cortex (Lewis et al. 2005), particularly in parvalbumin-containing neurons (Hashimoto et al. 2003). Deficits in GABA reuptake have also been reported (Lewis and Gonzalez-Burgos 2006). Importantly, these alterations in the GABA system may contribute to the working memory abnormalities seen in individuals with schizophrenia (Lewis and Gonzalez-Burgos 2006). Finally, there appears to be a compensatory upregulation of GABA-A receptors and GABA-A alpha2-subunit-immunoreactivity in the dorsolateral prefrontal cortex (Lewis et al. 2005).

There is evidence that suggests abnormalities of the nicotinic-cholinergic system in schizophrenia. Individuals with schizophrenia have increased rates of smoking and smoke more heavily than others (Goff et al. 1992; Kelly and McCreadie 1999; Lasser et al. 2000; Martin et al. 2004). Nicotine regulates a number of neurotransmitter systems, including dopamine and acetylcholine, and nicotine administration improves the performance of individuals with schizophrenia in a variety of neuropsychiatric domains (Martin et al. 2004; Smith et al. 2006). Decreased levels of muscarinic receptors (Crook et al. 2001) and abnormal regulation of nicotine receptors (Breese et al. 2000) have also been reported in individuals with schizophrenia. Finally, a recent SPECT study found decreased availability of muscarinic receptors in the basal ganglia, cortex and thalamus of individuals with schizophrenia (Raedler et al. 2003).

Abnormalities in serotonin levels, and levels of its receptors and metabolites, have also been observed in schizophrenia (Patel 2003). However, the relevance of these findings to the pathophysiology of schizophrenia remains unclear.

Neurophysiology

Neurophysiologic experiments generally measure the electrical activity of the brain at the scalp during experimental procedures or tasks (Keshavan et al. 2008). Mismatch negativity (MMN) is one such measure. In MMN, the patient focuses on one set of auditory stimuli and receives infrequent, deviant stimuli that represent the experimental stimuli (Keshavan et al. 2008). Studies have demonstrated that individuals with schizophrenia demonstrate deficits in MMN compared to control subjects (Catts et al. 1995; Umbricht and Krljes 2005), regardless of whether they are medicated (Catts et al. 1995). One study also found that bipolar patients were not different from controls, although individuals with schizophrenia were, suggesting that these deficits are relatively specific to schizophrenia (Catts et al. 1995). In addition, evidence suggests that MMN is at least in part heritable, making it a potentially valid endophenotype in this population (Hall et al. 2006).

The P50 response refers to the response of an individual fifty milliseconds after a second auditory stimulus occurring a short period of time after a first stimulus (Keshavan et al. 2008). Healthy individuals are inherently able to robustly inhibit their responses to the second stimulus, while individuals with schizophrenia are less able to inhibit their responses (Siegel et al. 1984; Bramon et al. 2004; de Wilde et al. 2007). Decreased ability to inhibit the 50-millisecond stimulus was also found in studies of family members of individuals with schizophrenia, supporting the relevance of these findings (Siegel et al. 1984; de Wilde et al. 2007).

Similar findings have been reported for the P300 ERP. This waveform is measured at 300 milliseconds after a stimulus (Keshavan et al. 2008). Individuals with schizophrenia demonstrated decreased amplitude of this waveform (Morstyn et al. 1983; Bramon et al. 2004) as well as increased latency (Bramon et al. 2004). Again, non-psychotic relatives of individuals with schizophrenia demonstrate similar neurophysiologic findings (Bramon et al. 2005).

Pre-pulse inhibition (PPI) is the phenomenon in which a relatively gentle stimulus precedes a more intense stimulus and dampens the response to the more intense stimulus (Braff et al. 2001). Numerous studies have been performed in individuals with schizophrenia and have demonstrated decreased PPI compared to healthy controls (Braff et al. 2001). These deficits also exist in patients with schizotypal personality disorder, characterized by odd or eccentric behaviour that is persistent throughout all facets of life, and in relatives of individuals with schizophrenia (Cadenhead et al. 2000). However, they are not specific to schizophrenia spectrum illnesses (Geyer 2006), and can be seen in other psychiatric disorders, such as panic disorder (Ludewig et al. 2002).

An ERP signal processing method (40Hz Gamma phase synchrony) focuses on a possible mechanism by which neural networks functionally interconnect, and has been found to be deficient in people with schizophrenia, pointing to a widespread brain neural network timing dysfunction.

Physical/Neurodevelopmental Findings

Individuals with schizophrenia have also been found to exhibit abnormalities on some gross neurologic measures. For example, individuals with schizophrenia demonstrate impaired smooth pursuit tracking compared to healthy controls (Mialet and Pichot 1981). *Smooth pursuit* refers to the flowing, slow, uninterrupted movement of the eyes as they follow a moving target. These are in contrast to *saccades*, which are rapid, jerky eye movements used to change gaze rapidly (Calkins and Iacono 2000). The impairments themselves and the underlying mechanisms appear to be familial (Chen et al. 1999).

Minor physical anomalies (MPAs) are gross examination findings considered to serve as a window into early neurodevelopment, given their persistence throughout life and their shared embryonic origin between the target anatomic areas and the nervous system (Green et al. 1994). They include abnormalities of the face, head, hands, and feet (e.g., abnormal distance between tear ducts, high steepled palate). Individuals with schizophrenia, as well as their siblings, have been shown to have more MPAs than healthy control subjects (Gualtieri et al. 1982; Guy et al. 1983; Green et al. 1994; Ismail et al. 1998). A greater number of MPAs may also signal a greater risk of developing schizophrenia (Schiffman et al. 2002). Similarly, neurological "soft signs," such as gaze deficits, incoordination, and abnormal reflexes, are thought to be nonspecific indicators of brain injury and occur to a greater degree in individuals with schizophrenia, compared to control subjects (Gupta et al. 1995).

Neurocognitive Indices

Approximately 75 percent of patients with schizophrenia have clinically meaningful deficits in at least two cognitive domains, and 90 percent have deficits in at least one domain (Palmer et al. 1997). These deficits span a range of neurocognitive domains and tests (Heinrichs and Zakzanis 1998). Attention, verbal memory, and working memory have been examined as potential endophenotypes in schizophrenia (Braff et al. 2008).

Attention, often measured by the continuous performance test (CPT), has been found to be frequently impaired in schizophrenia, and these abnormalities are not necessarily state-dependent (Cornblatt and Keilp 1994). These abnormalities are also found in unaffected family members of individuals with schizophrenia (Cornblatt and Keilp 1994; Chen et al. 1998).

Verbal memory is also impaired in schizophrenia (Saykin et al. 1991), including in first-episode patients (Saykin et al. 1994; Aleman et al. 1999). Healthy relatives of individuals with schizophrenia may also have deficits in this neurocognitive domain (Sitskoorn et al. 2004; Snitz et al. 2006).

Working memory is another neurocognitive domain in which individuals with schizophrenia demonstrate deficits (Gold et al. 1997). Working memory is an executive function aspect of short-term memory involved in manipulating, processing, and retrieving information. Deficits are also present in some non-psychotic family members of individuals with schizophrenia (Park et al. 1995), supporting the significance of this trait as a marker for schizophrenia.

Conclusion and the Future of Personalized Medicine in Schizophrenia

Over the course of the last several decades, there has been a burgeoning of new information about the genetics and pathobiology of schizophrenia, which serves as a foundation for discussion and a potential source of measures that can contribute to the personalization of medicine in the treatment of this illness. Genetic factors have been identified in numerous cohorts, associating several genetic markers with clinical response, response to specific antipsychotic medications, or specific side effects.

Such applications would revolutionize the way that psychiatry is practiced, giving clinicians the ability to take an informed and directed approach to the selection of psychiatric medications. Similarly, abnormalities in brain structure, function, and neurochemistry measured by magnetic resonance-based applications (sMRI, DTI, MRS, fMRI), along with findings of neurochemical abnormalities based on nuclear medical imaging (PET and SPECT), suggest that it may not be too long before imaging can be used as a clinical tool for the diagnosis of schizophrenia, prediction of progression from the prodromal stage to syndromal schizophrenia, or prediction of response to specific antipsychotic medications. Neurophysiological, neurodevelopmental, physical, and neurocognitive abnormalities in schizophrenia also suggest a potential role for these tests in the identification of endophenotypes, or as biomarkers of schizophrenia.

Much work remains to be done before the field can realize the potential of these genetic and pathobiological findings in schizophrenia. Genetic studies have so far been limited by sample size and multiple gene interactions, which are likely to have contributed to a failure of significant replication. Furthermore, genetic findings are often limited by a lack of reproducibility, or relevance to only a small or specific cohort of individuals with schizophrenia. And although many of the pathobiological and imaging findings are more established (e.g., hyperdopaminergic activity in the striatum, increased ventricular volume), the differences are often small, of unclear significance to the disease, and not always specific to schizophrenia. It may be that, with schizophrenia, clinicians will first assign patients "profiles" of a combination of the genetic and pathobiological markers described above. These markers may be more robust tools when used together, rather than individually, for the identification, diagnosis, and treatment of individuals with schizophrenia.

However, this discussion remains speculative. More solid outcomes may emerge from studies of sufficient size to explore replication, and from the organization of data into databases, to increase statistical power and allow systematic examination of interactions. Only with continued investigation into the genetics and pathobiology of schizophrenia will the field be able to one day to fulfill the promise of personalized medicine in schizophrenia.

References

Abdolmaleky, H. M., Thiagalingam, S., & Wilcox, M. (2005). Genetics and epigenetics in major psychiatric disorders: dilemmas, achievements, applications, and future scope. *Am J Pharmacogenomics, 5*(3), 149–160.

Abi-Dargham, A., & Laruelle, M. (2005). Mechanisms of action of second generation antipsychotic drugs in schizophrenia: insights from brain imaging studies. *Eur Psychiatry, 20*(1), 15–27.

Abi-Dargham, A., Mawlawi, O., Lombardo, I., Gil, R., Martinez, D., Huang, Y., Hwang, D. R., Keilp, J., Kochan, L., Van Heertum, R., Gorman, J. M., & Laruelle, M. (2002). Prefrontal dopamine D1 receptors and working memory in schizophrenia. *J Neurosci, 22*(9), 3708–3719.

Abi-Dargham, A., Rodenhiser, J., Printz, D., Zea-Ponce, Y., Gil, R., Kegeles, L. S., Weiss, R., Cooper, T. B., Mann, J. J., Van Heertum, R. L., Gorman, J. M., & Laruelle, M. (2000). Increased baseline occupancy of D2 receptors by dopamine in schizophrenia. *Proc Natl Acad Sci U S A, 97*(14), 8104–8109.

Achim, A. M., & Lepage, M. (2005). Episodic memory-related activation in schizophrenia: meta-analysis. *Br J Psychiatry, 187*, 500–509.

Al Hadithy, A. F., Ivanova, S. A., Pechlivanoglou, P., Semke, A., Fedorenko, O., Kornetova, E., Ryadovaya, L., Brouwers, J. R., Wilffert, B., Bruggeman, R., & Loonen, A. J. (2009). Tardive dyskinesia and DRD3, HTR2A and HTR2C gene polymorphisms in Russian psychiatric inpatients from Siberia. *Prog Neuropsychopharmacol Biol Psychiatry, 33*(3), 475–481.

Aleman, A., Hijman, R., de Haan, E. H., & Kahn, R. S. (1999). Memory impairment in schizophrenia: a meta-analysis. *Am J Psychiatry, 156*(9), 1358–1366.

Anttila, S., Illi, A., Kampman, O., Mattila, K. M., Lehtimaki, T., & Leinonen, E. (2004). Interaction between NOTCH4 and catechol-O-methyltransferase genotypes in schizophrenia patients with poor response to typical neuroleptics. *Pharmacogenetics, 14*(5), 303–307.

APA. (2000). *Diagnostic and Statistical Manual of Mental Disorders, Fourth Edition, Text Revision (DSM-IV-TR)*. Arlington, VA.

Arranz, M. J., & de Leon, J. (2007). Pharmacogenetics and pharmacogenomics of schizophrenia: a review of last decade of research. *Mol Psychiatry, 12*(8), 707–747.

Ashtari, M., Cottone, J., Ardekani, B. A., Cervellione, K., Szeszko, P. R., Wu, J., Chen, S., & Kumra, S. (2007). Disruption of white matter integrity in the inferior longitudinal fasciculus in adolescents with schizophrenia as revealed by fiber tractography. *Arch Gen Psychiatry, 64*(11), 1270–1280.

Barta, P. E., Pearlson, G. D., Powers, R. E., Richards, S. S., & Tune, L. E. (1990). Auditory hallucinations and smaller superior temporal gyral volume in schizophrenia. *Am J Psychiatry, 147*(11), 1457–1462.

Basile, V. S., Masellis, M., McIntyre, R. S., Meltzer, H. Y., Lieberman, J. A., & Kennedy, J. L. (2001). Genetic dissection of atypical antipsychotic-induced weight gain: novel preliminary data on the pharmacogenetic puzzle. *J Clin Psychiatry, 62 Suppl 23*, 45–66.

Basile, V. S., Masellis, M., Potkin, S. G., & Kennedy, J. L. (2002). Pharmacogenomics in schizophrenia: the quest for individualized therapy. *Hum Mol Genet, 11*(20), 2517–2530.

Bassett, A. S., Chow, E. W., & Weksberg, R. (2000). Chromosomal abnormalities and schizophrenia. *Am J Med Genet, 97*(1), 45–51.

Bertolino, A., Caforio, G., Blasi, G., Rampino, A., Nardini, M., Weinberger, D. R., Dallapiccola, B., Sinibaldi, L., & Douzgou, S. (2007). COMT Val158Met polymorphism predicts negative symptoms response to treatment with olanzapine in schizophrenia. *Schizophr Res, 95*(1–3), 253–255.

Bishop, J. R., Ellingrod, V. L., Moline, J., & Miller, D. (2006). Pilot study of the G-protein beta3 subunit gene (C825T) polymorphism and clinical response to olanzapine or olanzapine-related weight gain in persons with schizophrenia. *Med Sci Monit, 12*(2), BR47–50.

Blanc, O., Brousse, G., Meary, A., Leboyer, M., & Llorca, P. M. (2009). Pharmacogenetic of response efficacy to antipsychotics in schizophrenia: pharmacodynamic aspects. Review and implications for clinical research. *Fundam Clin Pharmacol*.

Braff, D. L., Freedman, R., Schork, N. J., & Gottesman, II. (2007). Deconstructing schizophrenia: an overview of the use of endophenotypes in order to understand a complex disorder. *Schizophr Bull, 33*(1), 21–32.

Braff, D. L., Geyer, M. A., & Swerdlow, N. R. (2001). Human studies of prepulse inhibition of startle: normal subjects, patient groups, and pharmacological studies. *Psychopharmacology (Berl), 156*(2–3), 234–258.

Braff, D. L., Greenwood, T. A., Swerdlow, N. R., Light, G. A., & Schork, N. J. (2008). Advances in endophenotyping schizophrenia. *World Psychiatry, 7*(1), 11–18.

Bramon, E., McDonald, C., Croft, R. J., Landau, S., Filbey, F., Gruzelier, J. H., Sham, P. C., Frangou, S., & Murray, R. M. (2005). Is the P300 wave an endophenotype for schizophrenia? A meta-analysis and a family study. *Neuroimage, 27*(4), 960–968.

Bramon, E., Rabe-Hesketh, S., Sham, P., Murray, R. M., & Frangou, S. (2004). Meta-analysis of the P300 and P50 waveforms in schizophrenia. *Schizophr Res, 70*(2–3), 315–329.

Breese, C. R., Lee, M. J., Adams, C. E., Sullivan, B., Logel, J., Gillen, K. M., Marks, M. J., Collins, A. C., & Leonard, S. (2000). Abnormal regulation of high affinity nicotinic receptors in subjects with schizophrenia. *Neuropsychopharmacology, 23*(4), 351–364.

Breier, A., Buchanan, R. W., Elkashef, A., Munson, R. C., Kirkpatrick, B., & Gellad, F. (1992). Brain morphology and schizophrenia. A magnetic resonance imaging study of limbic, prefrontal cortex, and caudate structures. *Arch Gen Psychiatry, 49*(12), 921–926.

Burns, J., Job, D., Bastin, M. E., Whalley, H., Macgillivray, T., Johnstone, E. C., & Lawrie, S. M. (2003). Structural disconnectivity in schizophrenia: a diffusion tensor magnetic resonance imaging study. *Br J Psychiatry, 182*, 439–443.

Cadenhead, K. S., Swerdlow, N. R., Shafer, K. M., Diaz, M., & Braff, D. L. (2000). Modulation of the startle response and startle laterality in relatives of schizophrenic patients and in subjects with schizotypal personality disorder: evidence of inhibitory deficits. *Am J Psychiatry, 157*(10), 1660–1668.

Calkins, M. E., & Iacono, W. G. (2000). Eye movement dysfunction in schizophrenia: a heritable characteristic for enhancing phenotype definition. *Am J Med Genet, 97*(1), 72–76.

Cardno, A. G., Marshall, E. J., Coid, B., Macdonald, A. M., Ribchester, T. R., Davies, N. J., Venturi, P., Jones, L. A., Lewis, S. W., Sham, P. C., Gottesman, II, Farmer, A. E., McGuffin, P., Reveley, A. M., & Murray, R. M. (1999). Heritability estimates for psychotic disorders: the Maudsley twin psychosis series. *Arch Gen Psychiatry, 56*(2), 162–168.

Catts, S. V., Shelley, A. M., Ward, P. B., Liebert, B., McConaghy, N., Andrews, S., & Michie, P. T. (1995). Brain potential evidence for an auditory sensory memory deficit in schizophrenia. *Am J Psychiatry, 152*(2), 213–219.

Chakos, M. H., Lieberman, J. A., Bilder, R. M., Borenstein, M., Lerner, G., Bogerts, B., Wu, H., Kinon, B., & Ashtari, M. (1994). Increase in caudate nuclei volumes of first-episode schizophrenic patients taking antipsychotic drugs. *Am J Psychiatry, 151*(10), 1430–1436.

Chen, S. F., Shen, Y. C., & Chen, C. H. (2009). HTR2A A-1438G/T102C polymorphisms predict negative symptoms performance upon aripiprazole treatment in schizophrenic patients. *Psychopharmacology (Berl), 205*(2), 285–292.

Chen, W. J., Liu, S. K., Chang, C. J., Lien, Y. J., Chang, Y. H., & Hwu, H. G. (1998). Sustained attention deficit and schizotypal personality features in nonpsychotic relatives of schizophrenic patients. *Am J Psychiatry, 155*(9), 1214–1220.

Chen, Y., Nakayama, K., Levy, D. L., Matthysse, S., & Holzman, P. S. (1999). Psychophysical isolation of a motion-processing deficit in schizophrenics and their relatives and its association with impaired smooth pursuit. *Proc Natl Acad Sci U S A, 96*(8), 4724–4729.

Cornblatt, B. A., & Keilp, J. G. (1994). Impaired attention, genetics, and the pathophysiology of schizophrenia. *Schizophr Bull, 20*(1), 31–46.

Corson, P. W., Nopoulos, P., Andreasen, N. C., Heckel, D., & Arndt, S. (1999). Caudate size in first-episode neuroleptic-naive schizophrenic patients measured using an artificial neural network. *Biol Psychiatry, 46*(5), 712–720.

Creese, I., Burt, D. R., & Snyder, S. H. (1976). Dopamine receptor binding predicts clinical and pharmacological potencies of antischizophrenic drugs. *Science, 192*(4238), 481–483.

Crook, J. M., Tomaskovic-Crook, E., Copolov, D. L., & Dean, B. (2001). Low muscarinic receptor binding in prefrontal cortex from subjects with schizophrenia: a study of Brodmann's areas 8, 9, 10, and 46 and the effects of neuroleptic drug treatment. *Am J Psychiatry, 158*(6), 918–925.

Davidson, L. L., & Heinrichs, R. W. (2003). Quantification of frontal and temporal lobe brain-imaging findings in schizophrenia: a meta-analysis. *Psychiatry Res, 122*(2), 69–87.

Davis, C. E., Jeste, D. V., & Eyler, L. T. (2005). Review of longitudinal functional neuroimaging studies of drug treatments in patients with schizophrenia. *Schizophr Res, 78*(1), 45–60.

Davis, K. L., Kahn, R. S., Ko, G., & Davidson, M. (1991). Dopamine in schizophrenia: a review and reconceptualization. *Am J Psychiatry, 148*(11), 1474–1486.

de Wilde, O. M., Bour, L. J., Dingemans, P. M., Koelman, J. H., & Linszen, D. H. (2007). A meta-analysis of P50 studies in patients with schizophrenia and relatives: differences in methodology between research groups. *Schizophr Res, 97*(1–3), 137–151.

Egan, M. F., Goldberg, T. E., Kolachana, B. S., Callicott, J. H., Mazzanti, C. M., Straub, R. E., Goldman, D., & Weinberger, D. R. (2001). Effect of COMT Val108/158 Met genotype on frontal lobe function and risk for schizophrenia. *Proc Natl Acad Sci U S A, 98*(12), 6917–6922.

Ende, G., Hubrich, P., Walter, S., Weber-Fahr, W., Kammerer, N., Braus, D. F., & Henn, F. A. (2005). Further evidence for altered cerebellar neuronal integrity in schizophrenia. *Am J Psychiatry, 162*(4), 790–792.

Foster, A., Miller del, D., & Buckley, P. F. (2007). Pharmacogenetics and schizophrenia. *Psychiatr Clin North Am, 30*(3), 417–435.

Gasparotti, R., Valsecchi, P., Carletti, F., Galluzzo, A., Liserre, R., Cesana, B., & Sacchetti, E. (2009). Reduced fractional anisotropy of corpus callosum in first-contact, antipsychotic drug-naive patients with schizophrenia. *Schizophr Res, 108*(1–3), 41–48.

Geyer, M. A. (2006). The family of sensorimotor gating disorders: comorbidities or diagnostic overlaps? *Neurotox Res, 10*(3–4), 211–220.

Glahn, D. C., Ragland, J. D., Abramoff, A., Barrett, J., Laird, A. R., Bearden, C. E., & Velligan, D. I. (2005). Beyond hypofrontality: a quantitative meta-analysis of functional neuroimaging studies of working memory in schizophrenia. *Hum Brain Mapp, 25*(1), 60–69.

Goff, D. C., Henderson, D. C., & Amico, E. (1992). Cigarette smoking in schizophrenia: relationship to psychopathology and medication side effects. *Am J Psychiatry, 149*(9), 1189–1194.

Gold, J. M., Carpenter, C., Randolph, C., Goldberg, T. E., & Weinberger, D. R. (1997). Auditory working memory and Wisconsin Card Sorting Test performance in schizophrenia. *Arch Gen Psychiatry, 54*(2), 159–165.

Goldman, A. L., Pezawas, L., Mattay, V. S., Fischl, B., Verchinski, B. A., Chen, Q., Weinberger, D. R., & Meyer-Lindenberg, A. (2009). Widespread reductions of cortical thickness in schizophrenia and spectrum disorders and evidence of heritability. *Arch Gen Psychiatry, 66*(5), 467–477.

Goto, N., Yoshimura, R., Moriya, J., Kakeda, S., Ueda, N., Ikenouchi-Sugita, A., Umene-Nakano, W., Hayashi, K., Oonari, N., Korogi, Y., & Nakamura, J. (2009). Reduction of brain gamma-aminobutyric acid (GABA) concentrations in early-stage schizophrenia patients: 3T Proton MRS study. *Schizophr Res, 112*(1–3), 192–193.

Gottesman, II, & Gould, T. D. (2003). The endophenotype concept in psychiatry: etymology and strategic intentions. *Am J Psychiatry, 160*(4), 636–645.

Gottestman, I. I., & Shields, J. (1972). *Schizophrenia and Genetics: A Twin Study Vantage Point.* New York and London: Academic Press.

Green, M. F. (1996). What are the functional consequences of neurocognitive deficits in schizophrenia? *Am J Psychiatry, 153*(3), 321–330.

Green, M. F., Satz, P., & Christenson, C. (1994). Minor physical anomalies in schizophrenia patients, bipolar patients, and their siblings. *Schizophr Bull, 20*(3), 433–440.

Greenwood, T. A., Braff, D. L., Light, G. A., Cadenhead, K. S., Calkins, M. E., Dobie, D. J., Freedman, R., Green, M. F., Gur, R. E., Gur, R. C., Mintz, J., Nuechterlein, K. H., Olincy, A., Radant, A. D., Seidman, L. J., Siever, L. J., Silverman, J. M., Stone, W. S., Swerdlow, N. R., Tsuang, D. W., Tsuang, M. T., Turetsky, B. I., & Schork, N. J. (2007). Initial heritability analyses of endophenotypic measures for schizophrenia: the consortium on the genetics of schizophrenia. *Arch Gen Psychiatry, 64*(11), 1242–1250.

Gualtieri, C. T., Adams, A., Shen, C. D., & Loiselle, D. (1982). Minor physical anomalies in alcoholic and schizophrenic adults and hyperactive and autistic children. *Am J Psychiatry, 139*(5), 640–643.

Gupta, S., Andreasen, N. C., Arndt, S., Flaum, M., Schultz, S. K., Hubbard, W. C., & Smith, M. (1995). Neurological soft signs in neuroleptic-naive and neuroleptic-treated schizophrenic patients and in normal comparison subjects. *Am J Psychiatry, 152*(2), 191–196.

Guy, J. D., Majorski, L. V., Wallace, C. J., & Guy, M. P. (1983). The incidence of minor physical anomalies in adult male schizophrenics. *Schizophr Bull, 9*(4), 571–582.

Hall, M. H., Schulze, K., Rijsdijk, F., Picchioni, M., Ettinger, U., Bramon, E., Freedman, R., Murray, R. M., & Sham, P. (2006). Heritability and reliability of P300, P50 and duration mismatch negativity. *Behav Genet, 36*(6), 845–857.

Hamdani, N., Tabeze, J. P., Ramoz, N., Ades, J., Hamon, M., Sarfati, Y., Boni, C., & Gorwood, P. (2008). The CNR1 gene as a pharmacogenetic factor for antipsychotics rather than a susceptibility gene for schizophrenia. *Eur Neuropsychopharmacol, 18*(1), 34–40.

Harris, J. M., Yates, S., Miller, P., Best, J. J., Johnstone, E. C., & Lawrie, S. M. (2004). Gyrification in first-episode schizophrenia: a morphometric study. *Biol Psychiatry, 55*(2), 141–147.

Harrison, P. J., Law, A. J., & Eastwood, S. L. (2003). Glutamate receptors and transporters in the hippocampus in schizophrenia. *Ann N Y Acad Sci, 1003*, 94–101.

Harrison, P. J., & Weinberger, D. R. (2005). Schizophrenia genes, gene expression, and neuropathology: on the matter of their convergence. *Mol Psychiatry, 10*(1), 40–68; image 45.

Hashimoto, T., Volk, D. W., Eggan, S. M., Mirnics, K., Pierri, J. N., Sun, Z., Sampson, A. R., & Lewis, D. A. (2003). Gene expression deficits in a subclass of GABA neurons in the prefrontal cortex of subjects with schizophrenia. *J Neurosci, 23*(15), 6315–6326.

Heckers, S. (2009). Who is at risk for a psychotic disorder? *Schizophr Bull, 35*(5), 847–850.

Heinrichs, R. W., & Zakzanis, K. K. (1998). Neurocognitive deficit in schizophrenia: a quantitative review of the evidence. *Neuropsychology, 12*(3), 426–445.

Hietala, J., Syvalahti, E., Vuorio, K., Rakkolainen, V., Bergman, J., Haaparanta, M., Solin, O., Kuoppamaki, M., Kirvela, O., Ruotsalainen, U., & et al. (1995). Presynaptic dopamine function in striatum of neuroleptic-naive schizophrenic patients. *Lancet, 346*(8983), 1130–1131.

Hill, K., Mann, L., Laws, K. R., Stephenson, C. M., Nimmo-Smith, I., & McKenna, P. J. (2004). Hypofrontality in schizophrenia: a meta-analysis of functional imaging studies. *Acta Psychiatr Scand, 110*(4), 243–256.

Ikeda, M., Yamanouchi, Y., Kinoshita, Y., Kitajima, T., Yoshimura, R., Hashimoto, S., O'Donovan, M. C., Nakamura, J., Ozaki, N., & Iwata, N. (2008). Variants of dopamine and serotonin candidate genes as predictors of response to risperidone treatment in first-episode schizophrenia. *Pharmacogenomics, 9*(10), 1437–1443.

Irvin, M. R., Wiener, H. W., Perry, R. P., Savage, R. M., & Go, R. C. (2009). Genetic risk factors for type 2 diabetes with pharmacologic intervention in African-American patients with schizophrenia or schizoaffective disorder. *Schizophr Res, 114*(1–3), 50–56.

Ismail, B., Cantor-Graae, E., & McNeil, T. F. (1998). Minor physical anomalies in schizophrenic patients and their siblings. *Am J Psychiatry, 155*(12), 1695–1702.

Jablensky, A. (2000). Epidemiology of schizophrenia: the global burden of disease and disability. *Eur Arch Psychiatry Clin Neurosci, 250*(6), 274–285.

Jakary, A., Vinogradov, S., Feiwell, R., & Deicken, R. F. (2005). N-acetylaspartate reductions in the mediodorsal and anterior thalamus in men with schizophrenia verified by tissue volume corrected proton MRSI. *Schizophr Res, 76*(2–3), 173–185.

Jayakumar, P. N., Gangadhar, B. N., Subbakrishna, D. K., Janakiramaiah, N., Srinivas, J. S., & Keshavan, M. S. (2003). Membrane phospholipid abnormalities of basal ganglia in never-treated schizophrenia: a 31P magnetic resonance spectroscopy study. *Biol Psychiatry, 54*(4), 491–494.

Jensen, J. E., Miller, J., Williamson, P. C., Neufeld, R. W., Menon, R. S., Malla, A., Manchanda, R., Schaefer, B., Densmore, M., & Drost, D. J. (2004). Focal changes in brain energy and phospholipid metabolism in first-episode schizophrenia: 31P-MRS chemical shift imaging study at 4 Tesla. *Br J Psychiatry, 184*, 409–415.

Johnstone, E. C., Crow, T. J., Frith, C. D., Husband, J., & Kreel, L. (1976). Cerebral ventricular size and cognitive impairment in chronic schizophrenia. *Lancet*, 2(7992), 924–926.

Karayiorgou, M., Morris, M. A., Morrow, B., Shprintzen, R. J., Goldberg, R., Borrow, J., Gos, A., Nestadt, G., Wolyniec, P. S., Lasseter, V. K., & et al. (1995). Schizophrenia susceptibility associated with interstitial deletions of chromosome 22q11. *Proc Natl Acad Sci U S A, 92*(17), 7612–7616.

Karlsson, P., Farde, L., Halldin, C., & Sedvall, G. (2002). PET study of D(1) dopamine receptor binding in neuroleptic-naive patients with schizophrenia. *Am J Psychiatry, 159*(5), 761–767.

Kelly, C., & McCreadie, R. G. (1999). Smoking habits, current symptoms, and premorbid characteristics of schizophrenic patients in Nithsdale, Scotland. *Am J Psychiatry, 156*(11), 1751–1757.

Kelsoe, J. R., Jr., Cadet, J. L., Pickar, D., & Weinberger, D. R. (1988). Quantitative neuroanatomy in schizophrenia. A controlled magnetic resonance imaging study. *Arch Gen Psychiatry, 45*(6), 533–541.

Keshavan, M. S., Rosenberg, D., Sweeney, J. A., & Pettegrew, J. W. (1998). Decreased caudate volume in neuroleptic-naive psychotic patients. *Am J Psychiatry, 155*(6), 774–778.

Keshavan, M. S., Tandon, R., Boutros, N. N., & Nasrallah, H. A. (2008). Schizophrenia, "just the facts": what we know in 2008 Part 3: neurobiology. *Schizophr Res, 106*(2–3), 89–107.

Kessler, R. C., Amminger, G. P., Aguilar-Gaxiola, S., Alonso, J., Lee, S., & Ustun, T. B. (2007). Age of onset of mental disorders: a review of recent literature. *Curr Opin Psychiatry, 20*(4), 359–364.

Kim, E., Levy, R., & Pikalov, A. (2007). Personalized treatment with atypical antipsychotic medications. *Adv Ther, 24*(4), 721–740.

Kohlrausch, F. B., Gama, C. S., Lobato, M. I., Belmonte-de-Abreu, P., Callegari-Jacques, S. M., Gesteira, A., Barros, F., Carracedo, A., & Hutz, M. H. (2008). Naturalistic pharmacogenetic study of treatment resistance to typical neuroleptics in European-Brazilian schizophrenics. *Pharmacogenet Genomics, 18*(7), 599–609.

Kohlrausch, F. B., Salatino-Oliveira, A., Gama, C. S., Lobato, M. I., Belmonte-de-Abreu, P., & Hutz, M. H. (2008). G-protein gene 825C>T polymorphism is associated with response to clozapine in Brazilian schizophrenics. *Pharmacogenomics, 9*(10), 1429–1436.

Konick, L. C., & Friedman, L. (2001). Meta-analysis of thalamic size in schizophrenia. *Biol Psychiatry, 49*(1), 28–38.

Kubicki, M., McCarley, R., Westin, C. F., Park, H. J., Maier, S., Kikinis, R., Jolesz, F. A., & Shenton, M. E. (2007). A review of diffusion tensor imaging studies in schizophrenia. *J Psychiatr Res, 41*(1–2), 15–30.

Kubicki, M., Westin, C. F., Maier, S. E., Frumin, M., Nestor, P. G., Salisbury, D. F., Kikinis, R., Jolesz, F. A., McCarley, R. W., & Shenton, M. E. (2002). Uncinate fasciculus findings in schizophrenia: a magnetic resonance diffusion tensor imaging study. *Am J Psychiatry, 159*(5), 813–820.

Kuperberg, G. R., Broome, M. R., McGuire, P. K., David, A. S., Eddy, M., Ozawa, F., Goff, D., West, W. C., Williams, S. C., van der Kouwe, A. J., Salat, D. H., Dale, A. M., & Fischl, B. (2003). Regionally localized thinning of the cerebral cortex in schizophrenia. *Arch Gen Psychiatry, 60*(9), 878–888.

Kuroki, N., Kubicki, M., Nestor, P. G., Salisbury, D. F., Park, H. J., Levitt, J. J., Woolston, S., Frumin, M., Niznikiewicz, M., Westin, C. F., Maier, S. E., McCarley, R. W., & Shenton, M. E. (2006). Fornix integrity and hippocampal volume in male schizophrenic patients. *Biol Psychiatry, 60*(1), 22–31.

Kwon, J. S., Shenton, M. E., Hirayasu, Y., Salisbury, D. F., Fischer, I. A., Dickey, C. C., Yurgelun-Todd, D., Tohen, M., Kikinis, R., Jolesz, F. A., & McCarley, R. W. (1998). MRI study of cavum septi pellucidi in schizophrenia, affective disorder, and schizotypal personality disorder. *Am J Psychiatry, 155*(4), 509–515.

Lachman, H. M., Papolos, D. F., Saito, T., Yu, Y. M., Szumlanski, C. L., & Weinshilboum, R. M. (1996). Human catechol-O-methyltransferase pharmacogenetics: description of a functional polymorphism and its potential application to neuropsychiatric disorders. *Pharmacogenetics, 6*(3), 243–250.

Lane, H. Y., Lee, C. C., Liu, Y. C., & Chang, W. H. (2005). Pharmacogenetic studies of response to risperidone and other newer atypical antipsychotics. *Pharmacogenomics, 6*(2), 139–149.

Laruelle, M., Abi-Dargham, A., van Dyck, C. H., Gil, R., D'Souza, C. D., Erdos, J., McCance, E., Rosenblatt, W., Fingado, C., Zoghbi, S. S., Baldwin, R. M., Seibyl, J. P., Krystal, J. H., Charney, D. S., & Innis, R. B. (1996). Single photon emission computerized tomography imaging of amphetamine-induced dopamine release in drug-free schizophrenic subjects. *Proc Natl Acad Sci U S A, 93*(17), 9235–9240.

Lasser, K., Boyd, J. W., Woolhandler, S., Himmelstein, D. U., McCormick, D., & Bor, D. H. (2000). Smoking and mental illness: A population-based prevalence study. *JAMA, 284*(20), 2606–2610.

Lavedan, C., Licamele, L., Volpi, S., Hamilton, J., Heaton, C., Mack, K., Lannan, R., Thompson, A., Wolfgang, C. D., & Polymeropoulos, M. H. (2009). Association of the NPAS3 gene and five other loci with response to the antipsychotic iloperidone identified in a whole genome association study. *Mol Psychiatry, 14*(8), 804–819.

Lavedan, C., Volpi, S., Polymeropoulos, M. H., & Wolfgang, C. D. (2008). Effect of a ciliary neurotrophic factor polymorphism on schizophrenia symptom improvement in an iloperidone clinical trial. *Pharmacogenomics, 9*(3), 289–301.

Lee, K., Yoshida, T., Kubicki, M., Bouix, S., Westin, C. F., Kindlmann, G., Niznikiewicz, M., Cohen, A., McCarley, R. W., & Shenton, M. E. (2009). Increased diffusivity in superior temporal gyrus in patients with schizophrenia: a Diffusion Tensor Imaging study. *Schizophr Res, 108*(1–3), 33–40.

Lewis, D. A., & Gonzalez-Burgos, G. (2006). Pathophysiologically based treatment interventions in schizophrenia. *Nat Med, 12*(9), 1016–1022.

Lewis, D. A., Hashimoto, T., & Volk, D. W. (2005). Cortical inhibitory neurons and schizophrenia. *Nat Rev Neurosci, 6*(4), 312–324.

Lieberman, J. A., Kane, J. M., & Alvir, J. (1987). Provocative tests with psychostimulant drugs in schizophrenia. *Psychopharmacology (Berl), 91*(4), 415–433.

Lieberman, J. A., Stroup, T. S., McEvoy, J. P., Swartz, M. S., Rosenheck, R. A., Perkins, D. O., Keefe, R. S., Davis, S. M., Davis, C. E., Lebowitz, B. D., Severe, J., & Hsiao, J. K. (2005). Effectiveness of antipsychotic drugs in patients with chronic schizophrenia. *N Engl J Med, 353*(12), 1209–1223.

Liou, Y. J., Lai, I. C., Lin, M. W., Bai, Y. M., Lin, C. C., Liao, D. L., Chen, J. Y., Lin, C. Y., & Wang, Y. C. (2006). Haplotype analysis of endothelial nitric oxide synthase (NOS3) genetic variants and tardive dyskinesia in patients with schizophrenia. *Pharmacogenet Genomics, 16*(3), 151–157.

Ludewig, S., Ludewig, K., Geyer, M. A., Hell, D., & Vollenweider, F. X. (2002). Prepulse inhibition deficits in patients with panic disorder. *Depress Anxiety, 15*(2), 55–60.

Lutkenhoff, E. S., van Erp, T. G., Thomas, M. A., Therman, S., Manninen, M., Huttunen, M. O., Kaprio, J., Lonnqvist, J., O'Neill, J., & Cannon, T. D. (2008). Proton MRS in twin pairs discordant for schizophrenia. *Mol Psychiatry.*

Martin, L. F., Kem, W. R., & Freedman, R. (2004). Alpha-7 nicotinic receptor agonists: potential new candidates for the treatment of schizophrenia. *Psychopharmacology (Berl), 174*(1), 54–64.

McCarley, R. W., Shenton, M. E., O'Donnell, B. F., Faux, S. F., Kikinis, R., Nestor, P. G., & Jolesz, F. A. (1993). Auditory P300 abnormalities and left posterior superior temporal gyrus volume reduction in schizophrenia. *Arch Gen Psychiatry, 50*(3), 190–197.

McClay, J. L., Adkins, D. E., Aberg, K., Stroup, S., Perkins, D. O., Vladimirov, V. I., Lieberman, J. A., Sullivan, P. F., & van den Oord, E. J. (2009). Genome-wide pharmacogenomic analysis of response to treatment with antipsychotics. *Mol Psychiatry.*

McGuffin, P., Farmer, A. E., Gottesman, II, Murray, R. M., & Reveley, A. M. (1984). Twin concordance for operationally defined schizophrenia. Confirmation of familiality and heritability. *Arch Gen Psychiatry, 41*(6), 541–545.

Meary, A., Brousse, G., Jamain, S., Schmitt, A., Szoke, A., Schurhoff, F., Gavaudan, G., Lancon, C., Macquin-Mavier, I., Leboyer, M., & Llorca, P. M. (2008). Pharmacogenetic study of atypical antipsychotic drug response: involvement of the norepinephrine transporter gene. *Am J Med Genet B Neuropsychiatr Genet, 147B*(4), 491–494.

Mialet, J. P., & Pichot, P. (1981). Eye-tracking patterns in schizophrenia. An analysis based on the incidence of saccades. *Arch Gen Psychiatry, 38*(2), 183–186.

Miller, T. J., McGlashan, T. H., Woods, S. W., Stein, K., Driesen, N., Corcoran, C. M., Hoffman, R., & Davidson, L. (1999). Symptom assessment in schizophrenic prodromal states. *Psychiatr Q, 70*(4), 273–287.

Minzenberg, M. J., Laird, A. R., Thelen, S., Carter, C. S., & Glahn, D. C. (2009). Meta-analysis of 41 functional neuroimaging studies of executive function in schizophrenia. *Arch Gen Psychiatry, 66*(8), 811–822.

Moghaddam, B., & Homayoun, H. (2008). Divergent plasticity of prefrontal cortex networks. *Neuropsychopharmacology, 33*(1), 42–55.

Morstyn, R., Duffy, F. H., & McCarley, R. W. (1983). Altered P300 topography in schizophrenia. *Arch Gen Psychiatry, 40*(7), 729–734.

Mossner, R., Schuhmacher, A., Kuhn, K. U., Cvetanovska, G., Rujescu, D., Zill, P., Quednow, B. B., Rietschel, M., Wolwer, W., Gaebel, W., Wagner, M., & Maier, W. (2009). Functional serotonin 1A receptor variant influences treatment response to atypical antipsychotics in schizophrenia. *Pharmacogenet Genomics, 19*(1), 91–94.

Muller, D. J., & Kennedy, J. L. (2006). Genetics of antipsychotic treatment emergent weight gain in schizophrenia. *Pharmacogenomics, 7*(6), 863–887.

Need, A. C., Keefe, R. S., Ge, D., Grossman, I., Dickson, S., McEvoy, J. P., & Goldstein, D. B. (2009). Pharmacogenetics of antipsychotic response in the CATIE trial: a candidate gene analysis. *Eur J Hum Genet, 17*(7), 946–957.

Nnadi, C. U., & Malhotra, A. K. (2007). Individualizing antipsychotic drug therapy in schizophrenia: the promise of pharmacogenetics. *Curr Psychiatry Rep, 9*(4), 313–318.

Ohrmann, P., Siegmund, A., Suslow, T., Pedersen, A., Spitzberg, K., Kersting, A., Rothermundt, M., Arolt, V., Heindel, W., & Pfleiderer, B. (2007). Cognitive impairment and in vivo metabolites in first-episode neuroleptic-naive and chronic medicated schizophrenic patients: a proton magnetic resonance spectroscopy study. *J Psychiatr Res, 41*(8), 625–634.

Ohrmann, P., Siegmund, A., Suslow, T., Spitzberg, K., Kersting, A., Arolt, V., Heindel, W., & Pfleiderer, B. (2005). Evidence for glutamatergic neuronal dysfunction in the prefrontal cortex in chronic but not in first-episode patients with schizophrenia: a proton magnetic resonance spectroscopy study. *Schizophr Res, 73*(2–3), 153–157.

Okubo, Y., Suhara, T., Suzuki, K., Kobayashi, K., Inoue, O., Terasaki, O., Someya, Y., Sassa, T., Sudo, Y., Matsushima, E., Iyo, M., Tateno, Y., & Toru, M. (1997). Decreased prefrontal dopamine D1 receptors in schizophrenia revealed by PET. *Nature, 385*(6617), 634–636.

Okugawa, G., Nobuhara, K., Minami, T., Takase, K., Sugimoto, T., Saito, Y., Yoshimura, M., & Kinoshita, T. (2006). Neural disorganization in the superior cerebellar peduncle and cognitive abnormality in patients with schizophrenia: A diffusion tensor imaging study. *Prog Neuropsychopharmacol Biol Psychiatry, 30*(8), 1408–1412.

Olney, J. W., Newcomer, J. W., & Farber, N. B. (1999). NMDA receptor hypofunction model of schizophrenia. *J Psychiatr Res, 33*(6), 523–533.

Ozdemir, V., Aklillu, E., Mee, S., Bertilsson, L., Albers, L. J., Graham, J. E., Caligiuri, M., Lohr, J. B., & Reist, C. (2006). Pharmacogenetics for off-patent antipsychotics: reframing the risk for tardive dyskinesia and access to essential medicines. *Expert Opin Pharmacother, 7*(2), 119–133.

Ozdemir, V., Basile, V. S., Masellis, M., & Kennedy, J. L. (2001). Pharmacogenetic assessment of antipsychotic-induced movement disorders: contribution of the dopamine D3 receptor and cytochrome P450 1A2 genes. *J Biochem Biophys Methods, 47*(1–2), 151–157.

Palmer, B. W., Heaton, R. K., Paulsen, J. S., Kuck, J., Braff, D., Harris, M. J., Zisook, S., & Jeste, D. V. (1997). Is it possible to be schizophrenic yet neuropsychologically normal? *Neuropsychology, 11*(3), 437–446.

Park, S., Holzman, P. S., & Goldman-Rakic, P. S. (1995). Spatial working memory deficits in the relatives of schizophrenic patients. *Arch Gen Psychiatry, 52*(10), 821–828.

Patel, J. K., Pinals, D.A., Breier, A. (2003a). *Schizophrenia and Other Psychoses* (2 ed. Vol. 2). Chichester: John Wiley & Sons, LTD.

Patel, J. K., Pinals, D.A., Breier, A. (2003b). Schizophrenia and Other Psychoses. In A. Tasman, Kay, J., and Lieberman, J.A. (Ed.), *Psychiatry* (2nd ed., Vol. 2, pp. 1131–1206). Chichester: John Wiley & Sons, LTD.

Raedler, T. J., Knable, M. B., Jones, D. W., Urbina, R. A., Gorey, J. G., Lee, K. S., Egan, M. F., Coppola, R., & Weinberger, D. R. (2003). In vivo determination of muscarinic acetylcholine receptor availability in schizophrenia. *Am J Psychiatry, 160*(1), 118–127.

Ragland, J. D., Gur, R. C., Valdez, J., Turetsky, B. I., Elliott, M., Kohler, C., Siegel, S., Kanes, S., & Gur, R. E. (2004). Event-related fMRI of frontotemporal activity during word encoding and recognition in schizophrenia. *Am J Psychiatry, 161*(6), 1004–1015.

Ragland, J. D., Laird, A. R., Ranganath, C., Blumenfeld, R. S., Gonzales, S. M., & Glahn, D. C. (2009). Prefrontal activation deficits during episodic memory in schizophrenia. *Am J Psychiatry, 166*(8), 863–874.

Rajarethinam, R., Upadhyaya, A., Tsou, P., Upadhyaya, M., & Keshavan, M. S. (2007). Caudate volume in offspring of patients with schizophrenia. *Br J Psychiatry, 191*, 258–259.

Reddy, R., & Keshavan, M. S. (2003). Phosphorus magnetic resonance spectroscopy: its utility in examining the membrane hypothesis of schizophrenia. *Prostaglandins Leukot Essent Fatty Acids, 69*(6), 401–405.

Rege, S. (2008). Antipsychotic induced weight gain in schizophrenia:mechanisms and management. *Aust N Z J Psychiatry, 42*(5), 369–381.

Reynolds, G. P. (2007). The impact of pharmacogenetics on the development and use of antipsychotic drugs. *Drug Discov Today, 12*(21–22), 953–959.

Reynolds, G. P., Arranz, B., Templeman, L. A., Fertuzinhos, S., & San, L. (2006). Effect of 5-HT1A receptor gene polymorphism on negative and depressive symptom response to antipsychotic treatment of drug-naive psychotic patients. *Am J Psychiatry, 163*(10), 1826–1829.

Rietschel, M., Krauss, H., Muller, D. J., Schulze, T. G., Knapp, M., Marwinski, K., Maroldt, A. O., Paus, S., Grunhage, F., Propping, P., Maier, W., Held, T., & Nothen, M. M. (2000). Dopamine D3 receptor variant and tardive dyskinesia. *Eur Arch Psychiatry Clin Neurosci, 250*(1), 31–35.

Rossler, W., Salize, H. J., van Os, J., & Riecher-Rossler, A. (2005). Size of burden of schizophrenia and psychotic disorders. *Eur Neuropsychopharmacol, 15*(4), 399–409.

Rusch, N., Tebartz van Elst, L., Valerius, G., Buchert, M., Thiel, T., Ebert, D., Hennig, J., & Olbrich, H. M. (2008). Neurochemical and structural correlates of executive dysfunction in schizophrenia. *Schizophr Res, 99*(1–3), 155–163.

Sallet, P. C., Elkis, H., Alves, T. M., Oliveira, J. R., Sassi, E., Campi de Castro, C., Busatto, G. F., & Gattaz, W. F. (2003). Reduced cortical folding in schizophrenia: an MRI morphometric study. *Am J Psychiatry, 160*(9), 1606–1613.

Saykin, A. J., Gur, R. C., Gur, R. E., Mozley, P. D., Mozley, L. H., Resnick, S. M., Kester, D. B., & Stafiniak, P. (1991). Neuropsychological function in schizophrenia. Selective impairment in memory and learning. *Arch Gen Psychiatry, 48*(7), 618–624.

Saykin, A. J., Shtasel, D. L., Gur, R. E., Kester, D. B., Mozley, L. H., Stafiniak, P., & Gur, R. C. (1994). Neuropsychological deficits in neuroleptic naive patients with first-episode schizophrenia. *Arch Gen Psychiatry, 51*(2), 124–131.

Schiffman, J., Ekstrom, M., LaBrie, J., Schulsinger, F., Sorensen, H., & Mednick, S. (2002). Minor physical anomalies and schizophrenia spectrum disorders: a prospective investigation. *Am J Psychiatry, 159*(2), 238–243.

Schobel, S. A., Lewandowski, N. M., Corcoran, C. M., Moore, H., Brown, T., Malaspina, D., & Small, S. A. (2009). Differential targeting of the CA1 subfield of the hippocampal formation by schizophrenia and related psychotic disorders. *Arch Gen Psychiatry, 66*(9), 938–946.

Scott, L. J. (2009). Iloperidone: in schizophrenia. *CNS Drugs, 23*(10), 867–880.

Seeman, P., & Lee, T. (1975). Antipsychotic drugs: direct correlation between clinical potency and presynaptic action on dopamine neurons. *Science, 188*(4194), 1217–1219.

Shenton, M. E., Dickey, C. C., Frumin, M., & McCarley, R. W. (2001). A review of MRI findings in schizophrenia. *Schizophr Res, 49*(1–2), 1–52.

Siegel, C., Waldo, M., Mizner, G., Adler, L. E., & Freedman, R. (1984). Deficits in sensory gating in schizophrenic patients and their relatives. Evidence obtained with auditory evoked responses. *Arch Gen Psychiatry, 41*(6), 607–612.

Sitskoorn, M. M., Aleman, A., Ebisch, S. J., Appels, M. C., & Kahn, R. S. (2004). Cognitive deficits in relatives of patients with schizophrenia: a meta-analysis. *Schizophr Res, 71*(2–3), 285–295.

Smesny, S., Rosburg, T., Nenadic, I., Fenk, K. P., Kunstmann, S., Rzanny, R., Volz, H. P., & Sauer, H. (2007). Metabolic mapping using 2D 31P-MR spectroscopy reveals frontal and thalamic metabolic abnormalities in schizophrenia. *Neuroimage, 35*(2), 729–737.

Smith, R. C., Warner-Cohen, J., Matute, M., Butler, E., Kelly, E., Vaidhyanathaswamy, S., & Khan, A. (2006). Effects of nicotine nasal spray on cognitive function in schizophrenia. *Neuropsychopharmacology, 31*(3), 637–643.

Snitz, B. E., Macdonald, A. W., 3rd, & Carter, C. S. (2006). Cognitive deficits in unaffected first-degree relatives of schizophrenia patients: a meta-analytic review of putative endophenotypes. *Schizophr Bull, 32*(1), 179–194.

Souza, R. P., De Luca, V., Muscettola, G., Rosa, D. V., de Bartolomeis, A., Romano Silva, M., & Kennedy, J. L. (2008). Association of antipsychotic induced weight gain and body mass index with GNB3 gene: a meta-analysis. *Prog Neuropsychopharmacol Biol Psychiatry, 32*(8), 1848–1853.

Spoletini, I., Cherubini, A., Banfi, G., Rubino, I. A., Peran, P., Caltagirone, C., & Spalletta, G. (2009). Hippocampi, Thalami, and Accumbens Microstructural Damage in Schizophrenia: A Volumetry, Diffusivity, and Neuropsychological Study. *Schizophr Bull.*

Srivastava, V., Deshpande, S. N., Nimgaonkar, V. L., Lerer, B., & Thelma, B. (2008). Genetic correlates of olanzapine-induced weight gain in schizophrenia subjects from north India: role of metabolic pathway genes. *Pharmacogenomics, 9*(8), 1055–1068.

Srivastava, V., Varma, P. G., Prasad, S., Semwal, P., Nimgaonkar, V. L., Lerer, B., Deshpande, S. N., & Bk, T. (2006). Genetic susceptibility to tardive dyskinesia among schizophrenia subjects: IV. Role of dopaminergic pathway gene polymorphisms. *Pharmacogenet Genomics, 16*(2), 111–117.

Steen, R. G., Hamer, R. M., & Lieberman, J. A. (2005). Measurement of brain metabolites by 1H magnetic resonance spectroscopy in patients with schizophrenia: a systematic review and meta-analysis. *Neuropsychopharmacology, 30*(11), 1949–1962.

Stefansson, H., Ophoff, R. A., Steinberg, S., Andreassen, O. A., Cichon, S., Rujescu, D., Werge, T., Pietilainen, O. P., Mors, O., Mortensen, P. B., Sigurdsson, E., Gustafsson, O., Nyegaard, M., Tuulio-Henriksson, A., Ingason, A., Hansen, T., Suvisaari, J., Lonnqvist, J., Paunio, T., Borglum, A. D., Hartmann, A., Fink-Jensen, A., Nordentoft, M., Hougaard, D., Norgaard-Pedersen, B., Bottcher, Y., Olesen, J., Breuer, R., Moller, H. J., Giegling, I., Rasmussen, H. B., Timm, S., Mattheisen, M., Bitter, I., Rethelyi, J. M., Magnusdottir, B. B., Sigmundsson, T., Olason, P., Masson, G., Gulcher, J. R., Haraldsson, M., Fossdal, R., Thorgeirsson, T. E., Thorsteinsdottir, U., Ruggeri, M., Tosato, S., Franke, B., Strengman, E., Kiemeney, L. A., Melle, I., Djurovic, S., Abramova, L., Kaleda, V., Sanjuan, J., de Frutos, R., Bramon, E., Vassos, E., Fraser, G., Ettinger, U., Picchioni, M., Walker, N., Toulopoulou, T., Need, A. C., Ge, D., Yoon, J. L., Shianna, K. V., Freimer, N. B., Cantor, R. M., Murray, R., Kong, A., Golimbet, V., Carracedo, A., Arango, C., Costas, J., Jonsson, E. G., Terenius, L., Agartz, I., Petursson, H., Nothen, M. M., Rietschel, M., Matthews, P. M., Muglia, P., Peltonen, L., St Clair, D., Goldstein, D. B., Stefansson, K., & Collier, D. A. (2009). Common variants conferring risk of schizophrenia. *Nature, 460*(7256), 744–747.

Stefansson, H., Rujescu, D., Cichon, S., Pietilainen, O. P., Ingason, A., Steinberg, S., Fossdal, R., Sigurdsson, E., Sigmundsson, T., Buizer-Voskamp, J. E., Hansen, T., Jakobsen, K. D., Muglia, P., Francks, C., Matthews, P. M., Gylfason, A., Halldorsson, B. V., Gudbjartsson, D., Thorgeirsson, T. E., Sigurdsson, A., Jonasdottir, A., Bjornsson, A., Mattiasdottir, S., Blondal, T., Haraldsson, M., Magnusdottir, B. B., Giegling, I., Moller, H. J., Hartmann, A., Shianna, K. V., Ge, D., Need, A. C., Crombie, C., Fraser, G., Walker, N., Lonnqvist, J., Suvisaari, J., Tuulio-Henriksson, A., Paunio, T., Toulopoulou, T., Bramon, E., Di Forti, M., Murray, R., Ruggeri, M., Vassos, E., Tosato, S., Walshe, M., Li, T., Vasilescu, C., Muhleisen, T. W., Wang, A. G., Ullum, H., Djurovic, S., Melle, I., Olesen, J., Kiemeney, L. A., Franke, B., Sabatti, C., Freimer, N. B., Gulcher, J. R., Thorsteinsdottir, U., Kong, A., Andreassen, O. A., Ophoff, R. A., Georgi, A., Rietschel, M., Werge, T., Petursson, H., Goldstein, D. B., Nothen, M. M., Peltonen, L., Collier, D. A., St Clair, D., & Stefansson, K. (2008). Large recurrent microdeletions associated with schizophrenia. *Nature, 455*(7210), 232–236.

Stroup, T. S. (2007). Heterogeneity of treatment effects in schizophrenia. *Am J Med, 120*(4 Suppl 1), S26–31.

Sussmann, J. E., Lymer, G. K., McKirdy, J., Moorhead, T. W., Maniega, S. M., Job, D., Hall, J., Bastin, M. E., Johnstone, E. C., Lawrie, S. M., & McIntosh, A. M. (2009). White matter abnormalities in bipolar disorder and schizophrenia detected using diffusion tensor magnetic resonance imaging. *Bipolar Disord, 11*(1), 11–18.

Takesada, M., Miyamoto, E., Kakimoto, Y., Sano, I., & Kaneko, Z. (1965). Phenolic and indole amines in the urine of schizophrenics. *Nature, 207*(5002), 1199–1200.

Templeman, L. A., Reynolds, G. P., Arranz, B., & San, L. (2005). Polymorphisms of the 5-HT2C receptor and leptin genes are associated with antipsychotic drug-induced weight gain in Caucasian subjects with a first-episode psychosis. *Pharmacogenet Genomics, 15*(4), 195–200.

Tezenas Du Montcel, S., Mendizabai, H., Ayme, S., Levy, A., & Philip, N. (1996). Prevalence of 22q11 microdeletion. *J Med Genet, 33*(8), 719.

Thelma, B., Srivastava, V., & Tiwari, A. K. (2008). Genetic underpinnings of tardive dyskinesia: passing the baton to pharmacogenetics. *Pharmacogenomics, 9*(9), 1285–1306.

Tiwari, A. K., Deshpande, S. N., Rao, A. R., Bhatia, T., Lerer, B., Nimgaonkar, V. L., & Thelma, B. K. (2005). Genetic susceptibility to tardive dyskinesia in chronic schizophrenia subjects: III. Lack of association of CYP3A4 and CYP2D6 gene polymorphisms. *Schizophr Res, 75*(1), 21–26.

Umbricht, D., & Krljes, S. (2005). Mismatch negativity in schizophrenia: a meta-analysis. *Schizophr Res, 76*(1), 1–23.

Volpi, S., Potkin, S. G., Malhotra, A. K., Licamele, L., & Lavedan, C. (2009). Applicability of a genetic signature for enhanced iloperidone efficacy in the treatment of schizophrenia. *J Clin Psychiatry, 70*(6), 801–809.

Wang, F., Sun, Z., Cui, L., Du, X., Wang, X., Zhang, H., Cong, Z., Hong, N., & Zhang, D. (2004). Anterior cingulum abnormalities in male patients with schizophrenia determined through diffusion tensor imaging. *Am J Psychiatry, 161*(3), 573–575.

Wang, L., Fang, C., Zhang, A., Du, J., Yu, L., Ma, J., Feng, G., Xing, Q., & He, L. (2008). The --1019 C/G polymorphism of the 5-HT(1)A receptor gene is associated with negative symptom response to risperidone treatment in schizophrenia patients. *J Psychopharmacol, 22*(8), 904–909.

Wei, Z., Wang, L., Xuan, J., Che, R., Du, J., Qin, S., Xing, Y., Gu, B., Yang, L., Li, H., Li, J., Feng, G., He, L., & Xing, Q. (2009). Association analysis of serotonin receptor 7 gene (HTR7) and risperidone response in Chinese schizophrenia patients. *Prog Neuropsychopharmacol Biol Psychiatry, 33*(3), 547–551.

Weinberger, D. R. (1987). Implications of normal brain development for the pathogenesis of schizophrenia. *Arch Gen Psychiatry, 44*(7), 660–669.

Wirshing, D. A., Boyd, J. A., Meng, L. R., Ballon, J. S., Marder, S. R., & Wirshing, W. C. (2002). The effects of novel antipsychotics on glucose and lipid levels. *J Clin Psychiatry, 63*(10), 856–865.

Wood, S. J., Berger, G. E., Lambert, M., Conus, P., Velakoulis, D., Stuart, G. W., Desmond, P., McGorry, P. D., & Pantelis, C. (2006). Prediction of functional outcome 18 months after a first psychotic episode: a proton magnetic resonance spectroscopy study. *Arch Gen Psychiatry, 63*(9), 969–976.

Wood, S. J., Yucel, M., Wellard, R. M., Harrison, B. J., Clarke, K., Fornito, A., Velakoulis, D., & Pantelis, C. (2007). Evidence for neuronal dysfunction in the anterior cingulate of patients with schizophrenia: a proton magnetic resonance spectroscopy study at 3 T. *Schizophr Res, 94*(1–3), 328–331.

Wright, I. C., Rabe-Hesketh, S., Woodruff, P. W., David, A. S., Murray, R. M., & Bullmore, E. T. (2000). Meta-analysis of regional brain volumes in schizophrenia. *Am J Psychiatry, 157*(1), 16–25.

Yamanouchi, Y., Iwata, N., Suzuki, T., Kitajima, T., Ikeda, M., & Ozaki, N. (2003). Effect of DRD2, 5-HT2A, and COMT genes on antipsychotic response to risperidone. *Pharmacogenomics J, 3*(6), 356–361.

Zakzanis, K. K., Poulin, P., Hansen, K. T., & Jolic, D. (2000). Searching the schizophrenic brain for temporal lobe deficits: a systematic review and meta-analysis. *Psychol Med, 30*(3), 491–504.

Zhang, X. Y., Tan, Y. L., Zhou, D. F., Haile, C. N., Cao, L. Y., Xu, Q., Shen, Y., Kosten, T. A., & Kosten, T. R. (2007). Association of clozapine-induced weight gain with a polymorphism in the leptin promoter region in patients with chronic schizophrenia in a Chinese population. *J Clin Psychopharmacol, 27*(3), 246–251.

7 Personalized Integrative Markers for Attention Deficit/Hyperactivity Disorder in Children and Adolescents

Michael R. Kohn, MD, Simon D. Clarke, MD, and Leanne M. Williams, PhD

Introduction

There is a "translational" gap between research findings and their implementation in clinical practice when it comes to disorders of brain health in adolescence and childhood (this is often the case more broadly with adult disorders as well). Here we focus on Attention Deficit Hyperactivity Disorder (ADHD), the most common disorder of child–adolescent brain health. We include examples from the associated conditions of anxiety, depression, and learning difficulties, to demonstrate the utility of a personalized and integrative neuroscience approach. Research studies have identified candidate markers to support diagnostic, functional assessment, and personalized treatment decisions in clinical practice, but there is little consensus about these markers. There is also little attention given to how to translate these markers into assessment platforms that may be readily used in a clinical setting. Identifying the objective brain-based markers with the most clinical utility is an important first step in implementing a personalized medicine program that uses increasingly predictive markers to tailor treatment decisions to the individual. Here, we outline an integrative neuroscience framework for ADHD that brings together measures of thinking, emotion, and feeling, along with their brain correlates. From this framework, markers sensitive and specific to ADHD are identified, together with the ways in which such objective markers may support personalized clinical decisions for ADHD and its allied conditions.

The Clinical Picture of ADHD

ADHD is considered to be the most common neurodevelopmental and chronic health condition affecting school-aged children. It has been described as "a serious disability with long-term consequences" (Ballard et al. 1997). It has severe consequences in social, individual, and family settings, often resulting in significant financial costs due to required treatments, psychological implications due to familial stress and breakdown, difficulties in academic and vocational areas, and increased risk of drug abuse (Faraone and Biederman 1998).

Using DSM-IV criteria, diagnosis of ADHD is currently founded on a classical triad of symptoms: inattention, hyperactivity, and impulsivity (Jensen, Martin, and Cantwell 1997). Diagnostic features include inappropriate and disruptive levels of inattention and/or hyperactivity with impulsivity. ADHD is typically identified in childhood or early adolescence. It is also considered a heterogenous condition that carries a high risk of comorbidity (Biederman 1998). Comorbid conditions frequently associated with ADHD may be broadly classified into three categories: learning problems, externalizing disorders (such as conduct disorder), and internalizing disorders (such as anxiety and depression). Both defining symptoms and comorbid features commonly persist into adulthood. ADHD is more prevalent in males, although underreporting in females may contribute to this perceived difference (Biederman 1998).

The Past

The diagnostic origin of ADHD can be traced back to the 1950s, when it was termed Minimal Brain Dysfunction (MBD). Subsequently, the diagnostic labels of Hyperactive Syndrome, Hyperkinesis, Hyperactivity Disorder of Childhood and variations of the term "Attention Deficit Disorder" were used. These changes may reflect the uncertainty about the cause and defining criteria for ADHD (Barkley 1998).

The most recent edition the *Diagnostic and Statistical Manual of Mental Disorders*, or DSM-IV (American Psychiatric Association 1994) has provided the most explicit recognition of the heterogeneity of symptoms. For example, diagnostic criteria could be met if an individual is inattentive in some situations but not others. While the historical evolution of these criteria reflects changing emphases in the conceptions of ADHD, there has been consistency in presuming an underlying brain disturbance in this condition.

The Present

The DSM-IV has based the diagnosis of ADHD on two symptom dimensions: inattention and hyperactivity/impulsivity. Each symptom dimension is defined by nine criteria. Meeting six or more of these criteria on either dimension will confirm a diagnosis of ADHD (Inattentive subtype) or ADHD (Hyperactive/Impulsive subtype), while meeting six or more on both dimensions will produce an ADHD (Combined subtype) diagnosis. The DSM-IV also requires that symptoms be present before the age of seven, in two or more settings, causing social and academic problems and not related to anxiety, mood, or stress (American Psychiatric Association 1994). A generally similar pattern has been observed in adult samples (Murphy, Barkley, and Bush 2002).

The Future

DSM-IV, following its predecessors, relies on subjective clinical assessment to classify behavior. In this regard, there may not always be a direct match between the "round hole" of diagnostic signs and symptoms and the "square pegs" of the brain–gene mechanisms underlying the pathophysiology of ADHD.

The focus on signs and symptoms has provided a pragmatic and consistent nomenclature among clinicians. Rapidly emerging findings for the brain–gene basis of psychopathological features has begun to be incorporated into the template for the next version of the DSM (DSM-V), which is expected to be published in 2013. While unrealistically optimistic expectations about single gene–brain dysfunction linkages still remain, the essential plan for converging biology with signs and symptoms into the DSM-V moves toward the concept of integration: "It is our goal to translate basic and clinical neuroscience research relating brain structure, brain function, and behavior into a classification of psychiatric disorders based on etiology and pathophysiology" (Charney et al. 2002).

The Need for Objective Markers for ADHD

In the move toward DSM-V, the importance of objective measures linked to underlying brain function has been highlighted (Castellanos and Tannock 2002; Insel et al. 2004; Gordon, 2007; Gordon et al. 2007b. Objective measures will enhance the reliability of clinical decisions, and provide concrete benchmarks for monitoring progress. These benchmarks offer the additional benefit of engaging both patient and family, and providing them with explicit feedback (Monastra 2005).

The search for objective markers within the framework of personalized medicine has so far focused on the use of genomic markers as support for diagnosis and treatment decisions at the individual level. The success of genomic markers for physical conditions has provided a reasonable proof of concept for personalized medicine (Gordon et al. 2007b). The focus on genomics has been driven by the breakthrough innovations of the Human Genome Project, and the more recent focus on genome-wide association studies (GWAS). This genomic focus is likely to produce important outcomes for those few disorders that involve a single genomic susceptibility or mutation. But the complexity of the brain is likely to require a shift from a single genetic marker focus to a more integrated focus, in which additional brain-related information is taken into account, in order to objectively diagnose patients and accurately predict their treatment response.

The brain has a complex neuronal ecosystem that is inherently different from that of other organs. This complexity is reflected in the highly interconnected nature of its 100 billion neurons and over one hundred neurotransmitters (Gordon 2000; Monastra 2005). At least 80 percent of each person's approximately 20,500 protein-coding genes (Clamp et al. 2007) are expressed in the brain to some extent. Many interacting genes confer susceptibility to psychiatric disorders, leading to heterogeneous phenotypic profiles. Given the expression of genetics in the brain, measures of brain structure and function provide important intermediate markers that link genetic susceptibility to behavioral symptoms. The term "endophenotypes" is sometimes used to describe these intermediate markers.

Thus, a personalized medicine for complex disorders of brain health, such as ADHD, will most likely be best facilitated by a combination of brain-based markers from genomics, as well as brain structure, brain function, and cognition. Objective measures of brain structure are provided by Magnetic Resonance Imaging (MRI), real time measures of brain function (including electroencephalography [EEG], event-related potential [ERP], magnetoencephalography [MEG], functional MRI [fMRI], and positron emission tomography [PET]), and cognitive performance measures of both general and emotional cognitive functions.

Identifying personalized integrative neuroscience markers will provide important steps toward establishing:

1. Diagnostic sensitivity and specificity,
2. Profiles of comorbidity, and
3. Treatment prediction, treatment efficacy, and side effect monitoring.

Such markers will provide the necessary first steps toward more precise diagnostic classification of ADHD, a better understanding of how treatments work at the individual level, the evidence-based ability to match the right drug to the right person, and objective evaluation of the efficacy of that treatment at the right time in the patient's clinical treatment plan.

The Real-World Benefits of Brain-Based Markers for ADHD

The real-world benefits of identifying personalized markers for ADHD include:

1. Providing *objective* indicators to support diagnostic and treatment-oriented decision-making in clinical practice. These markers complement information from clinical history and symptom criteria.
2. Reducing the trial and error of choosing treatments suited to individual patients, including those with a subtype of ADHD or comorbidities.
3. Enhancing the focus on the profile of each individual patient with ADHD rather than employing the "one size fits all" approach, and supporting the ability to make more precise decisions at once, such as the capacity to rule in or rule out stimulant medication. The personalized approach will reduce costs and potential over-servicing.
4. Providing objective indicators for developing new treatments that are optimized for the individual patient and the heterogeneity of ADHD across patients. These markers will improve the cost–benefit efficiency of clinical trials by enabling a focus on particular patients exhibiting candidate markers. Markers will also support the emerging emphasis on comparative effectiveness trials (which compare relative benefits of treatments), and provide new objective benchmarks for comparing pharmaceutical to non-pharmaceutical treatments.

The current focus of markers for ADHD emphasizes general levels of impairment. However, the rapid development of brain and cognitive measures provides the opportunity to match treatments to objective information based on individual strengths, as well as weaknesses.

Identifying Brain-Based Markers for ADHD

Toward this goal, theories of ADHD have seen an evolution from those focused on cognitive concepts to those focused on brain systems or genetic susceptibility. Progressively integrating these theories—from genes to brain, cognition, and clinical symptoms—aids in the overall understanding of ADHD.

A consensus is emerging that ADHD is characterized by disturbances across the spectrum of emotion, thinking, feeling, and self-regulation (Bush, Valera, and Seidman 2005). Yet no similar theoretical consensus exists regarding their cause and underlying brain–gene origin. An integrative neuroscience approach provides a framework to draw together theories that emphasize the involvement of multiple and dynamic brain pathways in the development of ADHD. It also offers a framework-promoting consensus regarding the objective brain-based markers that have utility for assessing ADHD and for predicting and evaluating the treatment response. An integrative neuroscience framework encompasses:

1. Integrative theoretical models,
2. Assessing multiple levels of organization,
3. Standardization of all data, and
4. Databases and large subject numbers.

An Integrative Neuroscience Framework for ADHD

The INTEGRATE Model is an example of an integrative neuroscience framework that may be applied to ADHD, to bring together levels of organization across timescale (when disturbances happen), brain systems (where they happen), and disposition (genetic susceptibility).

The INTEGRATE Model considers how four core processes may capture the inordinate detail of the brain: emotion, thinking, feeling and self-regulation (Gordon et al., 2008; Williams et al., 2008a; for detailed review, see Chapter 3). A fundamental organizing principle that integrates these four processes across all levels of function is the core motivation to "minimize danger and maximize reward" (see Figure 7-1). Susceptibility to impaired emotion, thinking, feeling, and/or self-regulation is contributed by one's genetic disposition, and influenced by ongoing interactions with the environment, such as family or life stressors.

The overt manifestation of ADHD is characterized by a loss of function that is most pronounced at the timescale of "thinking." However, it impacts the earlier timescale of "emotion," the associated timescale of "feeling," and the later timescale of "self-regulation." The following sections provide an overview of the current consensus for brain-based markers of ADHD, its comorbidities, and treatment across these domains. These markers are summarized in Tables 7-1a, 7-1b, and 7-1c.

Brain-Based Markers for Thinking and Emotion in ADHD

Cognitive Thinking Markers

Core thinking markers for ADHD encompass sustained attention, inhibition, impulsivity, errors of intrusion from irrelevant information, and response variability. Together, these markers have a sensitivity of 86 percent and a specificity of 91 percent (Table 7-1a).

The thinking timescale that is altered in ADHD spans the timescale from several hundred milliseconds to a few seconds (see Figure 7-1 and Chapter 3). This timescale encompasses disturbances in sustained attention, a "'thinking'" process vital to maintaining focus and concentration over time (Williams et al. 2008a). Without the capacity to sustain attention, nonsignificant or irrelevant information may intrude on thinking. Moreover, the capacity to react effectively to significant stimuli and withhold reactions to nonsignificant stimuli may be diminished, producing impulsive responses. These three interrelated processes are hypothesized to underlie the triad signs of inattention, hyperactivity, and impulsivity. In addition, the combination of poor attention, intrusions, and impulsive responding is likely to produce inconsistent reactions to tasks, as reflected in an excessive variability in response times.

On their own, sustained attention disturbances are present in about 68 percent of individuals with ADHD, consistent with evidence from the Test of Variable Attention (T.O.V.A.®) and Conners' Continuous Performance Test (CPT), which assess alterations in sustained attention (Williams et al. 2010a). When the marker of sustained attention is considered in combination with markers of impulsivity, errors of intrusion from irrelevant information, and response variability, sensitivity increases to 86 percent and specificity to 91 percent (Table 7-1a; Williams et al. 2010a).

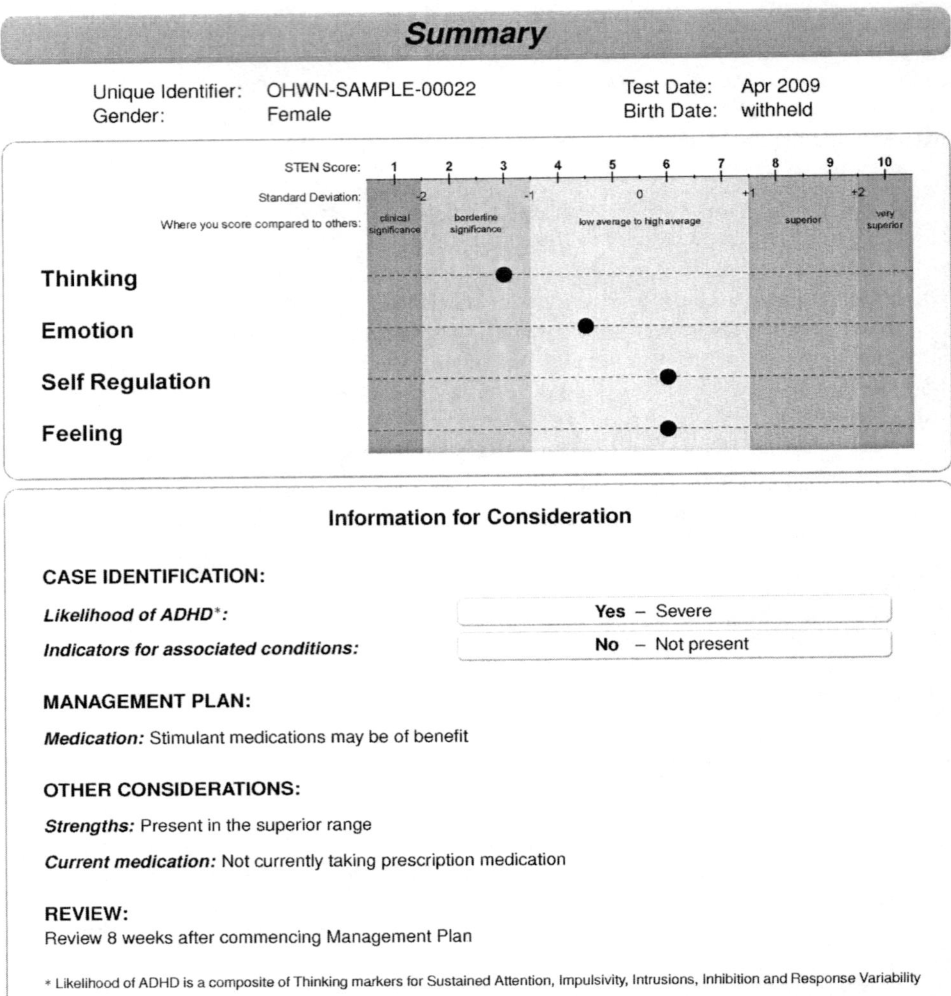

Summary

Unique Identifier: OHWN-SAMPLE-00022 — Test Date: Apr 2009
Gender: Female — Birth Date: withheld

Information for Consideration

CASE IDENTIFICATION:

Likelihood of ADHD*:	**Yes** – Severe
Indicators for associated conditions:	**No** – Not present

MANAGEMENT PLAN:

Medication: Stimulant medications may be of benefit

OTHER CONSIDERATIONS:

Strengths: Present in the superior range

Current medication: Not currently taking prescription medication

REVIEW:
Review 8 weeks after commencing Management Plan

* Likelihood of ADHD is a composite of Thinking markers for Sustained Attention, Impulsivity, Intrusions, Inhibition and Response Variability

Figure 7-1 Summary of an example of a report showing the profile of scores across cognitive and self-report measures within "emotion," "thinking," "feeling," and self-regulation" domains as summarized in Tables.

These findings build on the seminal theories of ADHD that have highlighted the cognitive basis of symptoms, and the role of interactions between cortical (top down feedback) and subcortical (bottom up feedforward) brain circuitry (Barkley 1997; Quay, 1997; Sergeant 2000; Sergeant et al. 2003).

Brain Imaging "Thinking" Markers

> *Brain imaging research shows a basis for cognitive "thinking" markers in raised EEG theta, slowed and reduced ERPs during cognitive tasks, reduced activation of prefrontal-anterior cingulate-parietal and basal ganglia circuitry in functional neuroimaging during thinking tasks, and reduced autonomic arousal during these tasks (Table 7-1a).*

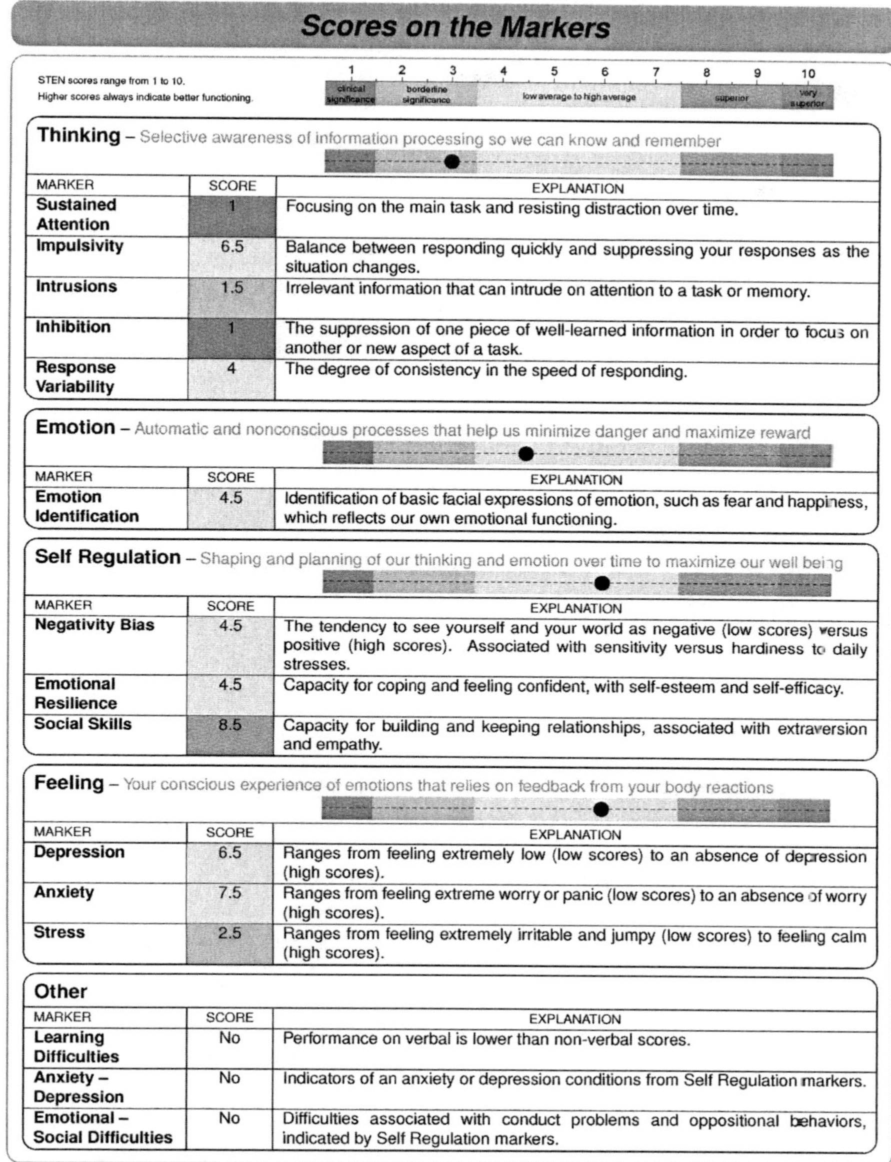

Scores on the Markers

STEN scores range from 1 to 10.
Higher scores always indicate better functioning.

Thinking – Selective awareness of information processing so we can know and remember

MARKER	SCORE	EXPLANATION
Sustained Attention	1	Focusing on the main task and resisting distraction over time.
Impulsivity	6.5	Balance between responding quickly and suppressing your responses as the situation changes.
Intrusions	1.5	Irrelevant information that can intrude on attention to a task or memory.
Inhibition	1	The suppression of one piece of well-learned information in order to focus on another or new aspect of a task.
Response Variability	4	The degree of consistency in the speed of responding.

Emotion – Automatic and nonconscious processes that help us minimize danger and maximize reward

MARKER	SCORE	EXPLANATION
Emotion Identification	4.5	Identification of basic facial expressions of emotion, such as fear and happiness, which reflects our own emotional functioning.

Self Regulation – Shaping and planning of our thinking and emotion over time to maximize our well being

MARKER	SCORE	EXPLANATION
Negativity Bias	4.5	The tendency to see yourself and your world as negative (low scores) versus positive (high scores). Associated with sensitivity versus hardiness to daily stresses.
Emotional Resilience	4.5	Capacity for coping and feeling confident, with self-esteem and self-efficacy.
Social Skills	8.5	Capacity for building and keeping relationships, associated with extraversion and empathy.

Feeling – Your conscious experience of emotions that relies on feedback from your body reactions

MARKER	SCORE	EXPLANATION
Depression	6.5	Ranges from feeling extremely low (low scores) to an absence of depression (high scores).
Anxiety	7.5	Ranges from feeling extreme worry or panic (low scores) to an absence of worry (high scores).
Stress	2.5	Ranges from feeling extremely irritable and jumpy (low scores) to feeling calm (high scores).

Other

MARKER	SCORE	EXPLANATION
Learning Difficulties	No	Performance on verbal is lower than non-verbal scores.
Anxiety – Depression	No	Indicators of an anxiety or depression conditions from Self Regulation markers.
Emotional – Social Difficulties	No	Difficulties associated with conduct problems and oppositional behaviors, indicated by Self Regulation markers.

Figure 7-1 (continued).

Brain imaging studies have yielded candidate markers for ADHD that may underlie alterations in thinking functions (Table 7-1a), including:

1. Raised EEG slow wave activity, particularly in the theta band (Loo et al. 2004; Hermens 2005a; Hermens et al. 2005a,b; Williams et al. 2010). Raised theta has been associated preferentially with impulsivity and lack of inhibition in ADHD (Williams et al. 2010a). EEG theta is typically reduced with maturation and development (Whitford et al. 2007), such that raised theta in ADHD is consistent with a deviation in maturation.

Table 7-1a Summary of Diagnostic Markers Implicated in ADHD

	Diagnostic Markers		
Type of marker	*Thinking*	*References*	*Next steps*
Cognition/ Self report	Sustained Attention[1,2] Impulsivity[1,2] Intrusions[1,2] Inhibition[1,2] Response Variability[1,2]	19 – 22	Confirmation of sensitivity and specificity to ADHD in larger samples with standardized assessments.
Brain imaging	Thinking task frontal-cingulate-parietal activation EEG Theta[1,2] Thinking ERPs[1] Arousal[1]	23 – 37	Confirmation of sensitivity and specificity to ADHD in larger samples with standardized assessments.
Gene	*DAT1* (10R allele)[3], *DRD4* (7R allele)[3]	39	Confirmation of sensitivity and specificity to ADHD in larger samples.

[1]Distinguish ADHD from healthy controls in group comparisons.

[2]Distinguish ADHD from healthy controls with sensitivity and specificity at individual subject level.

[3]Associated with ADHD, replicated and observed in meta-analyses. Also correlated with brain-based markers of cognition and brain imaging. DAT1 is related to poor attention, and DRD4 to poor sustained attention and impulsivity in particular.

2. Slowed and reduced ERPs are activated, which is most apparent in the period of 200–500 ms post-stimulus during sustained attention, selective attention, and inhibition tasks. These slowed and reduced ERPs are prominent over the frontal and parietal cortices (Callaway, Halliday, and Naylor 1983; Johnstone and Barry 1996; Yong-Liang et al. 2000; Hermens et al. 2005c). Reductions in the P450 ERP elicited by a continuous performance test of sustained attention have been associated with the sustained attention cognitive marker, and with markers of intrusion errors and response variability (Williams et al. 2010a).
3. On the basis of cognitive alterations in ADHD, brain imaging studies have focused on brain regions implicated in sustained attention, related working memory functions, inhibition, response organization, and impulsivity. These regions have preferentially encompassed the dorsolateral prefrontal cortex (DLPFC), ventrolateral prefrontal cortex (VLPFC), anterior cingulate (ACC, particularly the dorsal and mid-dorsal portions), and parietal cortex. The thalamus has also been implicated, consistent with the role of cortico-thalamic connections in attentional circuitry. Across activation tasks, activation in these regions has been reduced relative to healthy controls (Rubia et al. 1999; Durston et al. 2003; Dickstein et al. 2006; Tamm, Menon, and Reiss 2006; Vance et al. 2007; Bush 2009). Notably, the VLPFC and ACC, but not the DLPFC, are activated by an inhibition task, capturing impulsivity (lack of inhibition) in ADHD, suggesting that each thinking marker may preferentially engage a sub-network of cortical regions.
4. Reduced autonomic arousal—reflected in both lack of skin conductance arousal during resting conditions and reduced heart rate during cognitive tasks— is associated with disruptions in sustained attention, intrusive errors, and response variability (Hermens et al. 2005a,b; Williams et al. 2010a). These arousal systems implicate the brainstem and ascending reticular activating system.

Table 7-1b Summary of Markers of Comorbidity Related to ADHD

Comorbidity Markers			
Type of marker	*Emotion*	*References*	*Next steps*
Cognition/ Self report	Emotion Identification[4]	55 – -61	Confirmation of whether marker is specific to particular emotions in larger samples with standardized assessments.
Brain imaging	Emotion task ERPs[4] Emotion task frontal-posterior cingulate activation	61, 62	Replication in larger samples with standardized assessments.
Gene	COMT Val-Val[5a]	63	Replication in an independent study.
Type of marker	*Feeling*	*References*	*Next steps*
Cognition/ Self report	Anxiety	65 – -69	Confirmation in larger samples with standardized assessments.
Brain imaging	Emotion task ERPs[4]	61	Replication in an independent study.
Gene	Absence of DAT1and DRD4 7R[5b]	70	Replication in an independent study.
Type of marker	*Self Regulation*	*References*	*Next steps*
Cognition/Self report	Emotional Resilience (self-esteem, self-efficacy)[1] Social Skills[1] Theory of Mind (empathy)[1]	75 – -78	Confirmation of which markers are the most sensitive and specific to which comorbid conditions in larger samples with standardized assessments.
Brain imaging	-		Identify candidate brain imaging markers for comorbid conditions.
Gene	-		Identify candidate genomic markers for comorbid conditions.

[1]Distinguish ADHD from healthy controls in group comparisons.

[4]Distinguish ADHD from healthy controls in group comparisons, and are associated with comorbid anxiety, depression, and symptoms of emotional lability in particular.

[5]a. Associated with comorbid features of conduct disorder in ADHD. b. Associated with ADHD with anxiety and depression.

Genetic Variants Related to Thinking Markers

The most consistent evidence to date for genetic susceptibility to ADHD comes from polymorphisms of the dopamine transporter (DAT1 10-repeat homozygosity) and DRD4 (7-repeat allele). These variants have been related to both cognitive and brain imaging measures of "thinking" markers (Table 7-1a).

Table 7-1c **Summary of Markers Implicated in Prediction of Treatment Response in ADHD**

Treatment Markers			
Type of marker	*Thinking*	*References*	*Next steps*
Cognition/ Self report	Sustained Attention[7] Impulsivity[7] Intrusions[7] Inhibition[7] Response Variability[7]	40–43	Large randomized trials using standardized assessments.
Brain imaging	EEG Theta[7] Thinking ERPs[7] Arousal[7]	44–51	Large randomized trials using standardized assessments.
Gene	DAT1 10R[7] DAT1 9R[6]	39, 52,83	Large randomized trials.
Type of marker	*Emotion*	*References*	*Next steps*
Cognition/Self report	Emotion Identification6	61	Replication in an independent study
Brain imaging	Emotion ERPs[6]	61	Replication in an independent study
Gene	-		Identify candidate genomic markers for treatment related to emotion processes.
Type of marker	*Feeling*	*References*	*Next steps*
Cognition/Self report	Anxiety[8]	71,72	Confirmation in trials with standardized assessments.
Brain imaging	-		Identify candidate brain imaging markers for treatment related to anxiety.
Gene	DRD4 (4R, absence of 7R allele)[6]	70	Replication in an independent study.
Type of marker	*Self Regulation*	*References*	*Next steps*
Cognition/Self report	Emotional Resilience (Self-esteem)[8] Social Skills[8]	79–82	Confirmation in trials with standardized assessments.
Brain imaging	-		Identify candidate brain imaging markers for treatment related to self regulation.
Gene	-		Identify candidate genomic markers for treatment related to self regulation.

[6]Associated with poor response to stimulant medication.

[7]Associated with positive response to stimulant medication.

[8]Indicate response to behavioral therapies, and associated with comorbid behavioral problems (linked to Oppositional Defiance and Conduct Disorders).

Evidence for the role of polymorphisms of the dopamine transporter (*DAT1*) in genetic susceptibility to ADHD comes from linkage, association, and meta-analysis studies (for review, Sharp, McQuillin, and Gurling 2009). In addition, the *DAT1* marker polymorphisms have been correlated with both cognitive and brain imaging markers, including poor attention, EEG theta activity, fronto-striatal functional activation and brain structure (Sharp, McQuillin, and Gurling 2009).

Thinking Markers and Treatment

Cognitive and brain imaging markers for "thinking" also indicate a positive response to stimulants—particularly "sustained attention," and the correlated markers of P450 ERP and EEG theta. DAT1 *10R has been associated with positive response, and 9R homozygosity with poorer response (Table7-1b).*

Based on decades of research with stimulant medications, we know that the mechanisms of action implicate the catecholamines dopamine (DA) and norepinephrine (NE). Performance on cognitive tasks of sustained attention has been shown to normalize with stimulants in a number of studies (Sostek, Buchsbaum, and Rapoport 1980; Weingartner et al. 1980; Losier, McGrath, and Klein 1996; Bedard and Tannock 2008). On its own, the sustained attention marker captures 86 percent of individuals who have a positive response to stimulants (defined by a 25 percent or greater reduction in symptoms). Additional cognitive markers of impulsivity, intrusions, inhibition, and response variability boost sensitivity to 89 percent, indicating they are largely confirmatory (Williams et al. 2010a).

EEG, ERPs, and measures of autonomic arousal have each shown some potential as candidate markers of response to stimulant medication in ADHD (Klorman et al. 1990; Taylor et al. 1993; Sunohara et al. 1997; Loo, Teale, and Reite 1999; Clarke et al.2003). Normalization of EEG theta is a particularly robust marker, consistent with its correlation with cognitive markers of "thinking" (Williams et al. 2010a). In terms of sensitivity, it captures 86 percent of ADHD individuals who show a positive response to stimulants, defined by a 25 percent or greater reduction in symptom severity (Williams et al., 2006). The P450 ERP related to cognitive "thinking" markers also shows a normalization following stimulants (Keage et al. 2008), with a similar sensitivity of 82 percent. Similarly, reductions in arousal improve with stimulants (Satterfield, Cantwell, and Satterfield 1974), which may explain how stimulants may be effective for an "under-aroused" brain (Loo et al. 2004).

In ADHD, individuals with *DAT1* 10-repeat (10R) homozygosity have been found to show a relatively positive response to stimulant medications, while those with *DAT* 1 9-repeat (9R) homozygosity have been found to show a poorer response (Bellgrove et al., 2005; Cheon et al., 2007; Kirley et al. 2003; Gruber et al. 2009; Sharpe et al. 2009; Stein et al., 2005 Table 1b). Other *DAT1* variants may moderate different types of side effects.

Cognitive Emotion Markers

There have been comparatively few studies of "emotion" markers, but evidence to date points to impairments in identifying facial expressions of emotion in ADHD, which may be associated in particular with comorbid features (Tables 7-1a and 7-1c).

Emotion-related difficulties have been found to co-occur with particular comorbidities, including anxiety and conduct and oppositional defiance disorders (Table 7-1c). These findings highlight the feasibility of identifying markers for profiles of comorbidity in ADHD.

Complementary studies have focused on ADHD impairments in processing delayed rewards and reward-related motivational cues, also implicating fronto-subcortical systems (Nigg and Casey 2005; Sonuga-Barke 2005).

Markers of emotion processing disturbances in ADHD (Tables 7-1a and 7-1c) include:

1. Impairments in identifying facial expressions of emotion, associated with comorbid anxiety and symptoms of emotional lability in particular (Williams et al., 2008b); and
2. Deficits in matching emotional prosody to content and facial emotions, and identifying emotion in sentences (Shapiro et al. 1993; Cadesky, Mota, and Schachar 2000; Corbett and Glidden, 2000; Downs and Smith, 2004; Pelc et al. 2006).

Brain Imaging Emotion Markers

Candidate brain imaging markers for "emotion" in ADHD include a reduction in early electrical brain activity and excessive activation of frontal-posterior cingulate and limbic circuitry during processing of facial emotions (Table7-1a and 7-1c).

1. Markers from ERP in unmedicated children and adolescents with ADHD measure a reduction of the P120 potential, peaking around 120 post-stimulus for anger as well as for fear. This reduction is most apparent over the bilateral occipital cortex (Williams et al. 2008b). The P120 is generally associated with early, automatic appraisal of emotion cues. P120 reductions were also correlated with poorer identification of anger and fear, and with comorbid anxiety and emotional lability (Williams et al. 2008b); and
2. Markers from fMRI reflect an excess of activation of the frontal and posterior cingulate cortex, with reduction in limbic amygdala response to facial expressions of anger and fear in children and adolescents with ADHD, compared to healthy controls (Marsh et al., 2008; see Figure 7-1). This marker for excessive activation may reflect a neural hypersensitivity to emotion cues that interferes with the capacity to accurately identify them.

Genomic Markers Related to Emotion

The val108/158 COMT *variant has been associated with emotion disturbances in ADHD, particularly in externalizing comorbidities (Tables 7-1a and 7-1c).*

The *val108/158* variation of the catechol-O-methyl transferase (COMT) gene has specifically been associated with comorbid conduct disorder in ADHD linked to emotional disturbances (Thapar et al. 2005). This association also interacts with environmental stress. The *COMT Met* allele has been associated with alterations in emotional brain function (Williams et al, 2010b).

Emotion Markers and Treatment

> *While there has been little research into "emotion" markers for treatment in ADHD, the emerging evidence suggests that these markers indicate poorer response to stimulants.*

There has been little research into "emotion" markers of treatment in ADHD. Yet, markers for emotion disturbances in ADHD have been found to persist relative to healthy controls following stimulant medication, despite some normalization (Williams et al. 2008). The possibility that these markers indicate a comparatively poorer response to stimulants suggests that alternatives (non-stimulants, behavioral therapies) warrant further study.

Brain-Based Markers for Feeling in ADHD

Self-Report Feeling Markers

> *Markers of anxiety are important for identifying the anxiety and affective disorders that most commonly co-occur with ADHD.*

Anxiety disorders, in both children and adults, are among the disorders that most commonly co-occur with ADHD (Biederman, Newcorn, and Sprich 1991; MTA Cooperative Group 1999; Culpepper 2006). There is a high comorbidity of anxiety with ADHD of up to 33.5 percent (Biederman, Mick, and Faraone 2002; Culpepper 2006). There is also an associated high comorbidity of ADHD and affective disorders (Biederman et al. 1991).

In ADHD, intensity of experienced emotion, or feelings, has been inversely associated with identification of emotion cues, while this relationship is positive in healthy controls (Rapport et al. 2002). This inverse relationship correlates with the possibility that excessive levels of experienced emotion may interfere with identifying emotions.

Brain Imaging Feeling Markers

> *Feeling markers, such as "anxiety," have not been the focus of brain imaging studies in ADHD. The available evidence suggests an association between "anxiety" and "emotion" markers that disrupt earlier brain activity.*

To date, comorbid alterations in anxiety and depression from the "feeling" domain have not been the focus of brain imaging studies. The "emotion" marker of reduced P120 ERP has specifically been associated with severity of anxiety and depression (Williams et al. 2008b; Table 1c). This finding suggests that research into "feeling" markers associated with comorbid features of ADHD and with disruptions to emotional brain activity is a promising avenue for study.

Genomic Markers Related to Feeling

As outlined earlier, both the *DAT1* (homozygous for 10R) and *DRD4* (7R) variants have been associated with susceptibility to ADHD (Sharp et al. 2009). Yet, compared to controls,

A lower frequency of the DAT1 *10/10R and* DRD4 *7R allele variants that confer susceptibility to core features of ADHD may distinguish the subset that have internalizing comorbidities (anxiety, depression) with ADHD.*

ADHD patients with comorbid internalizing features of anxiety and depression have a lower frequency of the *DAT1* 10/10R and *DRD4* 7R allele variants (Table 7-1c). The combined absence of these variants was found to predict an association with the internalized comorbidities of ADHD, anxiety, and depression (Martinex-Levy et al. 2009).

Feeling Markers and Treatment

Anxiety with ADHD is indicative of a poorer response to stimulant medication, and supports consideration of alternatives, such as non-stimulants and behavior therapies (Table 7-1b).

Anxiety, as a comorbid feature of ADHD, may be an important marker for treatment outcomes. For instance, self-reported anxiety did not decline following stimulant medication in a study of emotional brain function in ADHD (Williams et al. 2008b). Comorbid anxiety has also been linked to poorer outcome with stimulants (Pliszka 1998). Non-stimulant medications may have a more beneficial effect on ADHD with anxiety. These findings were incorporated into the Texas Children's Medication Algorithm (Pliszka et al. 2006).

Given the highly comorbid nature of ADHD and anxiety, the identification of brain imaging markers related to treatment of ADHD with anxiety is an area for increasing future research. The findings have clinical implications for identifying markers that support decisions about when to consider non-stimulant medications, or when to combine medication with behavioral therapy.

Brain-Based Markers for Self-Regulation in ADHD

ADHD has been characterized as "a disorder in key aspects of self-regulation" (Nigg 2005). Children with ADHD have difficulties with self-regulation of emotion, thinking, and feeling functions. ADHD models have suggested a primary deficit in behavioral inhibition, causing secondary deficits in executive functions that impact emotional regulation. Recent evidence suggests that deficits in self-regulation are not simply a consequence of poor thinking functions (Perner, Kain, and Barchfeld 2002).

Self-Report Self-Regulation Markers

Markers of poor self-regulation of emotion and feeling in ADHD include measures of emotional resilience, social function (including empathy), theory of mind, and management of motivational goal-directed behavior (Tables 7-1a and 7-1c).

Emotional resilience and social skills are key aspects of self-regulation (Williams et al. 2008a). These aspects are associated with self control and empathy, respectively.

In self-reported assessments of these aspects, children and adolescents with ADHD perform at least two standard deviations below the normative mean (Tables 7-1a and 7-1c). The few studies conducted to date suggest that candidate markers for deficits in self-regulation of emotion and feeling in ADHD include empathy, capacity to anticipate feelings about future events, management of goal-directed behavior, and associated maladaptive reliance on external cues for motivation and arousal to persist in goal-directed behaviors (Braaten and Rosén 2000; Crundwell 2005).

These candidate markers may be most relevant to assessing comorbidity in ADHD. Loss of self-control has been associated with the behavioral disruptions characteristic of conduct and oppositional defiance disorders. Poor self-control has also been linked to loss of inhibition, or impulsivity. Related research indicates moderate associations between poor social functions and the aggressive behaviors characteristic of these comorbid conditions (Gilmour et al. 2004; Mikami and Pfiffner 2008).

Brain Imaging Self-Regulation Markers

Brain imaging markers of self-regulation have not been examined, but are hypothesized to involve complex feedback–feedforward (top down–bottom up) interactions between prefrontal and subcortical basal ganglia–limbic systems.

To date, the focus of brain-based markers for self-regulation in ADHD has centered on regulation of thinking functions, particularly in the prefrontal cortex. In contrast, brain imaging markers of disturbed self-regulation of emotion and feeling have not been examined in ADHD. Cortical-subcortical circuitry implicated in other features of ADHD has also been linked to self-regulation. For instance, connections between striatal circuits of the basal ganglia and medial prefrontal and anterior cingulate cortices may promote self-regulation of emotion and feeling, while connections with the lateral prefrontal areas may support regulation of thinking functions. With the input of important new information, feedback–feedforward interactions in these systems will allow encourage consideration of subcortical-cortical interactions in treatment decisions.

Genomic Markers for Self-Regulation

The COMT *variant linked to conduct-related "emotion" problems in ADHD may also contribute to poor self-regulation.*

As outlined earlier, the *COMT* variants have been linked to emotional disturbances and conduct problems commonly seen in conjunction with ADHD. Future studies are warranted to determine the potential role of this and other variants as markers for self-regulation.

Self-Regulation Markers and Treatment

Self-regulation markers may indicate the benefit of behavioral therapies targeting emotion management, and social and communication skills, particularly for those with comorbid externalizing problems (Tables 7-1b and 7-1c).

Self-regulation difficulties are a candidate marker for supporting clinical decisions concerning non-pharmaceutical treatment. While stimulants reduce the core thinking disturbances of ADHD, cognitive-behavioral therapy has been more effective in ameliorating self-regulation problems and enhancing coping strategies (Hinshaw, Henker, and Whalen 1984). There remains little support for the use of such behavioral treatment for ADHD as a group. However, findings point to the value of a personalized approach that identifies children experiencing comorbid self-regulation problems of poor resilience and self-esteem who will benefit from cognitive-behavioral therapy.

Children with self-regulation problems characteristic of conduct disorder may benefit from non-pharmaceutical therapies that target training of social and communication skills (Reid et al. 1999; Webster-Stratton, Reid, and Hammond 2001). Similarly, a program targeting social cognitive processing, in particular, has been found to increase effective social self-regulation and control, and decrease aggressive behavior, in a randomized controlled trial of children defined by aggressive conduct problems (Van Manen, Prins, and Emmelkamp 2004).

An Example of Clinical Application of Markers for ADHD

Translating markers for ADHD into the clinical setting will require a few key elements. First, we need identification of straightforward and brief marker assessments that can be readily used in clinical practice, or completed at home by the patient and brought in to the doctor. Next, cost-effective, brief, and straightforward assessments to measure the markers are necessary. More complex brain imaging assessments are only likely to be considered clinically in treatment-resistant and more complex cases. Finally, we need to develop straightforward assessments that can be used in routine practice, but have been shown to have direct correlates in a brain–gene basis. It is clinically feasible to undertake assessments of markers using brief computerized cognitive and self-report measures. Establishing the linkages among these markers, and results from brain imaging and genomics, on the same subjects would provide an evidence base for their basis in the brain. Comparatively simple cognitive and self-report assessments are then able to provide "proxy" markers for underlying brain systems.

One example of this approach is the association of cognitive and self-report markers with brain imaging markers from EEG and arousal assessments (Williams et al. 2010a). These markers are assessed using an Internet platform, WebNeuro (or the touchscreen equivalent, IntegNeuro). Results from the marker assessment are provided immediately to the clinician in the form of a summary report that includes considerations for clinical decisions based on the existing evidence base. The report provided by each assessment may therefore be considered a clinical decision support system (CDSS), since it adds actionable information (based on scientific evidence), to the results each patient receives on the assessment tool.

Figure 7-1, of this chapter, shows an example of a non-complex case of ADHD in which diagnosis is confirmed against DSM-IV criteria. In this situation, impairments in thinking are the most prominent, due in part to extremely poor sustained attention, inhibition, and below-average errors of intrusion (in Figure 7-1, lower scores indicate poorer performance). Given the severity of these impairments, and the absence of comorbid difficulties in anxiety or other difficulties of emotion or self-regulation, the profile indicates that stimulant medication may be considered. Of course, the goal of these suggestions is to provide additional objective information to aid in the clinician's decision-making process.

This example serves to illustrate the way that the systematic integration of information with brain-based markers may be implemented in a clinically feasible manner, offering the clinician objective markers for consideration in their diagnostic and treatment decisions. Such markers and Clinical Decision Support Systems are common in other areas of medicine, and with the web applications may also now be called 'Electronic Decision Support Systems; EDDS).

Conclusion

This chapter has summarized brain-based and gene markers of ADHD that can be considered in clinical practice to support diagnostic, comorbidity, and treatment-related decisions. Replicated studies using such markers will facilitate a personalized medicine approach to ADHD, in which the right treatment may be matched to each individual ADHD profile. The key points are highlighted in the following paragraphs.

In most cases, clinical management of ADHD is currently based on symptomatic criteria. The next edition of the diagnostic and statistical manual (DSM-V) aims to incorporate objective brain-based information from cognitive, brain imaging, and genomic studies.

There is reasonable consensus that ADHD is characterized primarily by "thinking" dysfunctions, which also affect "emotion," "feeling," and "self-regulation" processes.

Core "thinking" markers for ADHD encompass sustained attention, impulsivity, and associated inhibition—errors of intrusive irrelevant information and variability of responding. These markers are increasingly being used in clinical practice.

ADHD "thinking" markers are correlated with alterations in cortical-subcortical brain systems, electrical brain activity (EEG, ERPs), and autonomic arousal. They have also been associated with specific genetic variants (e.g., *DAT1, DRD4*).

"Emotion" markers (such as poor identification of facial emotion), "feeling" markers (such as anxiety) and "self-regulation" markers (such as poor resilience and theory of mind), may characterize distinctive profiles of ADHD linked to internalizing (anxiety-depression) and externalizing (conduct, oppositional defiance disorder) comorbidities. They may also relate to other genetic variants (*COMT*, and other *DAT1* and *DRD4* variants).

"Thinking" markers with moderate-to-severe impairments generally indicate a positive response to stimulants in ADHD. Emerging evidence suggests that "emotion," "feeling," and "self-regulation" markers indicate poor response to stimulants, and alternatives, such as non-stimulant medication or behavioral therapy (alone or in conjunction with medication) may be warranted.

An integrative neuroscience approach provides a way to translate markers into real-world clinical settings. Clinical feasibility relies on using markers that are assessed most effectively in the clinic or at home, using computerized assessments of cognition and self-report. By establishing the brain–gene correlates of these markers in the same subjects, it is possible for clinicians to make decisions based on objective information from straightforward assessments that they know have an established basis in the brain.

To date, few studies have assessed self-report, cognitive, brain, and genomic markers in the same subjects. Understanding the linkages among these levels of organization will be key to consolidating the evidence base of markers that may be used in the clinic. This evidence base would also enable a step-wise process, by which more severe and complex cases may require more extensive assessment of genomic and brain imaging markers. The goal is a set of consensus markers to help support personalized clinical decisions about diagnosis, comorbid conditions, and the treatment most suited to each individual child with ADHD.

References

American Psychiatric Association. 1994. *Diagnostic and statistical manual of mental disorders*, 4th ed. American Psychiatric Press: Washington, DC.

Ballard, S., M. Bolan, M. Burton, S. Snyder, C. Pasterczyk-Seabolt, and D. Martin. 1997. The neurological basis of attention deficit hyperactivity disorder. *Adolescence* 32: 855–862.

Barkley, R. A. 1997. *ADHD and the nature of self-control.* New York: Guilford Press.

Barkley, R. A. 1998. Attention-deficit hyperactivity disorder. *Scientific American*. 279: 66–71.

Bedard, A-C., and R. Tannock. 2008. Anxiety, methylphenidate response, and working memory in children with ADHD. *Journal of Affective Disorder* 11(5): 546–557.

Bellgrove, M. A., Z. Hawi, A. Kirley, M. Fitzgerald, M. Gill, and I. H. Robertson. 2005. Association between Dopamine Transporter (DAT1) Genotype, left-sided inattention, and an enhanced response to methylphenidate in Attention-Deficit Hyperactivity Disorder. *Neuropsychopharmacology* 30: 2290–97.

Biederman, J. 1998. Attention-deficit/hyperactivity disorder: a life-span perspective. *Journal of Clinical Psychiatry* 59: 4–16.

Biederman, J., E. Mick, S. V.Faraone, E. Braaten, A. Doyle, T. Spencer, T.E. Wilens, E. Frazier and M.A. Johnson 2002. Influence of gender on attention deficit hyperactivity disorder in children referred to a psychiatric clinic. *American Journal of Psychiatry* 159: 36–42.

Biederman, J., J. Newcorn, and S. Sprich. 1991. Comorbidity of attention deficit hyperactivity disorder with conduct, depressive, anxiety, and other disorders. *Am J Psychiatry*. 148: 564–77.

Braaten, E. B., and L. A. Rosén. 2000. Elf-regulation of affect in attention deficit-hyperactivity disorder (ADHD) and non-ADHD boys: Differences in empathic responding. *Journal of Consulting and Clinical Psychology* 68(2): 313–21.

Bush, G. 2009. Dorsal anterior midcingulate cortex: Roles in normal cognition and disruption in attention-deficit/hyperactivity disorder. In *Cingulate Neurobiology and Disease*, ed. B. A. Vogt, 207–218. New York: Oxford University Press.

Bush, G., E. M. Valera, and L. J.Seidman. 2005. Functional neuroimaging of attention-deficit/hyperactivity disorder: A review and suggested future directions. *Biological Psychiatry* 57: 1273–84.

Cadesky, E. B., V. L. Mota, and R. J. Schacha. 2000. Beyond words: How do children with ADHD and/or conduct problems process nonverbal information about affect?. *J. Am. Acad. Child Adolesc. Psychiatry* 39 (9): 1160–67.

Callaway, E., R. Halliday, and H. Naylor. 1983. Hyperactive children's event-related potentials fail to support underarousal and maturational-lag theories. *Archives of General Psychiatry* 40: 1243–48.

Castellanos, F. X., and R. Tannock. 2002. Neuroscience of attention deficit/hyperactivity disorder: The search for endophenotypes. *Nat Rev Neurosci*. 3: 617–28.

Charney, D. S., D. H. Barlow, K. Botteron, J.D. Cohen, D. Goldman, R.E. Gur, K.M. Lin, J.H. Meador-Woodruff, S.O. Moldin, E.J. Nestler, S.J. Watson, S.J., Zalcman 2002. Neuroscience research agenda to guide development of a pathophysiologically based classification system In *Research Agenda for DSM-V,* ed. D. J. Kupfer, M. B. First, and D. A. Regier, 31–38. Washington D.C.: The American Psychiatric Association.

Cheon K-A, K. Boong-Nyun, C. Soo-Churl. 2007. Association of 4-repeat allele of the dopamine d4 receptor gene exon iii polymorphism and response to methylphenidate treatment in Korean ADHD Children. *Neuropsychopharmacology* 32: 1377–1383.

Clamp, M., B. Fry, M. Kamal, X. Xie, J. Cuff, M. F. Lin, M. Kellis, K. Lindblad-Toh, and E. S. Lander. 2007. Distinguishing protein-coding and noncoding genes in the human genome. *Proc Natl Acad Sci USA*. 104(49): 19428–33.

Clarke, A. R., R. J. Barry, R. McCarthy, M. Selikowitz, C. R. Brown, and R. J. Croft. 2003. Effects of stimulant medications on the EEG of children with Attention-Deficit/Hyperactivity Disorder Predominantly Inattentive type. *International Journal of Psychophysiology* 47: 129–37.

Corbett, B., and H. Glidden. 2000. Processing affective stimuli in children with attention-deficit hyperactivity disorder. Child Neuropsychol. 6(2): 144–55.

Crundwell, M. R. A. 2005. An initial investigation of the impact of self-regulation and emotionality on behavior problems in children with ADHD. *Canadian Journal of School Psychology* 20: 62.

Culpepper, L. 2006. Primary care treatment of attention-deficit/hyperactivity disorder. *J Clin Psychiatry* 67: 51–58.

Dickstein, S. G., K. Bannon, X. F. Castellanos, and M. P. Milham. 2006. The neural correlates of attention and hyperactivity disorder: An ALE meta-analysis. *Journal of Child Psychology and Psychiatry* 47: 1051–62.

Downs, A., and T. Smith. 2004. Emotional understanding, cooperation, and social behavior in high-functioning children with autism. *J. Autism Dev. Disord.* 34 (6): 625–35.

Durston, S., M. C. Davidson, K. M. Thomas, M. S. Worden, N. Tottenham, A. Martinez, R. Watts, A.M. Ulug and B.J. Casey. 2003. Parametric manipulation of conflict and response competition using rapid mixed-trial event-related fMRI. *Neuroimage* 20: 2135–41.

Faraone, S. V., and J. Biederman. 1998. Neurobiology of attention-deficit hyperactivity disorder. *Biological Psychiatry* 44: 951–58.

Gilmour, J., B. Hill, M. Place, and D. H. Skuse. 2004. Social communication deficits in conduct disorder: A clinical and community survey. *Journal of Child Psychology & Psychiatry & Allied Disciplines* 45(5): 967–978.

Gordon, E. 2000. Integrative neuroscience: Bringing together biological, psychological and clinical models of the human brain. London: Harwood Academic.

Gordon, E., 2007a. Integrative genomics and neuromarkers for the era of brain-related personalized medicine. *Personalized Medicine* 4(2): 201–15.

Gordon, E., B.J. Liddell, K.J. Brown, R.A. Bryant, C.R. Clark, P. Das, C Dobson-Stone, E. Falconer, K. Felmingham, G. Flynn, J.M. Gatt, A. Harris, D.F. Hermens, P.J. Hopkinson, A.H. Kemp, S.A. Kuan, I. Lazzarro, J. Moyle, R.H. Paul, C.J. Rennie, P. Schofield, T.J. Whitford and L.M. Williams. 2007b. Integrating objective gene-brain-behavior markers of psychiatric disorder. *Journal of Integrative Neuroscience* 6: 11–34.

Gordon, E., K. J. Barnett, N. J. Cooper, N. Tran, and L. M. Williams. 2008. An "'integrative neuroscience'" platform: Application to profiles of negativity and positivity bias. *J Integr Neurosci.* 7: 345–66.

Gruber, R., R. Joober, N. Grizenko, B. L. Leventhal, E. H. Cook, and M. A. Stein. 2009. Dopamine transporter genotype and stimulant side effect factors in youth diagnosed with attention- deficit/ hyperactivity disorder. *Journal of Child and Adolescent Psychopharmacology* 19(3): 233–39.

Hermens, D. F., M. R. Kohn, S. Clarke, E. Gordon, and L. M. Williams. 2005a. Sex differences in adolescent ADHD: Findings from concurrent EEG and EDA. *Clin Neurophysiol.* 116: 1455–63.

Hermens, D. F., E. Soei, S. D. Clarke, M. R. Kohn, E. Gordon, and L. M. Williams. 2005b. Resting EEG theta activity predicts cognitive performance in ADHD. *Pediatric Neurol.* 32: 248–56.

Hermens, D. F., L. M. Williams, S. Clarke, M. Kohn, N. Cooper, and E. Gordon. 2005c. Responses to methylphenidate in adolescent AD/HD: Evidence from concurrently recorded autonomic (EDA) and central (EEG and ERP) measures. *Int J Psychophysiol.* 58: 21–33.

Hinshaw, S. P., B. Henker, and C. K. Whalen. 1984. Self-control in hyperactive boys in anger-inducing situations: Effects of cognitive-behavioral training and of methylphenidate. *J Abnorm Child Psychol.* 12(1): 55–77.

Insel, T. R., N. D. Volkow, S. C. Landis, T. K. Li, J. F. Battey, and P. Sieving. 2004. Limits to growth: Why neuroscience needs large-scale science. *Nat Neurosci.* 7: 426–27.

Jensen, P. S., D. Martin, and D. P. Cantwell. 1997. Comorbidity in ADHD: Implications for research, practice, and DSM-V. *Journal of the American Academy of Child & Adolescent Psychiatry* 36: 1065–79.

Johnstone, S. J., and R. J. Barry. 1996. Auditory event-related potentials to a two-tone discrimination paradigm in attention deficit hyperactivity disorder. *Psychiatry Research* 64: 179–92.

Keage, H. A. D., C. R. Clark, D. F. Hermens, L. M. Williams, M. R. Kohn, S. Clarke, C. Lamb, D. Crewther, and E. Gordon. 2008. ERP indices of working memory updating in AD/HD: Differential aspects of development, subtype and medication. *Journal of Clinical Neurophysiology* 25: 1–10.

Kirley, A., N. Lowe, Z. Hawi, C. Mullins, G. Daly, I. Waldman, M. McCarron, D. O'Donnell, M. Fitzgerald, and M. Gill. 2003. Association of the 480 bp DAT1 allele with methylphenidate response in a sample of Irish children with ADHD. *American Journal of Medical Genetics* 121B: 50–54.

Klorman, R., J. T. Brumaghim, L. F. Salzman, J. Strauss, A. D. Borgstedt, M. C. McBride, and S. Loeb. 1990. Effects of methylphenidate on processing negativities in patients with attention-deficit hyperactivity disorder. *Psychophysiology* 27: 328–37.

Loo, S. K., C. Hopfer, P. D. Teale, M. L. Reite. 2004. EEG correlates of methylphenidate response in ADHD: Association with cognitive and behavioral measures. *J Clin Neurophysiol.* 21(6): 457–64.

Loo, S. K., P. D. Teale, and M. L. Reite. 1999. EEG correlates of methylphenidate response among children with ADHD: A preliminary report. *Biological Psychiatry* 45: 1657–60.

Losier, B. J., P. J. McGrath, and R. M. Klein. 1996. Error patterns of the continuous performance test in non-medicated and medicated samples of children with and without ADHD: A meta-analytic review. *J Child Psychol Psychiatry* 37: 971–87.

Marsh, A. A., E. C. Finger, D. G. Mitchell, M. E. Reid, C. Sims, D. S. Kosson, K. E. Towbin, E. Leibenluft, D. S. Pine, and R. J. Blair. 2008. Reduced amygdala response to fearful expressions in children and adolescents with callous-unemotional traits and disruptive behavior disorders. *American Journal of. Psychiatry* 165 (6): 712–20.

Martinez-Levy, G., J. Diaz-Galvis, M. Briones-Velasco, A. Gomez-Sanchez, F. De le Pena-Olvera, L. Sosa-Mora, L. Palacios-Cruz, J. Ricardo-Garcell, E. Reyes-Zamorano and C. Cruz-Fuentes. 2009. Genetic interaction analysis for DRD4 and DAT1 genes in a group of Mexican ADHD patients. *Neuroscience Letters* 451(3): 257–60.

Mikami, A. Y., and L. J. Pfiffner. 2008. Sibling relationships among children with ADHD. *J. Atten. Disord.* 11 (4): 482–92.

Monastra, V. J. 2005. Overcoming the barriers to effective treatment for attention-deficit/hyperactivity disorder: A neuro-educational approach. *Int J Psychophysiol.* 58: 71–78.

MTA Cooperative Group. 1999. A 14-month randomized clinical trial of treatment strategies for attention deficit hyperactivity disorder. *Archives of General Psychiatry* 56: 1073–86.

Murphy, K. R., R. A. Barkley, and T. Bush. 2002. Young adults with attention deficit hyperactivity disorder: Subtype differences in comorbidity, educational, and clinical history. *Journal of Nervous & Mental Disease* 190: 147–57.

Nigg, J. T. 2005. Neuropsychologic theory and findings in attention-deficit/hyperactivity disorder: The state of the field and salient challenges for the coming decade. *Biological Psychiatry* 57: 1424–35.

Nigg, J. T., and B. J. Casey. 2005. An integrative theory of attention-deficit/hyperactivity disorder based on the cognitive and affective neurosciences. *Development and Psychopathology* 17(3): 785–806.

Pelc, K., C. Kornreich, M. L. Foisy, and B. Dan. 2006. Recognition of emotional facial expressions in attention-deficit hyperactivity disorder. *Pediatriatric Neurology*. 35 (2): 93–97.

Perner, J., W. Kain, and P. Barchfeld. 2002. Executive control and higher-order theory of mind in children at risk of ADHD. *Infant Child Development.* 11: 141–58.

Pliszka, S. R., M. L. Crismon, C. W. Hughes, C. K. Corners, G. J. Emslie, P. S. Jensen, J. T. McCracken, J. M. Swanson, and M. Lopez. 2006. Texas Consensus Conference Panel on Pharmacotherapy of Childhood Attention Deficit Hyperactivity Disorder.: The Texas Children's Medication Algorithm Project: revision of the algorithm for pharmacotherapy of attention-deficit/hyperactivity disorder. *Journal of the American Academy of Child and Adolescent Psychiatry* 45: 642–657.

Pliszka, S. R. 1998. Comorbidity of attention-deficit/hyperactivity disorder with psychiatric disorder: An overview. *Journal of Clinical Psychiatry* 59: Suppl-8.

Quay, H. C: 1997. Inhibition and attention deficit hyperactivity disorder. *Journal of Abnormal Child Psychology*. 25: 7–13.

Rapport, L. J., S. R. Friedman, A. Tzelepis, and A. Van Voorhis. 2002. Experienced emotion and affect recognition in adult attention-deficit hyperactivity disorder. *Neuropsychology* 16(1): 102–10.

Reid, J. B., J. M. Eddy, R. A. Fetrow, and M. Stoolmiller. 1999. Description and immediate impacts of a preventive intervention for conduct problems. *American Journal of Community Psychology* 27(4): 483–517.

Rubia, K., S. Overmeyer, E. Taylor, M. Brammer, S. C. Williams, A. Simmons and E.T. Bullmore. 1999. Hypofrontality in attention deficit hyperactivity disorder during higher-order motor control: A study with functional MRI. *American Journal of Psychiatry* 156: 891–96.

Satterfield, J. H., D. P. Cantwell, and B. T. Satterfield. 1974. Pathophysiology of the hyperactive child syndrome. *Archives of General Psychiatry* 31: 839–44.

Sergeant, J. 2000. The cognitive-energetic model: An empirical approach to attention-deficit hyperactivity disorder. *Neuroscience & Biobehavioral Reviews* 24: 7–12.

Sergeant, J. A., H. Geurts, S. Huijbregts, A. Scheres, and J. Oosterlaan. 2003. The top and the bottom of ADHD: A neuropsychological perspective. *Neurosci Biobehav Rev*. 27: 583–92.

Shapiro, E. G., S. J. Hughes, G. J. August, and M. L. Bloomquist. 1993. Processing of emotional information in children with attention-deficit hyperactivity disorder. *Dev. Neuropsychol.* 9: 207–24.

Sharp, S. I., A. McQuillin, and H. M. D. Gurling. 2009. Genetics of attention-deficit hyperactivity disorder (ADHD). *Neuropharmacology* 57: 590–600.

Sonuga-Barke, E. J. S. 2005. Causal models of attention-deficit/hyperactivity disorder: From common simple deficits to multiple developmental pathways. *Biol Psychiatry* 57: 1231–38.

Sostek, A. J., M. S. Buchsbaum, and J. L. Rapoport. 1980. Effects of amphetamine on vigilance performance in normal and hyperactive children. *J Abnorm Child Psychol.* 8: 491–500.

Stein M.A., I.D. Waldman, C.S. Sarampote, K.E. Seymour, A.S. Robb, C. Conlon, S-J. Kim, E.H. Cook Jr. 2005. Dopamine transporter genotype and methylphenidate dose response in children with ADHD. *Neuropsychopharmacology* 30: 1374–1382.

Sunohara, G. A., J. G. Voros, M. A. Malone, and M. J. Taylor. 1997. Effects of methylphenidate in children with attention deficit hyperactivity disorder: A comparison of event-related potentials between medication responders and non-responders. *International Journal of Psychophysiology* 27: 9–14.

Tamm, L., V. Menon, and A. L. Reiss. 2006. Parietal attentional system aberrations during target detection in adolescents with Attention Deficit Hyperactivity Disorder: Event-related fMRI evidence. *American Journal of Psychiatry* 163: 1033–43.

Taylor, M. J., J. G. Voros, W. J. Logan, and M. A. Malone. 1993. Changes in event-related potentials with stimulant medication in children with attention deficit hyperactivity disorder. *Biological Psychology* 36: 139–56.

Thapar, A., K. Langley, T. Fowler, F. Rice, D. Turic, N. Whittinger, J. Aggleton, M. Van den Bree, M. Owen, and M O'Donovan. 2005. Catechol O-methyltransferase gene variant and birth weight predict early-onset antisocial behavior in children with attention-deficit/hyperactivity disorder. *Arch. Gen. Psychiatry* 62: 1275–78.

Van Manen, T. G., P. J. Prins, and P. M. Emmelkamp. 2004. Reducing aggressive behavior in boys with a social cognitive group treatment: Results of a randomized, controlled trial. *Journal of the American Academy of Child & Adolescent Psychiatry* 43(12): 1478–87.

Vance, A., T. J. Silk, M. Casey, N. J. Rinehart, J. L. Bradshaw, M. A. Bellgrove, and R. Cunnington 2007. Right parietal dysfunction in children with attention deficit hyperactivity disorder combined type: A functional MRI study. *Molecular Psychiatry* 12: 793, 826–32.

Webster-Stratton, C., J. Reid, and M. Hammond. 2001. Social skills and problem-solving training for children with early-onset conduct problems: Who benefits? *Journal of Child Psychology & Psychiatry & Allied Disciplines* 42(7): 943–52.

Weingartner, H., 1980. Cognitive processes in normal and hyperactive children and their response to amphetamine treatment. *J Abnorm Psychol.* 89: 25–37.

Whitford, T. J., C. J. Rennie, S. M. Grieve, C.R. Clark, E. Gordon, L.M. Williams 2007. Brain maturation in adolescence: Concurrent changes in neuroanatomy and neurophysiology. *Human Brain Mapping*. 28: 228–37.

Williams L,M., D.F. Hermens, M. Kohn, S. Clarke and E. Gordon. 2006. ADHD across the lifespan: from the laboratory to the clinic. Integrating Cognitive and Affective Markers of ADHD. *Neuropsychiatric Disease and Treatment*, 2: 74.

Williams, L. M., J. M. Gatt, S. M. Grieve, C. Dobson-Stone, R. H. Paul, E. Gordon, and P. R. Schofield. 2010b. COMT Val108/158Met polymorphism effects on emotional brain function and negativity bias. Neuroimage (online early view; doi:10.1016/j.neuroimage.2010.01.084)

Williams, L. M., J. M. Gatt, A. Hatch,D.M. Palmer, M. Nagy, C. Rennie, N.J. Cooper, C. Morris, S. Grieve, C. Dobson-Stone, P. Schofield, C.R. Clark, E. Arns, R.H. Paul, E. Gordon. 2008a. The INTEGRATE model of emotion, thinking and self regulation: Application to the "paradox of aging." *Journal of Integrative Neuroscience.* 7: 367–404.

Williams, L. M., D. F. Hermens, T. Thein, C. R. Clark, N. Cooper, S.D. Clarke, C. Lamb, E. Gordon and M.R. Kohn 2010a. Using brain-based cognitive measures to support clinical decisions in ADHD. *Pediatric Neurology* 42: 118–126.

Williams, L. M., D. F. Hermens, D. Palmer, M. Kohn, S. Clarke, H. Keage, C. R. Clark, and E. Gordon. 2008b. Misinterpreting emotional expressions in attention-deficit/hyperactivity disorder: Evidence for a neural marker and stimulant effects. *Biol. Psychiatry.* 63(10): 917–26.

Yong-Liang, G., P. Robaey, F. Karayanidis, M. Bourassa, G. Pelletier, and G. Geoffroy. 2000. ERPs and behavioral inhibition in a Go/No-go task in children with attention-deficit hyperactivity disorder. *Brain Cogn.* 43: 215–20.

Section 3

Personalized Medicine and Other Brain Disorders

8

The Role of Neuroimaging Biomarkers in Personalized Medicine for Neurodegenerative and Psychiatric Disorders

Ellen M. Migo, PhD, Steve C.R. Williams, PhD, William R. Crum, DPhil, Matthew J. Kempton, PhD, and Ulrich Ettinger, PhD

Introduction

Personalizing medicine in order to accurately predict treatment response in patients with psychiatric disorders relies on accurate diagnosis. This requires the exclusion of disorders with overlapping symptoms, as well as the differential diagnosis of conditions with similar superficial presentations. Most psychiatric illnesses involve a protracted and highly subjective diagnosis, primarily based on "signs and symptoms." Therefore, the need for rapid and objective diagnostic markers would be a timely complement to this approach. The hope is that identification of biomarkers of disease will aid in making a correct diagnosis, prescribing drugs that will be effective, and monitoring the point where the disease is controlled, whether in remission or entirely cured.

Neuroimaging is among the modalities under study to aid in identification of biomarkers of psychiatric disorders. Neuroimaging covers a wide range of complementary methodologies, but newer technologies allow neuroimaging to act as the bridging point between behavioral measures and molecular mechanisms. It therefore provides an indication of a pathological process and, by capturing this information, is seen as a major hope for biomarker development for psychiatry (Agid et al. 2007; Lucignani 2007). An accurate prognostic technique, such as neuroimaging, may well help stratify patients, predict response to treatment, and enhance patient compliance with treatment.

This chapter is focused on the use of neuroimaging to identify biomarkers for three conditions: dementia, schizophrenia, and mood disorders (major depressive disorder [MDD] and bipolar disorder [BP]).

Much effort is being invested in the search for reliable biomarkers, which are reproducible measurements correlating with at least some aspects of the molecular or brain pathophysiology of the disease in question. In this context, biomarkers link measurable physical or functional quantities with a psychiatric disease process and, in principle, add objectivity to the identification and staging of the disorder. In general, biomarkers should have some or all of the following properties:

1. Can be reproducibly identified;
2. Reliable;
3. Inexpensive to identify;
4. Identifiable by noninvasive or minimally invasive procedures;
5. Can distinguish healthy subjects from those with disease or those likely to develop disease;
6. Can distinguish between diseases;
7. Reflect the severity of the disease (both in terms of physical/structural progression and symptomatic changes);
8. Serve as predictors of treatment response; and
9. Reflect the effects of successful therapeutic intervention.

In practice, we have no biomarkers in neuropsychiatry that currently fulfill all of these criteria. However, there have been many identified biomarkers that meet at least some of these criteria for specific disorders.

Many psychiatric disorders feature a preclinical or prodromal stage in which the patient is unaware of the disorder. A screening test for at-risk individuals to predict future disease onset, or a test that can identify early stages of a disorder, would allow early intervention to prevent, or at least delay, the onset of disruptive symptoms. This may be achievable via a single biomarker or combination of biomarkers. When the disease has a physical substrate, such as dementia, early diagnosis and treatment could preserve cognitive function. Biomarkers could also aid differential diagnosis among disorders with similar symptom profiles, or improve the predictability of treatment response. This is particularly important because, for many disorders, there are a variety of pharmacological and behavioral treatment options available. These are all clinical benefits to the patient, but there is also an important cost implication for care providers in targeting appropriate therapies to appropriate patients most likely to respond.

Biomarkers are also required as surrogate outcome measures in research studies—notably in drug discovery—to reduce the follow-up period and identify participants who are more likely to respond to a given treatment. They can take various forms, from gene combinations to different functional connectivity in the brain (see Mayeux 2004; Singh and Rose 2009 for a wider discussion). Although much research has focused on drug treatments and the role of biomarkers in drug development, there are also important therapeutic interventions that are purely behavioral, such as cognitive-behavioral therapy (CBT). CBT has been shown to be beneficial for a variety of neuropsychiatric disorders (e.g., DeRubeis and Crits-Christoph 1998; Butler et al. 2006), but has a longer window, compared to medication-based treatment, before benefits can be seen. Using neuroimaging to predict CBT response in patients with a variety of disorders is an emerging research area (e.g., McClure et al. 2007; Fu et al. 2008; Konarski et al. 2009; Premkumar et al. 2009).

Neuroimaging provides an indication of a pathological process and, by capturing this information, is seen as a major hope for biomarker development for psychiatry (Agid et al. 2007; Lucignani 2007). In this unique position, neuroimaging biomarkers have an important role, either alone or in conjunction with biomarkers derived from cheaper but invasive genomic methods, such as cerebrospinal fluid (CSF) or blood plasma analyses.

Imaging Methodologies

Early research into psychiatric disorders using neuroimaging centered on techniques using radiotracers, such as single photon emission computed tomography (SPECT) or positron

Table 8-1 Summary of Abbreviations Used in this Chapter

Imaging term	*Measure*
Blood-Oxygenation-Level Dependent (BOLD)	Changes in local blood oxygenation levels as a measure of brain activity
Diffusion Tensor Imaging (DTI)	Constrained movement of water molecules
Functional MRI (fMRI)	Using MRI to study the active brain
Structural MRI (sMRI)	Non-invasive images of anatomy
Magnetic Resonance Spectroscopy (MRS)	Chemical concentration and distribution
Pharmacological MRI (phMRI)	fMRI with a pharmacological agent
Positron Emission Tomography (PET)	Dynamic distribution of radio-labelled positron emitting molecules
Single Photon Emission Computed Tomography (SPECT)	Distribution of radio-labelled photon emitting molecules

emission tomography (PET) (see Abou-Saleh 2006 and Capote 2009 for reviews of their applications to psychiatry). Radiotracers are given to participants and these then emit energy in the form of photons (SPECT) or positrons (PET). These outputs are then detected, although PET gives much better spatial and temporal resolution than SPECT. PET can allow functional changes to be seen within an experiment, but the longer half-lives of the radioactive isotopes used in SPECT do not make that possible.

The improved spatial resolution of PET over SPECT is itself limited, compared to what can be achieved using magnetic resonance imaging (MRI). This is a noninvasive technique that does not require administration of radiotracers and can give information on function (functional MRI or fMRI) or structure (structural MRI or sMRI). Functional MRI measures the blood-oxygenation-level dependent (BOLD) change in levels of blood deoxyhemoglobin, an indicator of neural activity. The BOLD measurement depends upon the magnetic qualities of oxyhemoglobin and deoxyhemoglobin; the latter is slightly more paramagnetic than oxyhemoglobin. Paramagnetism is magnetism that is manifest only in response to an externally applied magnetic field. Blood flow to the brain varies with the neurons' activation and use of oxygen. It is that variation in the amount of deoxyhemoglobin and oxyhemoglobin that is measured in BOLD. It is widely used to investigate a broad range of psychiatric disorders (e.g., Mitterschiffthaler et al. 2006).

MRI scanners can also obtain information on tissue metabolism using magnetic resonance spectroscopy (MRS), which measures the levels of metabolites within different brain regions. This is informative for many psychiatric conditions (e.g., Lyoo and Renshaw 2002). MRS can show abnormalities in tissue function in the absence of any gross structural changes. Other techniques used within psychiatric research can measure the white matter integrity and connectivity between regions, such as diffusion tensor imaging (DTI), which is based on diffusion characteristics of water molecules (Taylor et al. 2004). Applying fMRI to psychopharmacology allows investigation of how drugs can alter brain activity. This use is termed pharmacological MRI or phMRI (Leslie and James 2000; Honey and Bullmore 2004).

There is clearly a broad range of imaging techniques available to researchers and clinicians within psychiatry which covers many characteristics of brain structure and function. All of these different imaging modalities can provide information on psychiatric disorders at a single time point, or can be used in longitudinal studies to assess longer-term changes.

The key outputs from neuroimaging modalities all present possible biomarkers of disease status. Here, we will discuss the role of neuroimaging biomarkers in relation to three main groups of disorders: dementia, schizophrenia, and mood disorders (which include MDD and BD).

Dementia

Background

Dementia is the progressive loss of brain function in at least two cognitive domains, one of which is usually memory. The lifetime risk of dementia is approaching one in four for people in the developed world, with those affected usually over the age of 60; age is the most important risk factor, with known genetic factors significant in a small number of cases with early onset. Alzheimer's disease (AD) is the most common cause of dementia (Alzheimer's Association 2009). At the cellular level, AD is associated with abnormal beta-amyloid deposits, "plaques," and tau-protein "tangles."

The second most common form of dementia is vascular dementia (VaD), sometimes known as multi-infarct dementia, which is caused by a succession of micro-strokes affecting brain blood flow. Symptoms can overlap with AD, and many sufferers may have concurrent microscopic features of both diseases or of other, less common forms, such as fronto-temporal dementia and dementia with Lewy bodies.

Dementia sufferers place a continuous and increasing burden on the health care system. Some studies estimate the cost at up to three times the cost of treating other disorders (Alzheimer's Association 2009). Dementia sufferers can survive from four to twenty years after diagnosis with worsening symptoms and comorbidities. There are currently no cures; some drugs offer symptomatic relief in some patients, often only with short-term improvements (Laks and Engelhardt 2008), making current treatment a trial-and-error process. There is a real need for the development of more effective medications with more predictable patient response. Candidate therapeutic agents are currently being developed. The high risk of dementia in the elderly, which will increase alongside the predicted higher life expectancy among the population, makes biomarker development important for diagnosis, treatment prediction, and drug discovery.

Biomarkers that can (a) detect and differentiate between dementia subtypes at an early (even preclinical) stage and (b) correlate with physical and cognitive treatment response are vital for future diagnosis and treatment strategies. Much recent work has focused on patients with mild cognitive impairment (MCI), who experience memory deficits greater than is expected in normal aging without other cognitive impairments. MCI is considered to be a prodromal stage of dementia (Petersen and Knopman 2006), and while conversion rates vary across studies, the annual conversion rate is approximately 10 percent (see Bruscoli and Lovestone 2004).

Probable diagnosis of dementia is usually on clinical grounds and based on cognitive impairment by the time substantial physical damage has already occurred. Supporting investigations include blood tests, brain scans, and neuropsychological tests, and are often used to exclude other treatable conditions (e.g., depression). Currently, definitive diagnosis requires postmortem confirmation, because diagnostic accuracy is generally low in the early stages of the disease. Criteria for diagnostic biomarkers in AD (where the most research effort has been devoted) have specified diagnostic sensitivity and specificity against other dementias of over 80 percent (Davies et al. 1998). To be useful in the current climate, biomarkers in dementia should also dynamically reflect the effects of successful therapeutic intervention (Frank et al. 2003).

Risk for Dementia

A large body of research has looked at the genetic predictors or risk factors for dementia. There are known genetic mutations associated with familial predisposition to AD (in the *APP*, *PSEN1*, and *PSEN2* genes) and susceptibility to sporadic AD (associated with the ε4-allele of *APOE*; see Rademakers and Rovelet-Lecrux 2009). However, many patients with sporadic AD do not carry *APOE-ε4* and some familial AD cohorts do not have known genetic mutations (Frey et al. 2005). This means that significant genetic factors in dementia remain unidentified.

Imaging Biomarkers for Dementia

Noninvasive imaging has been used extensively to study dementia and to serve as a possible source of disease-related biomarkers (Kantarci and Jack Jr. 2003). The most widely applied imaging modalities are MRI (Kantarci and Jack Jr. 2004; Mueller et al. 2006), SPECT, and PET; X-Ray computed tomography (CT) is often used diagnostically to exclude other disorders.

Diagnosis of Dementia

The most obvious macroscopic changes in brain morphology that are confirmed by post-mortem studies and correlate with cognitive decline are global (Fox et al. 1999; Schott et al. 2008) and local (Rusinek et al. 2003; Jack Jr. et al. 2004; Ridha et al. 2008) measures of atrophy determined from MRI. The most widely reported structural indicators of disease progression alongside diffuse whole-brain atrophy are hippocampal atrophy (Fox et al. 1996; Kaye et al. 1997; Jack Jr. et al. 2000), which is thought to precede clinical symptoms of AD, and ventricular expansion.

While MRI produces images of gross tissue structure, MRS measures the quantity of constituent metabolites, most commonly N-acetyl aspartate (NAA), myoinositol (mI), creatine (Cr), and choline (Cho) (Lin et al. 2005). The quantity, anatomical distribution, and change in metabolites have potential to identify early disease and differentiate between dementia variants (Frank et al. 2003; Jones and Waldman 2004). Raised mI levels have been measured in both AD (Firbank et al. 2002) and MCI (Kantarci et al. 2002) and compared with normal controls; an inverse correlation of mI levels has been observed with memory impairment in the MCI group. Longitudinal studies have found that NAA reduction in gray matter is more pronounced in AD than in normal controls (Adalsteinsson et al. 2000) and that reduction in NAA and Cr can predict cognitive decline and conversion to dementia in patients initially classified as MCI (e.g., Metastasio et al. 2006; Fayed et al. 2008; Pilatus et al. 2009).

Nuclear medicine techniques have been widely applied to study regional perfusion (HMPAO-SPECT) and glucose metabolism (fludeoxyglucose [FDG]-PET) in dementia. Studies have consistently found reduced perfusion and metabolism in regions (e.g., frontal, temporoparietal) known to be affected by dementia (Dougall et al. 2004; Foster et al. 2007). Both SPECT (Pimlott and Ebmeier 2007) and PET (Herholz et al. 2007) have become the foci of renewed interest due to recent investigations with new tracer agents that can target specific neurotransmitter systems. Novel SPECT agents that target muscarinic and nicotinic acetylcholine receptors and some PET agents with similar targets have also been tested (Paterson and Nordberg 2000; Horti and Villemagne 2006; Ogawa et al. 2009).

The most widely reported new PET ligands, of which the best known is Pittsburgh Compound B (PIB) (Klunk et al. 2004; Nordberg 2004), are believed to label amyloid plaques. These agents have already shown promise in group studies, but are currently still some distance away from playing a role in personalized medicine.

Fluid Biomarkers for Dementia

The prospect of identifying neuroimaging biomarkers is complemented by research into fluid biomarkers, with some indication that a combination of these objective measures will have a future clinical function (De Leon et al. 2007). This need for using neuroimaging and fluid markers together means that although this chapter is focused on neuroimaging, a brief review of fluid-based biomarkers is important.

Enormous effort has been invested in searching for biomarkers of dementia in CSF (Blennow 2004) and plasma (Irizarry 2004). CSF is not routinely collected, but its composition should be strongly related to the extracellular environment in the brain. Conversely, plasma is straightforward to collect but likely to be less specific to dementia pathology, and the measurements may not be as specific for the brain as they are for other areas of the body.

As a key constituent of the characteristic plaques found in AD, amyloid beta (Aβ) peptides have been widely investigated as biomarkers in both CSF and plasma. In particular, the prevalent isoform, Aβ42, is significantly reduced in CSF in AD, when compared with age-matched controls, and there is evidence that a reduction of Aβ42 in CSF in MCI is associated with progression to AD (e.g., Hansson et al. 2006; Frankfort et al. 2008). In addition, although studies have shown an inverse correlation of CSF Aβ42 with both plaque numbers (Strozyk et al. 2003) and results of PIB imaging (Fagan et al. 2006), specificity for other dementias is low (Hulstaert et al. 1999).

Although there are currently limited treatment options for dementia, if future drugs can target particular dementia types, differential diagnosis will be important. Although increases in plasma Aβ42 have been reported in some familial AD and Down syndrome cases, no consistent findings are reported in the more common sporadic AD. Soluble tau proteins are the other widely investigated CSF biomarker, as elevated levels are associated with neurofibrillar pathology (Buerger et al. 2006). Studies often combine total tau measurements with Aβ42 to improve specificity of predicting MCI to dementia conversion (e.g., Hansson et al. 2006).

Many more candidate CSF biomarkers, such as ubiquitin, have been investigated recently, including in sophisticated studies using proteomic methods (Simonsen et al. 2007; Kovacech et al. 2009; Zellner et al. 2009). Broader investigations of plasma biomarkers, which reflect effects such as oxidation and inflammation, have found differences in AD compared with controls, and these may also contribute to diagnosis in the future (Flirski and Sobow 2005).

Biomarkers for Dementia Progression, and Future Prospects

Currently, the most sensitive techniques for measuring disease progression in the individual involve serial structural MRI scanning with intervals of at least six months (e.g., Fox and Freeborough 1997).

Research into biomarkers for dementia is continuing on several broad fronts (Craig-Schapiro et al. 2009). Currently, no single candidate biomarker has the required specificity for disease prediction, or the required sensitivity to accurately track progression in the range of individuals that present clinically with or without treatment. Biomarkers may be useful individually to capture certain aspects of a complex disease mechanism, or to identify different facets in the aftermath of disease. They may also correlate to specific symptoms, such as specific memory problems, rather than attempting to represent the myriad of features of cognitive decline in dementia. This could prove useful in monitoring therapeutic interventions that target particular aspects of disease or for investigating interactions of complementary parts of the disease process. The understanding of the disease mechanism, the development of new therapeutics, and the identification of disease-specific biomarkers (Verbeek et al. 2003) are intimately linked.

The continuing development of hypothesis-led fluid analysis (Shaw et al. 2007; Craig-Schapiro et al. 2009) and high-throughput proteomic (Zellner et al. 2009) and genetic screening techniques will undoubtedly increase the number of candidate biomarkers. This has already resulted in panel approaches that deduce a multi-protein disease signature (Carrette et al. 2003; Finehout et al. 2007; Britschgi and Wyss-Coray 2009). Proteomics has also renewed interest in plasma biomarkers (Hye et al. 2006). Carefully chosen combinations of individual biomarkers (Hampel et al. 2008), possibly from complementary measurement modalities such as specific structural MRI and CSF measures (De Leon et al. 2006), may be required to achieve both disease sensitivity and specificity and robustness across the clinical population.

One particularly promising development in the field of dementia has been the establishment of the Alzheimer's Disease Neuroimaging Initiative (ADNI) (Mueller et al. 2005; Jack Jr. et al. 2008; Carrillo et al. 2009). This multicenter initiative (over fifty sites) recruited participants with MCI and AD, as well as elderly controls, across North America (and now also in Europe), aiming to improve neuroimaging methods and, once optimized, establish a standard for sMRI acquisition. Participants were followed longitudinally and additional information on cognitive performance and CSF measures was collected. Data collection has been completed and the database of MRI, CSF, and behavioral measures is providing a wealth of data to help understand disease diagnosis (e.g., Chupin et al. 2009) and disease progression (e.g., Risacher et al. 2009) using MRI. Other work has used the neuroimaging data in combination with genetic and CSF measures (e.g., Leow et al. 2009; Schuff et al. 2009; Shaw et al. 2009). There is still a wide range of data to be published from the ADNI cohort, which will help researchers understand which of the many potential biomarkers will be most appropriate for clinical or drug discovery uses.

This type of collaboration, between large numbers of academic and clinical sites, as well as pharmaceutical companies and charitable foundations, is a model of how different stakeholders can create a partnership to collect and share multimodal data from a large cohort of participants. Large-scale initiatives, like ADNI and the European Alzheimer Disease Consortium (EADC) (Winblad et al. 2008), should accelerate the development of biomarkers that have clinical utility. In this way, dementia research and Alzheimer's research, in particular, are leading the way in psychiatry in terms of collaborative methods contributing to cutting-edge research.

Schizophrenia

Background

Schizophrenia is a severe neuropsychiatric condition diagnosed in just under 1 percent of the population. It is characterized by positive psychotic symptoms such as hallucinations and delusions, thought disorder, and negative symptoms such as limited emotional expression and reduced motivation. Cognitive dysfunction is prominent and most, if not all, patients show significant loss of professional and social function. The course of schizophrenia is often chronic and, as an often lifelong condition, from adolescence to old age, differs greatly from the onset and duration of dementia disorders. Although it has a much lower prevalence in the population than dementia, the longevity of the condition makes the experience of schizophrenia highly distressing to sufferers, as well as to their relatives and friends. In addition to the emotional costs, financial costs to society are considerable (Knapp et al. 2004).

Currently, the best treatments are pharmacological; however, there is no cure, relapse is frequent, and there is considerable inter-individual variability in clinical response. There is a clear clinical need for biomarkers that could predict treatment response, so that patients could be given the most appropriate medication. All antipsychotic drugs currently in use act, at least in part, on the D2 dopamine receptor. This focus likely stems from the influential dopamine hypothesis of schizophrenia and its subsequent revisions (see Howes and Kapur 2009 for a recent review, including recent developments).

The original dopamine hypothesis, founded in early observations of the dopaminergic action of antipsychotics, as well as the psychotomimetic effects of amphetamine, assumed that psychosis was due to an excess of dopamine neurotransmission (termed hyperdopaminergia). The revised dopamine hypothesis, which still prevails, integrates more recent findings into a regionally specific differentiation between prefrontal hypodopaminergia and subcortical hyperdopaminergia.

Imaging Biomarkers for Schizophrenia

Schizophrenia is associated with changes in a large number of sMRI and fMRI measures (Tost et al. 2009). Based on a number of potential preclinical and clinical applications (Borsook et al. 2006), these changes may serve as biomarkers in two contexts. First, they might act as surrogate endpoints in treatment studies. Second, they might serve as baseline measures to predict inter-individual variability in drug response.

The biomarker approach to treatment prediction can be viewed in two ways. First, individual differences in a baseline measure may be used to predict the magnitude of treatment response. Second, patients can be classified into subgroups on the basis of biomarker data. The latter approach could lead to the grouping of patients into more homogenous samples with or without certain neural deficits. Treatment response may then differ by group, and novel treatments could be tailored to patients presenting with certain neural profiles. This approach thus aims to capture the heterogeneity of the clinical condition, where patients display different neural changes and different responses to a given treatment.

Despite the large number of potential neuroimaging biomarkers, none of the observed brain changes have been definitively confirmed as biomarkers of schizophrenia. However, the following are examples of potentially promising biological markers.

Prediction of Treatment Response

A recent example of neuroimaging having predictive value in a clinical setting has been seen in a group of first-episode psychosis patients. This study showed that the baseline (pre-treatment) volume of the pituitary gland, which may serve as a marker of hypothalamic-pituitary-adrenal (HPA) axis function, was a significant predictor of treatment response, where a larger pituitary volume led to less improvement in symptoms (Garner et al. 2009). However, earlier studies have found only mixed support for structural brain predictors of treatment outcome (Lawrie et al. 1995; Robinson et al. 1999; Arango et al. 2003; Molina et al. 2003). Interestingly, neuroimaging may be able to predict more than just a pharmacological treatment response. A recent fMRI study found that activity in the dorsolateral prefrontal cortex, and its connectivity with the cerebellum, predicted subsequent CBT response in schizophrenia, where a stronger signal and increased connectivity predicted a better response (Kumari et al. 2009).

Surrogate Endpoints for Treatment Response

The normalization of the fronto-striatal BOLD signal with clinically successful antipsychotic treatment has been described on several occasions. Patients frequently, but not always, show reduced BOLD signal in cortical networks during task performance (Tost et al. 2009). A normalization of this hypofunction has been observed in studies of second-generation antipsychotics, compared to first-generation compounds (Braus et al. 1999; Honey et al. 1999; Miller et al. 2001; Stephan et al. 2001; Jones et al. 2004; Lahti et al. 2004; Meisenzahl et al. 2006; Kumari et al. 2007). This normalization may thus represent a criterion biomarker against which to compare novel compounds.

Use of sMRI has also illustrated antipsychotic treatment effects in schizophrenia. The striatum is the target of prominent dopamine projections and, therefore, one of the key sites of action of currently available antipsychotics. sMRI studies have shown that treatment with first-generation antipsychotics leads to increased striatal volume, an effect that may reflect the well-known motor side effects of these compounds and is reversible with a switch to second-generation compounds (Scherk and Falkai 2006; Brandt and Bonelli 2008). Novel antipsychotics may thus be assessed for their ability to reduce or reverse striatal volume increases induced by older antipsychotics.

Future Prospects for Schizophrenia

While the prospect of obtaining reproducible neuroimaging biomarkers of schizophrenia is exciting and may revolutionize drug development, a number of important issues must first be considered to achieve these aims. First, none of the currently studied biomarkers is found in every patient (i.e., there is low sensitivity) and many are also seen in other psychiatric populations, as well as in apparently healthy individuals (i.e., there is low specificity). Further work is necessary to refine biomarker specificity, selectivity, and predictive ability. Until these characteristics are improved, biomarkers will not be able to translate to a clinical setting.

Secondly, fMRI biomarker validation needs to address the issue of dissociable drug effects at neural and behavioral levels. In fMRI experiments, drug effects may be observed at the level of task performance, in relation to BOLD, or both. The complexities of dissociations between the neural and performance levels have been described elsewhere (Wilkinson and Halligan 2004) but need to be considered in this context. Unfortunately, we do not know if BOLD fMRI will be useful in determining if a novel compound has its expected and desirable effect in a patient. For cognitive biomarkers, the desirable effect is typically an improvement in performance efficiency or effectiveness; however, it is unclear if BOLD fMRI will show the optimal drug effect, which will likely depend on a number of factors (e.g., task difficulty, baseline performance, neurotransmitter levels, and other trait and state factors).

Future comparative cross-species studies need to fully exploit the translational potential of imaging biomarkers. Specifically, it will be of considerable interest to employ activation tasks during pharmacological disease models that can be modeled in humans as well as in model animals, such as glutamatergic models of schizophrenia (Condy et al. 2005; Honey et al. 2005; Hutton and Ettinger 2006).

Experiments based on the ketamine and phencyclidine (PCP) models of psychosis administer NMDA receptor antagonists to healthy participants to model the symptoms of schizophrenia (Javitt and Zukin 1991; Krystal et al. 1994; Lahti et al. 1995). This centers on the theory that there is a glutamatergic deficiency in patients with schizophrenia (Kim et al. 1980; Wang and Yang 2005). This may be particularly relevant given that it is a major challenge to depart from dopamine-centred drug discovery, as argued (Carpenter and Koenig 2008) and implemented recently (Patil et al. 2007).

Finally, it was recently argued that the number of cognitive and neurophysiological biomarkers of antipsychotic response is already large without considering neuroimaging (De Visser et al. 2001). Therefore, great care should be taken not to prematurely add non-validated neuroimaging measures to the list of potential biomarkers. Instead, before application to preclinical or clinical studies, there should be an intense validation phase of potential biomarkers. Consortia involving both industry and academia (Green and Nuechterlein 2004; Borsook et al. 2008), as implemented in dementia research through programs such as ADNI, would be helpful in validating neuroimaging biomarkers against established cognitive and behavioral markers, and in piloting these in relation to different compounds across different sites.

Discussion

Despite focus on D2-acting antipsychotics, negative symptoms and cognitive deficits do not respond well to treatment. The predominance of the dopamine hypothesis and the primary focus on treating positive psychotic symptoms are, perhaps, a hindrance to the development of new treatments that could help with other aspects of the schizophrenia (Carpenter and Koenig 2008). Patients' variable and less-than-complete responses to pharmacological treatment, and the industry's slow progress toward diversifying compounds with alternative mechanisms of action, make the development of novel drugs of prime importance. Biomarkers may represent an important platform in this context (Thaker 2007; Borsook et al. 2008; Javitt et al. 2008).

Mood Disorders

Background

While everyone experiences low mood from time to time, patients with major depressive disorder (MDD) face prolonged periods of depression or anhedonia, that is, an inability to experience interest or pleasure in everyday activities. Patients may also experience changes in body weight, suffer from sleep disturbance and fatigue, and harbor thoughts of suicide. The lifetime prevalence of MDD is 16 percent, although the risk of MDD among women is 70 percent greater than in men (Kessler et al. 2003). The mean age of onset of MDD in patients in treatment is 29 years, although there is evidence that this has been decreasing over the last century (Judd 1998).

Patients with bipolar disorder (BD) suffer from both depression, as described above, and mania, which exhibits as an abnormally and persistently elevated mood. During an episode of mania, psychosis, in which the patient experiences delusions or even hallucinations, may also be present. BD is rarer than MDD, with a lifetime prevalence of 1 percent and a younger median age of onset of 18 years (Merikangas et al. 2007). Unlike MDD, men are as likely to be affected by the illness as women. Twin studies have shown that genetic influences are important in both affective disorders, although the genetic contribution in BD is somewhat greater than in MDD (Kendler and Prescott 1999; Smoller and Finn 2003). Environmental risk factors for both disorders include the postpartum period in women, and stressful life events, such as early parental loss (through death or divorce) and divorce of the patient (Agid et al. 1999; Tsuchiya et al. 2003).

The treatment of the two disorders is different, and this makes correct diagnosis essential. In practice, this is not always easy, as a patient with MDD may be indistinguishable from a depressed patient with BD. MDD patients are commonly treated with antidepressants, such as selective serotonin reuptake inhibitors (SSRIs) and serotonin-norepinephrine reuptake inhibitors (SNRIs). However, if a depressed BD patient is given an antidepressant, this may initiate a manic episode. Accurate diagnosis to disentangle the two disorders, separating the overlapping symptom profiles, is therefore critical for patient care. For BD patients, the first line of treatment is the use of mood stabilizers, such as lithium or sodium valproate, and antipsychotic medication, especially for patients with acute mania.

Response to antidepressants in MDD is far from optimal: 30 to 40 percent of depressed patients do not improve in response to the first antidepressant (Adli et al. 2006), with the remaining patients trying a number of medications before recovering, or, in a small proportion of cases, developing treatment-resistant depression. Treatment response is also relatively poor in BD; only 50 percent of patients respond to the initial medication (DelBello and Strakowski 2004). One of the most commonly used medications for BD is lithium, and although its effect in preventing manic episodes is well established, there is less evidence that it prevents depressive episodes (Geddes et al. 2004).

Identifying biomarkers for MDD and BD would enable clinicians to distinguish between the two disorders soon after onset, and would then facilitate the administration of appropriate medication. Biomarkers to predict treatment response within each disorder would allow clinicians to choose the best medication for the individual, significantly shortening mood episodes, improving compliance, and reducing the impairment and distress to the patient.

Imaging Biomarkers for Disease Diagnosis

Candidate biomarkers for BD and MDD span the breadth of neuroimaging modalities. Structural MRI, the most established of the MR imaging methods, has been used for the last twenty-five years to compare regional brain volumes in BD and MDD patients to those of control subjects.

In a meta-analysis carried out by our group of ninety-eight structural imaging studies of BD patients, we confirmed that patients with BD have enlarged lateral ventricles (17 percent larger) compared to matched healthy controls, and increased visibility of deep white matter hyperintensities (Kempton et al. 2008). Hyperintensities are small regions of increased signal observed on T_2 weighted MR images of the brain, typically in white matter (T_1 weighed MR images are particularly suited to cerebral gray/white matter contrasts, whereas T_2 weighted MR images are especially sensitive to water content). These lesions are thought to be caused by ischemic demyelination and are linked to cerebrovascular disease.

In MDD, increased hyperintensities have also been observed, especially in patients with a late age of onset. The link between hyperintensities, cardiovascular factors, and late-onset MDD has been considered to be so strong by some investigators that the term "vascular depression" was proposed to indicate a subtype of depression essentially caused by cardiovascular disease (Alexopoulos et al. 1997). An abnormality that appears to separate MDD from BD is decreased hippocampal volume, which is observed across the age range in MDD patients, while patients with BD have hippocampal volumes comparable to those of the controls (Videbech and Ravnkilde 2004).

Looking to fMRI studies, Blumberg and colleagues scanned BD patients during episodes of mania, depression, and euthymia (Blumberg et al. 2003). They identified a small region in the ventral prefrontal cortex that was consistently less active during the Stroop task (Stroop 1935) when compared to healthy controls, potentially identifying a trait marker of the illness (Blumberg et al. 2003). The Stroop task is a long-established task within psychology that measures brain inhibitory processing (MacLeod 1991). Participants read a list of the names of colors twice: once when the color name and font color match, and once when the color name and font color do not match. The ability to suppress the natural response to name the font color in the second reading is an established indicator of prefrontal functioning (Perret 1974). Lawrence et al. (2004) directly compared brain activations in response to facial expressions of affect in MDD and BD patients. They reported that while BD patients showed increased activations of the ventral prefrontal cortex and subcortical structures, MDD patients had diminished activations in these regions. Reviews of fMRI studies in BD generally agree that there is decreased activity in the prefrontal cortex and increased activation of the amygdala. Increased activity of the amygdala has also been observed in MDD, and while this has been suggested as an endophenotype for MDD by some researchers (Hasler et al. 2004), this finding does not appear to separate MDD from BD.

Studies investigating neurochemical abnormalities with MRS in patients with BD appear to be sensitive to both medication and mood state. N-acetyl aspartate (NAA), which has been described as a marker of neuronal content and integrity, is reduced in BD patients, but increases following lithium treatment (Moore et al. 2000; Silverstone et al. 2003), while phosphomonoesters (PME) appear to be increased in symptomatic patients compared to remitted BD patients (Yildiz et al. 2001). Overall, the most consistent MRS findings in BD are decreased intracellular pH, indicated by a frequency shift in metabolites, and increased choline and glutamate/glutamine (Glx) levels. These diverse findings have been integrated into a theory of mitochondrial dysfunction in BD by Stork and Renshaw (2005). In contrast to these findings in BD, no alteration in NAA and a decrease in Glx levels were found in a meta-analysis of MDD studies (Yildiz-Yesiloglu and Ankerst 2006).

There have been a relatively small number of DTI studies in patients with mood disorders. A study of medication-naive adolescents experiencing their first episode of mania (Adler et al. 2006) showed reduced fractional anisotropy (FA)—the degree to which water diffusion is directional—in prefrontal and posterior white matter. This result indicates that this abnormality occurred early in the illness and was not a direct result of medication. Indeed, a recent review reports that reduced FA is consistent in both BD and MDD patients, primarily in frontal and temporal white matter tracts (White et al. 2008).

Prediction of Treatment Response

The search for neuroimaging biomarkers for treatment response in affective disorders has primarily focused on predicting antidepressant response in MDD patients. An early finding was that MDD patients with white matter hyperintensities had a poorer response to pharmacological treatments (Hickie et al. 1995; Simpson et al. 1997). A longitudinal study has also suggested that patients who have a greater increase in white matter hyperintensities over time are less likely to respond to treatment (Taylor et al. 2003).

If these patients do not respond well to antidepressants, what other options are available? Taragano et al. (2005) conducted a randomized clinical trial in patients with 'vascular depression,' augmenting the antidepressant fluoxetine with nimodipine, which is used to treat hypertension. The combination of antidepressant and antihypertensive medication was more effective than the antidepressant alone, indicating that treating one of the putative causes of hyperintensities improved the patients' symptoms.

Using fMRI, Davidson et al. (2003) reported that greater anterior cingulate activation in response to affective visual stimuli was predictive of an improved response to the antidepressant venlafaxine. Alexopoulos et al. (2002) measured FA in MDD patients with DTI before initiating treatment with citalopram, and reported that high FA values predicted an improved response to the medication.

More recently, artificial intelligence classification algorithms, such as support vector machines (SVM), have been applied to neuroimaging data. Using this approach, Costafreda and colleagues (2009) were able to correctly predict the patients who would respond to antidepressant medication with a specificity and sensitivity of 88.9 percent from whole brain structural imaging data. The technique could predict treatment response for pharmacological (fluoxetine) but not behavioral (CBT) interventions. Similar machine classification has also shown some success using fMRI data to classify treatment response with a specificity of 70 percent and a sensitivity of 65 percent, using data from a working memory task (Marquand et al. 2008). This type of analysis using SVM may also be useful when applied to psychiatric conditions other than mood disorders.

Future Prospects for Mood Disorders

In summary, while no imaging biomarkers for MDD and BD diagnosis currently exist in clinical practice, either for unequivocally separating BD and MDD at diagnosis or for predicting treatment response, candidates are emerging from the neuroimaging literature. Due to the overlap between healthy controls and patients with mood disorders observed in neuroimaging measures, it is likely that no one biomarker will be used, but rather a number of biomarkers from which a probability of a diagnosis may be calculated. This value could be combined with genetic and neuropsychological measures to assign a diagnosis and select the best treatment course to follow. No matter what methods are used to calculate a value to assign a diagnosis, specificity, in particular, is and will remain important in mood disorder diagnosis.

Conclusions

Although there are some idiosyncrasies in different areas of mental health, there are some common themes available for the use of imaging biomarkers in neuro-degenerative and psychiatric disorders. Biomarkers can potentially contribute to personalized medicine by aiding diagnosis, predicting disease progression, and predicting treatment response.

In dementia, schizophrenia, and the two mood disorders (MDD and BD), there is a significant potential role for biomarkers, from neuroimaging in particular, to have a positive impact on clinical care. Neuroimaging methods have begun to play an important role in drug discovery (Borsook et al. 2008) and there is hope that this trend will increase in the future.

Although there are many biomarker candidates for different disorders, none, when used in isolation, are yet clinically unequivocal. Showing general group differences is not enough; a biomarker must be able to allow categorization or predict outcome on an individual basis. Since the search for new biomarkers tends to be driven by group studies of new treatments, it might be expected that further investigations could identify those sensitive and specific enough either alone, but most likely in combination, for personalized applications. Specificity and selectivity are most important, and issues related to these parameters are relevant to the application of neuroimaging biomarkers to all psychiatric disorders. There remains substantial variety in the types of biomarkers that can be used, and in how expensive or invasive it is to identify them. This leads to implications for how they can be used in clinical practice. The process of discovering a biomarker should not be completely removed from the reality of what will be financially and practically viable in a clinical setting.

Neuroimaging offers a broad range of options, and sMRI, in particular, is often carried out as part of routine clinical assessment. Any imaging biomarker will likely be required in combination with other modalities (e.g., -omics using blood, other tissue, or CSF analyses), but should be able to target treatment to benefit the care of patients, as well as being more cost-effective. There are also neuroimaging modalities that are still relatively novel in terms of applications to psychiatric research and could potentially offer benefits, such as arterial spin labeling (ASL), which measures cerebral blood flow (Brown et al. 2007).

It will be important to clarify the pharmacological mechanisms of observed BOLD signal changes induced by administration of different compounds using pharmacological interaction designs or additional imaging methods, such as SPECT or PET. A key feature of the BOLD response is that it remains an indirect measure of neural activity. As such, it is not necessarily informative about the underlying neurotransmitter or receptor systems. Thus, an integrative approach is important for assessing drug efficacy in all psychiatric disorders.

A particularly exciting prospect for future research in this field may be the use of biomarkers that are also endophenotypes, or genetic vulnerability markers (see Ettinger and Kumari 2003 and Thaker 2007 for a discussion of this in relation to schizophrenia). Adopting neurobiologically informative and specific endophenotypes that are heritable and linked to known genetic loci and polymorphisms adds a further and very informative dimension to biomarker use. Such endophenotypic markers of pathophysiology should represent important targets for correction through pharmacological treatment, as they are more "upstream" than the clinical phenotype or a complex cognitive biomarker. Eventually, the use of such endophenotypic biomarkers should allow the development of genetically informed compounds. Hypothetically, a biomarker that has reliably been associated with a polymorphism in a certain pharmacologically relevant gene (e.g., a receptor gene) might be sensitive to a compound targeting that receptor. It will be important to identify a battery of biomarkers that are statistically, neurophysiologically, and genetically independent of each other, in order to maximize the spectrum of neural deficits targetable by novel compounds that can be assessed.

Large-scale collaborations between industry and academia also have an important role to play in furthering knowledge and understanding. Sharing information across all types of psychiatric research is beneficial for development, but it is also apparent that different MRI scanners and sequences can dramatically alter sensitivity to a given neuroimaging biomarker, particularly on an individual level. This means that standardization of MRI and other imaging sequences is required, especially across manufacturers (see Deoni et al. 2008 and Kruggel et al. 2010 for examples of progress in this area). This approach is already being adopted in multicenter imaging studies, such as ADNI, and one can envisage that it will also be applied to other psychiatric disorders.

Imaging techniques are not yet ready to support truly personalized medicine, but there is much promise in recent research results and in the latest combined complementary (gene–brain) marker direction in which the field is moving.

References

Abou-Saleh, M. T. 2006. Neuroimaging in psychiatry: An update. *Journal of Psychosomatic Research* 61(3): 289–293.

Adalsteinsson, E., Sullivan, E. V., Kleinhans, N., Spielman, D. M., and Pfefferbaum, A. 2000. Longitudinal decline of the neuronal marker N-acetyl aspartate in Alzheimer's disease. *Lancet* 355(9216): 1696–1697.

Adler, C. M., Adams, J., DelBello, M. P., Holland, S. K., Schmithorst, V., Levine, A., Jarvis, K., and Strakowski, S. M. 2006. Evidence of white matter pathology in bipolar disorder adolescents experiencing their first episode of mania: A diffusion tensor imaging study. *American Journal of Psychiatry* 163(2): 322–324.

Adli, M., Bauer, M., and Rush, A. J. 2006. Algorithms and collaborative-care systems for depression: are they effective and why? A systematic review. *Biological Psychiatry* 59(11): 1029–1038.

Agid, O., Shapira, B., Zislin, J., Ritsner, M., Hanin, B., Murad, H., Troudart, T., Bloch, M., Heresco-Levy, U., and Lerer, B. 1999. Environment and vulnerability to major psychiatric illness: A case control study of early parental loss in major depression, bipolar disorder and schizophrenia. *Molecular Psychiatry* 4(2): 163–172.

Agid, Y., Buzsaki, G., Diamond, D. M., Frackowiak, R., Giedd, J., Girault, J.-A., Grace, A., Lambert, J. J., Manji, H., Mayberg, H., Popoli, M., Prochiantz, A., Richter-Levin, G., Somogyi, P., Spedding, M., Svenningsson, P., and Weinberger, D. 2007. How can drug discovery for psychiatric disorders be improved? *Nat Rev Drug Discov* 6(3): 189–201.

Alexopoulos, G. S., Kiosses, D. N., Choi, S. J., Murphy, C. F., and Lim, K. O. 2002. Frontal white matter microstructure and treatment response of late-life depression: A preliminary study. *American Journal of Psychiatry* 159(11): 1929–1932.

Alexopoulos, G. S., Meyers, B. S., Young, R. C., Campbell, S., Silbersweig, D., and Charlson, M. 1997. 'Vascular depression' hypothesis. *Archives of General Psychiatry* 54(10): 915–922.

Alzheimer's Association. 2009. Alzheimer's disease facts and figures. *Alzheimer's & Dementia* 5(3): 234–270.

Arango, C., Breier, A., McMahon, R., Carpenter, W. T., and Buchanan, R. W. 2003. The relationship of clozapine and haloperidol treatment response to prefrontal, hippocampal, and caudate brain volumes. *The American Journal of Psychiatry* 160(8): 1421–1427.

Blennow, K. 2004. Cerebrospinal Fluid Protein Biomarkers for Alzheimer's disease. *NeuroRx* 1(2): 213–225.

Blumberg, H. P., Leung, H. C., Skudlarski, P., Lacadie, C. M., Fredericks, C. A., Harris, B. C., Charney, D. S., Gore, J. C., Krystal, J. H., and Peterson, B. S. 2003. A functional magnetic resonance imaging study of bipolar disorder: State- and trait-related dysfunction in ventral prefrontal cortices. *Archives of General Psychiatry* 60(6): 601–609.

Borsook, D., Becerra, L., and Hargreaves, R. 2006. A role for fMRI in optimizing CNS drug development. *Nature Reviews Drug Discovery* 5(5): 411–424.

Borsook, D., Bleakman, D., Hargreaves, R., Upadhyay, J., Schmidt, K. F., and Becerra, L. 2008. A 'BOLD' experiment in defining the utility of fMRI in drug development. *Neuroimage* 42(2): 461–466.

Brandt, G. N., and Bonelli, R. M. 2008. Structural neuroimaging of the basal ganglia in schizophrenic patients: A review. *Wiener Medizinische Wochenschrift* 158(3–4): 84–90.

Braus, D. F., Ende, G., Weber-Fahr, W., Sartorius, A., Krier, A., Hubrich-Ungureanu, P., Ruf, M., Stuck, S., and Henn, F. A. 1999. Antipsychotic drug effects on motor activation measured by functional magnetic resonance imaging in schizophrenic patients. *Schizophrenia Research* 39(1): 19–29.

Britschgi, M., and Wyss-Coray, T. 2009. Blood protein signature for the early diagnosis of alzheimer disease. *Archives of Neurology* 66(2): 161–165.

Brown, G. G., Clark, C., and Liu, T. T. 2007. Measurement of cerebral perfusion with arterial spin labeling: Part 2. Applications. *Journal of the International Neuropsychological Society* 13(3): 526–538.

Bruscoli, M., and Lovestone, S. 2004. Is MCI really just early dementia? A systematic review of conversion studies. *International Psychogeriatrics* 16(2): 129–140.

Buerger, K., Ewers, M., Pirttilä, T., Zinkowski, R., Alafuzoff, I., Teipel, S. J., DeBernardis, J., Kerkman, D., McCulloch, C., Soininen, H., and Hampel, H. 2006. CSF phosphorylated tau protein correlates with neocortical neurofibrillary pathology in Alzheimer's disease. *Brain* 129(11): 3035–3041.

Butler, A. C., Chapman, J. E., Forman, E. M., and Beck, A. T. 2006. The empirical status of cognitive-behavioral therapy: A review of meta-analyses. *Clinical Psychology Review* 26(1): 17–31.

Capote, H. A. 2009. Neuroimaging in psychiatry. *Neurologic Clinics* 27(1): 237–249.

Carpenter, W. T., and Koenig, J. I. 2008. The evolution of drug development in schizophrenia: Past issues and future opportunities. *Neuropsychopharmacology* 33(9): 2061–2079.

Carrette, O., Demalte, I., Scherl, A., Yalkinoglu, O., Corthals, G., Burkhard, P., Hochstrasser, D. F., and Sanchez, J. C. 2003. A panel of cerebrospinal fluid potential biomarkers for the diagnosis of Alzheimer's disease. *Proteomics* 3(8): 1486–1494.

Carrillo, M. C., Sanders, C. A., and Katz, R. G. 2009. Maximizing the Alzheimer's Disease Neuroimaging Initiative II. *Alzheimer's and Dementia* 5(3): 271–275.

Chupin, M., Gérardin, E., Cuingnet, R., Boutet, C., Lemieux, L., Lehéricy, S., Benali, H., Garnero, L., and Colliot, O. 2009. Fully automatic hippocampus segmentation and classification in Alzheimer's disease and mild cognitive impairment applied on data from ADNI. *Hippocampus* 19(6): 579–587.

Condy, C., Wattiez, N., Rivaud-Pechoux, S., and Gaymard, B. 2005. Ketamine-induced distractibility: An oculomotor study in monkeys. *Biological Psychiatry* 57(4): 366–372.

Costafreda, S. G., Khanna, A., Mourao-Miranda, J., and Fu, C. H. Y. 2009. Neural correlates of sad faces predict clinical remission to cognitive behavioural therapy in depression. *NeuroReport* 20(7): 637–641.

Craig-Schapiro, R., Fagan, A. M., and Holtzman, D. M. 2009. Biomarkers of Alzheimer's disease. *Neurobiology of Disease* 35(2): 128–140.

Davidson, R. J., Irwin, W., Anderle, M. J., and Kalin, N. H. 2003. The neural substrates of affective processing in depressed patients treated with venlafaxine. *American Journal of Psychiatry* 160(1): 64–75.

Davies, P., Resnick, J., Resnick, B., Gilman, S., Growdon, J. H., Khachaturian, Z. S., Radebaugh, T. S., Roses, A. D., Selkoe, D. J., Trojanowski, J. Q., Blass, J. P., Gibson, G. E., Sheu, K. F. R., Blennow, K., Delacourte, A., Frisoni, G. B., Jefferies, W. A., McRae, A., Wisniewski, H. M., Mehta, P. D., Pirttla, T., Parshad, R., Scinto, L. F. M., Scheltens, P., Riekkinen, P. J., Soininen, H. S., Swanwick, G. R. J., Wahlund, L. O., Arnold, S. E., Winblad, B., Ihara, Y., Yamazaki, T., Arai, H., Clark, C. M., Ewbank, D. C., Takase, S., Higuchi, S., Mirua, M., Seki, H., Higuchi, M., Matsui, T., Lee, V. M. Y., Sasaki, H., Foster, N. L., Galasko, D., Hock, C., Hyman, B. T., St. George-Hyslop, P. H., Iwatsubo, T., Kennard, M., Klunk, W. E., Lannfelt, L., Litvan, I., Mayeux, R., Robles, A., Fabian, V. A., Jones, T. M., Wilton, S. D., Dench, J. E., Davis, M. R., Lim, L., Kakulas, B. A., Kennard, M. L., Feldman, H., Yamada, T., Sweet, R. A., and Zubenko, G. S. 1998. Consensus report of the working group on: 'Molecular and biochemical markers of Alzheimer's disease'. *Neurobiology of Aging* 19(2): 109–116.

De Leon, M. J., DeSanti, S., Zinkowski, R., Mehta, P. D., Pratico, D., Segal, S., Rusinek, H., Li, J., Tsui, W., Saint Louis, L. A., Clark, C. M., Tarshish, C., Li, Y., Lair, L., Javier, E., Rich, K., Lesbre, P., Mosconi, L., Reisberg, B., Sadowski, M., DeBernadis, J. F., Kerkman, D. J., Hampel, H., Wahlund, L. O., and Davies, P. 2006. Longitudinal CSF and MRI biomarkers improve the diagnosis of mild cognitive impairment. *Neurobiology of Aging* 27(3): 394–401.

De Leon, M. J., Mosconi, L., Blennow, K., DeSanti, S., Zinkowski, R., Mehta, P. D., Pratico, D., Tsui, W., Saint Louis, L. A., Sobanska, L., Brys, M., Li, Y., Rich, K., Rinne, J., and Rusinek, H.

2007. Imaging and CSF studies in the preclinical diagnosis of Alzheimer's disease. *Annals of the New York Academy of Sciences* 1097: 114–145.

De Visser, S. J., Van der Post, J., Pieters, M. S. M., Cohen, A. F., and Van Gerven, J. M. A. 2001. Biomarkers for the effects of antipsychotic drugs in healthy volunteers. *British Journal of Clinical Pharmacology* 51(2): 119–132.

DelBello, M. P., and Strakowski, S. M. 2004. Neurochemical predictors of response to pharmacologic treatments for bipolar disorder. *Current Psychiatry Reports* 6(6): 466–472.

Deoni, S. C. L., Williams, S. C. R., Jezzard, P., Suckling, J., Murphy, D. G. M., and Jones, D. K. 2008. Standardized structural magnetic resonance imaging in multicentre studies using quantitative T1 and T2 imaging at 1.5 T. *Neuroimage* 40(2): 662–671.

DeRubeis, R. J., and Crits-Christoph, P. 1998. Empirically supported individual and group psychological treatments for adult mental disorders. *Journal of Consulting and Clinical Psychology* 66(1): 37–52.

Dougall, N. J., Bruggink, S., and Ebmeier, K. P. 2004. Systematic review of the diagnostic accuracy of 99mTc-HMPAO- SPECT in dementia. *American Journal of Geriatric Psychiatry* 12(6): 554–570.

Ettinger, U., and Kumari, V. 2003. Pharmacological studies of smooth pursuit and antisaccade eye movements in schizophrenia: Current status and directions for future research. *Current Neuropharmacology* 1: 285–300.

Fagan, A. M., Mintun, M. A., Mach, R. H., Lee, S. Y., Dence, C. S., Shah, A. R., LaRossa, G. N., Spinner, M. L., Klunk, W. E., Mathis, C. A., DeKosky, S. T., Morris, J. C., and Holtzman, D. M. 2006. Inverse relation between in vivo amyloid imaging load and cerebrospinal fluid $A\beta_{42}$ in humans. *Annals of Neurology* 59(3): 512–519.

Fayed, N., Dávila, J., Oliveros, A., Castillo, J., and Medrano, J. J. 2008. Utility of different MR modalities in mild cognitive impairment and its use as a predictor of conversion to probable dementia. *Academic Radiology* 15(9): 1089–1098.

Finehout, E. J., Franck, Z., Choe, L. H., Relkin, N., and Lee, K. H. 2007. Cerebrospinal fluid proteomic biomarkers for Alzheimer's disease. *Annals of Neurology* 61(2): 120–129.

Firbank, M. J., Harrison, R. M., and O'Brien, J. T. 2002. A comprehensive review of proton magnetic resonance spectroscopy studies in dementia and Parkinson's disease. *Dementia and Geriatric Cognitive Disorders* 14(2): 64–76.

Flirski, M., and Sobow, T. 2005. Biochemical markers and risk factors of Alzheimer's disease. *Current Alzheimer Research* 2(1): 47–64.

Foster, N. L., Heidebrink, J. L., Clark, C. M., Jagust, W. J., Arnold, S. E., Barbas, N. R., DeCarli, C. S., Scott Turner, R., Koeppe, R. A., Higdon, R., and Minoshima, S. 2007. FDG-PET improves accuracy in distinguishing frontotemporal dementia and Alzheimer's disease. *Brain* 130(10): 2616–2635.

Fox, N. C., and Freeborough, P. A. 1997. Brain atrophy progression measured from registered serial MRI: Validation and application to Alzheimer's disease. *Journal of Magnetic Resonance Imaging* 7(6): 1069–1075.

Fox, N. C., Scahill, R. I., Crum, W. R., and Rossor, M. N. 1999. Correlation between rates of brain atrophy and cognitive decline in AD. *Neurology* 52(8): 1687–1689.

Fox, N. C., Warrington, E. K., Freeborough, P. A., Hartikainen, P., Kennedy, A. M., Stevens, J. M., and Rossor, M. N. 1996. Presymptomatic hippocampal atrophy in Alzheimer's disease: A longitudinal MRI study. *Brain* 119(6): 2001–2007.

Frank, R. A., Galasko, D., Hampel, H., Hardy, J., De Leon, M. J., Mehta, P. D., Rogers, J., Siemers, E., and Trojanowski, J. Q. 2003. Biological markers for therapeutic trials in Alzheimer's disease: Proceedings of the biological markers working group; NIA initiative on neuroimaging in Alzheimer's disease. *Neurobiology of Aging* 24(4): 521–536.

Frankfort, S. V., Tulner, L. R., van Campen, J. P. C. M., Verbeekl, M. M., Jansen, R. W. M. M., and Beijnen, J. H. 2008. Amyloid beta protein and tau in cerebrospinal fluid and plasma as biomarkers for dementia: A review of recent literature. *Current Clinical Pharmacology* 3(2): 123–131.

Frey, H. J., Mattila, K. M., Korolainen, M. A., and Pirttilä, T. 2005. Problems associated with biological markers of Alzheimer's disease. *Neurochemical Research* 30(12): 1501–1510.

Fu, C. H. Y., Williams, S. C. R., Cleare, A. J., Scott, J., Mitterschiffthaler, M. T., Walsh, N. D., Donaldson, C., Suckling, J., Andrew, C., Steiner, H., and Murray, R. M. 2008. Neural Responses to Sad Facial Expressions in Major Depression Following Cognitive Behavioral Therapy. *Biological Psychiatry* 64(6): 505–512.

Garner, B., Berger, G. E., Nicolo, J. P., Mackinnon, A., Wood, S. J., Pariante, C. M., Dazzan, P., Proffitt, T. M., Markulev, C., Kerr, M., McConchie, M., Phillips, L. J., Pantelis, C., and McGorry, P. D. 2009. Pituitary volume and early treatment response in drug-naive first-episode psychosis patients. *Schizophrenia Research* 113(1): 65–71.

Geddes, J. R., Burgess, S., Hawton, K., Jamison, K., and Goodwin, G. M. 2004. Long-Term Lithium Therapy for Bipolar Disorder: Systematic Review and Meta-Analysis of Randomized Controlled Trials. *American Journal of Psychiatry* 161: 217–222.

Green, M. F., and Nuechterlein, K. H. 2004. The MATRICS initiative: Developing a consensus cognitive battery for clinical trials. *Schizophrenia Research* 72(1): 1–3.

Hampel, H., Bürger, K., Teipel, S. J., Bokde, A. L. W., Zetterberg, H., and Blennow, K. 2008. Core candidate neurochemical and imaging biomarkers of Alzheimer's disease. *Alzheimer's and Dementia* 4(1): 38–48.

Hansson, O., Zetterberg, H., Buchhave, P., Londos, E., Blennow, K., and Minthon, L. 2006. Association between CSF biomarkers and incipient Alzheimer's disease in patients with mild cognitive impairment: A follow-up study. *Lancet Neurology* 5(3): 228–234.

Hasler, G., Drevets, W. C., Manji, H. K., and Charney, D. S. 2004. Discovering endophenotypes for major depression. *Neuropsychopharmacology* 29(10): 1765–1781.

Herholz, K., Carter, S. F., and Jones, M. 2007. Positron emission tomography imaging in dementia. *British Journal of Radiology* 80(SPEC. ISS. 2): S160–167.

Hickie, I., Scott, E., Mitchell, P., Wilhelm, K., Austin, M. P., and Bennett, B. 1995. Subcortical hyperintensities on magnetic resonance imaging: Clinical correlates and prognostic significance in patients with severe depression. *Biological Psychiatry* 37(3): 151–160.

Honey, G., and Bullmore, E. 2004. Human pharmacological MRI. *Trends in Pharmacological Sciences* 25(7): 366–374.

Honey, G. D., Bullmore, E. T., Soni, W., Varatheesan, M., Williams, S. C. R., and Sharma, T. 1999. Differences in frontal cortical activation by a working memory task after substitution of risperidone for typical antipsychotic drugs in patients with schizophrenia. *Proceedings of the National Academy of Sciences of the United States of America* 96(23): 13432–13437.

Honey, G. D., Honey, R. A. E., O'Loughlin, C., Sharar, S. R., Kumaran, D., Suckling, J., Menon, D. K., Sleator, C., Bullmore, E. T., and Fletcher, P. C. 2005. Ketamine disrupts frontal and hippocampal contribution to encoding and retrieval of episodic memory: an fMRI study. *Cerebral Cortex* 15(6): 749–759.

Horti, A. G., and Villemagne, V. L. 2006. The quest for Eldorado: Development of radioligands for in vivo imaging of nicotinic acetylcholine receptors in human brain. *Current Pharmaceutical Design* 12(30): 3877–3900.

Howes, O. D., and Kapur, S. 2009. The dopamine hypothesis of schizophrenia: Version III - The final common pathway. *Schizophrenia Bulletin* 35(3): 549–562.

Hulstaert, F., Blennow, K., Ivanoiu, A., Schoonderwaldt, H. C., Riemenschneider, M., De Deyn, P. P., Bancher, C., Cras, P., Wiltfang, J., Mehta, P. D., Iqbal, K., Pottel, H., Vanmechelen, E., and Vanderstichele, H. 1999. Improved discrimination of AD patients using β-amyloid(1–42) and tau levels in CSF. *Neurology* 52(8): 1555–1562.

Hutton, S. B., and Ettinger, U. 2006. The antisaccade task as a research tool in psychopathology: A critical review. *Psychophysiology* 43(3): 302–313.

Hye, A., Lynham, S., Thambisetty, M., Causevic, M., Campbell, J., Byers, H. L., Hooper, C., Rijsdijk, F., Tabrizi, S. J., Banner, S., Shaw, C. E., Foy, C., Poppe, M., Archer, N., Hamilton, G., Powell, J., Brown, R. G., Sham, P., Ward, M., and Lovestone, S. 2006. Proteome-based plasma biomarkers for Alzheimer's disease. *Brain* 129(11): 3042–3050.

Irizarry, M. C. 2004. Biomarkers of Alzheimer Disease in Plasma. *NeuroRx* 1(2): 226–234.

Jack Jr, C. R., Bernstein, M. A., Fox, N. C., Thompson, P., Alexander, G., Harvey, D., Borowski, B., Britson, P. J., Whitwell, J. L., Ward, C., Dale, A. M., Felmlee, J. P., Gunter, J. L., Hill, D. L. G.,

Killiany, R., Schuff, N., Fox-Bosetti, S., Lin, C., Studholme, C., DeCarli, C. S., Krueger, G., Ward, H. A., Metzger, G. J., Scott, K. T., Mallozzi, R., Blezek, D., Levy, J., Debbins, J. P., Fleisher, A. S., Albert, M., Green, R., Bartzokis, G., Glover, G., Mugler, J., and Weiner, M. W. 2008. The Alzheimer's Disease Neuroimaging Initiative (ADNI): MRI methods. *Journal of Magnetic Resonance Imaging* 27(4): 685–691.

Jack Jr, C. R., Petersen, R. C., Xu, Y., O'Brien, P. C., Smith, G. E., Ivnik, R. J., Boeve, B. F., Tangalos, E. G., and Kokmen, E. 2000. Rates of hippocampal atrophy correlate with change in clinical status in aging and AD. *Neurology* 55(4): 484–489.

Jack Jr, C. R., Shiung, M. M., Gunter, J. L., O'Brien, P. C., Weigand, S. D., Knopman, D. S., Boeve, B. F., Ivnik, R. J., Smith, G. E., Cha, R. H., Tangalos, E. G., and Petersen, R. C. 2004. Comparison of different MRI brain atrophy rate measures with clinical disease progression in AD. *Neurology* 62(4): 591–600.

Javitt, D. C., Spencer, K. M., Thaker, G. K., Winterer, G., and Hajos, M. 2008. Neurophysiological biomarkers for drug development in schizophrenia. *Nature Reviews Drug Discovery* 7(1): 68–83.

Javitt, D. C., and Zukin, S. R. 1991. Recent advances in the phencyclidine model of schizophrenia. *American Journal of Psychiatry* 148(10): 1301–1308.

Jones, H. M., Brammer, M. J., O'Toole, M., Taylor, T., Ohlsen, R. I., Brown, R. G., Purvis, R., Williams, S., and Pilowsky, L. S. 2004. Cortical effects of quetiapine in first-episode schizophrenia: a preliminary functional magnetic resonance imaging study. *Biological Psychiatry* 56(12): 938–942.

Jones, R. S., and Waldman, A. D. 2004. 1H-MRS evaluation of metabolism in Alzheimer's disease and vascular dementia. *Neurological Research* 26(5): 488–495.

Judd, L. L. 1998. A decade of antidepressant development: the SSRIs and beyond. *Journal of Affective Disorders* 51(3): 211–213.

Kantarci, K., and Jack Jr, C. R. 2003. Neuroimaging in Alzheimer disease: An evidence-based review. *Neuroimaging Clinics of North America* 13(2): 197–209.

Kantarci, K., and Jack Jr, C. R. 2004. Quantitative Magnetic Resonance Techniques as Surrogate Markers of Alzheimer's Disease. *NeuroRx* 1(2): 196–205.

Kantarci, K., Smith, G. E., Ivnik, R. J., Petersen, R. C., Boeve, B. F., Knopman, D. S., Tangalos, E. G., and Jack Jr, C. R. 2002. 1H magnetic resonance spectroscopy, cognitive function, and apolipoprotein E genotype in normal aging, mild cognitive impairment and Alzheimer's disease. *Journal of the International Neuropsychological Society* 8(7): 934–942.

Kaye, J. A., Swihart, T., Howieson, D., Dame, A., Moore, M. M., Karnos, T., Camicioli, R., Ball, M., Oken, B., and Sexton, G. 1997. Volume loss of the hippocampus and temporal lobe in healthy elderly persons destined to develop dementia. *Neurology* 48(5): 1297–1304.

Kempton, M. J., Geddes, J. R., Ettinger, U., Williams, S. C. R., and Grasby, P. M. 2008. Meta-analysis, database, and meta-regression of 98 structural imaging studies in bipolar disorder. *Archives of General Psychiatry* 65(9): 1017–1032.

Kendler, K. S., and Prescott, C. A. 1999. A population-based twin study of lifetime major depression in men and women. *Archives of General Psychiatry* 56(1): 39–44.

Kessler, R. C., Berglund, P., Demler, O., Jin, R., Koretz, D., Merikangas, K. R., Rush, A. J., Walters, E. E., and Wang, P. S. 2003. The Epidemiology of Major Depressive Disorder: Results from the National Comorbidity Survey Replication (NCS-R). *Journal of the American Medical Association* 289(23): 3095–3105.

Kim, J. S., Kornhuber, H. H., Schmid-Burgk, W., and Holzmuller, B. 1980. Low cerebrospinal fluid glutamate in schizophrenic patients and a new hypothesis on schizophrenia. *Neuroscience Letters* 20(3): 379–382.

Klunk, W. E., Engler, H., Nordberg, A., Wang, Y., Blomqvist, G., Holt, D. P., Bergström, M., Savitcheva, I., Huang, G. F., Estrada, S., Ausén, B., Debnath, M. L., Barletta, J., Price, J. C., Sandell, J., Lopresti, B. J., Wall, A., Koivisto, P., Antoni, G., Mathis, C. A., and Långström, B. 2004. Imaging Brain Amyloid in Alzheimer's Disease with Pittsburgh Compound-B. *Annals of Neurology* 55(3): 306–319.

Knapp, M., Mangalore, R., and Simon, J. 2004. The global costs of Schizophrenia. *Schizophrenia Bulletin* 30(2): 279–293.

Konarski, J. Z., Kennedy, S. H., Segal, Z. V., Lau, M. A., Bieling, P. J., McIntyre, R. S., and Mayberg, H. S. 2009. Predictors of nonresponse to cognitive behavioural therapy or venlafaxine using glucose metabolism in major depressive disorder. *Journal of Psychiatry and Neuroscience* 34(3): 175–180.

Kovacech, B., Zilka, N., and Novak, M. 2009. New age of neuroproteomics in Alzheimer's disease research. *Cellular and Molecular Neurobiology* 29(6–7): 799–805.

Kruggel, F., Turner, J., and Muftuler, L. T. 2010. Impact of scanner hardware and imaging protocol on image quality and compartment volume precision in the ADNI cohort. *Neuroimage* 49(3): 2123–2133.

Krystal, J. H., Karper, L. P., Seibyl, J. P., Freeman, G. K., Delaney, R., Bremner, J. D., Heninger, G. R., Bowers Jr, M. B., and Charney, D. S. 1994. Subanesthetic effects of the noncompetitive NMDA antagonist, ketamine, in humans: Psychotomimetic, perceptual, cognitive, and neuroendocrine responses. *Archives of General Psychiatry* 51(3): 199–214.

Kumari, V., Antonova, E., Geyer, M. A., Ffytche, D., Williams, S. C. R., and Sharma, T. 2007. A fMRI investigation of startle gating deficits in schizophrenia patients treated with typical or atypical antipsychotics. *International Journal of Neuropsychopharmacology* 10(4): 463–477.

Kumari, V., Peters, E. R., Fannon, D., Antonova, E., Premkumar, P., Anilkumar, A. P., Williams, S. C. R., and Kuipers, E. 2009. Dorsolateral Prefrontal Cortex Activity Predicts Responsiveness to Cognitive-Behavioral Therapy in Schizophrenia. *Biological Psychiatry* 66(6): 594–602.

Lahti, A. C., Holcomb, H. H., Weiler, M. A., Medoff, D. R., Frey, K. N., Hardin, M., and Tamminga, C. A. 2004. Clozapine but not Haloperidol Re-establishes Normal Task-Activated rCBF Patterns in Schizophrenia within the Anterior Cingulate Cortex. *Neuropsychopharmacology* 29(1): 171–178.

Lahti, A. C., Koffel, B., LaPorte, D., and Tamminga, C. A. 1995. Subanesthetic doses of ketamine stimulate psychosis in schizophrenia. *Neuropsychopharmacology* 13(1): 9–19.

Laks, J., and Engelhardt, E. 2008. Reports in pharmacological treatments in geriatric psychiatry: Is there anything new or just adding to old evidence? *Current Opinion in Psychiatry* 21(6): 562–567.

Lawrence, N. S., Williams, A. M., Surguladze, S., Giampietro, V., Brammer, M. J., Andrew, C., Frangou, S., Ecker, C., and Phillips, M. L. 2004. Subcortical and ventral prefrontal cortical neural responses to facial expressions distinguish patients with bipolar disorder and major depression. *Biological Psychiatry* 55(6): 578–587.

Lawrie, S. M., Ingle, G. T., Santosh, C. G., Rogers, A. C., Rimmington, J. E., Naidu, K. P., Best, J. J., O'Carroll, R. e., Goodwin, G. M., and Ebmeier, K. P. 1995. Magnetic resonance imaging and single photon emission tomography in treatment-responsive and treatment-resistant schizophrenia. *British Journal of Psychiatry* 167(2): 202–210.

Leow, A. D., Yanovsky, I., Parikshak, N., Hua, X., Lee, S., Toga, A. W., Jack Jr, C. R., Bernstein, M. A., Britson, P. J., Gunter, J. L., Ward, C. P., Borowski, B., Shaw, L. M., Trojanowski, J. Q., Fleisher, A. S., Harvey, D., Kornak, J., Schuff, N., Alexander, G. E., Weiner, M. W., and Thompson, P. M. 2009. Alzheimer's Disease Neuroimaging Initiative: A one-year follow up study using tensor-based morphometry correlating degenerative rates, biomarkers and cognition. *Neuroimage* 45(3): 645–655.

Leslie, R. A., and James, M. F. 2000. Pharmacological magnetic resonance imaging: A new application for functional MRI. *Trends in Pharmacological Sciences* 21(8): 314–318.

Lin, A., Ross, B. D., Harris, K., and Wong, W. 2005. Efficacy of proton magnetic resonance spectroscopy in neurological diagnosis and neurotherapeutic decision making. *NeuroRx* 2(2): 197–214.

Lucignani, G. 2007. Imaging biomarkers: From research to patient care - A shift in view. *European Journal of Nuclear Medicine and Molecular Imaging* 34(10): 1693–1697.

Lyoo, I. K., and Renshaw, P. F. 2002. Magnetic resonance spectroscopy: Current and future applications in psychiatric research. *Biological Psychiatry* 51(3): 195–207.

MacLeod, C. M. 1991. Half a century of research on the Stroop effect: An integrative review. *Psychological Bulletin* 109: 162–203.

Marquand, A. F., Mourão-Miranda, J., Brammer, M. J., Cleare, A. J., and Fu, C. H. Y. 2008. Neuroanatomy of verbal working memory as a diagnostic biomarker for depression. *NeuroReport* 19(15): 1507–1511.

Mayeux, R. 2004. Biomarkers: Potential Uses and Limitations. *NeuroRx* 1(2): 182–188.

McClure, E. B., Adler, A., Monk, C. S., Cameron, J., Smith, S., Nelson, E. E., Leibenluft, E., Ernst, M., and Pine, D. S. 2007. fMRI predictors of treatment outcome in pediatric anxiety disorders. *Psychopharmacology* 191(1): 97–105.

Meisenzahl, E. M., Scheuerecker, J., Zipse, M., Ufer, S., Wiesmann, M., Frodl, T., Koutsouleris, N., Zetzsche, T., Schmitt, G., Riedel, M., Spellmann, I., Dehning, S., Linn, J., Bruckmann, H., and Moller, H. J. 2006. Effects of treatment with the atypical neuroleptic quetiapine on working memory function: a functional MRI follow-up investigation. *European Archives of Psychiatry and Clinical Neuroscience* 256(8): 522–531.

Merikangas, K. R., Akiskal, H. S., Angst, J., Greenberg, P. E., Hirschfeld, R. M. A., Petukhova, M., and Kessler, R. C. 2007. Lifetime and 12-month prevalence of bipolar spectrum disorder in the national comorbidity survey replication. *Archives of General Psychiatry* 64(5): 543–552.

Metastasio, A., Rinaldi, P., Tarducci, R., Mariani, E., Feliziani, F. T., Cherubini, A., Pelliccioli, G. P., Gobbi, G., Senin, U., and Mecocci, P. 2006. Conversion of MCI to dementia: Role of proton magnetic resonance spectroscopy. *Neurobiology of Aging* 27(7): 926–932.

Miller, D. D., Andreasen, N. C., O'Leary, D. S., Watkins, G. L., Boles Ponto, L. L., and Hichwa, R. D. 2001. Comparison of the effects of risperidone and haloperidol on regional cerebral blood flow in schizophrenia. *Biological Psychiatry* 49(8): 704–715.

Mitterschiffthaler, M. T., Ettinger, U., Mehta, M. A., Mataix-Cols, D., and Williams, S. C. R. 2006. Applications of functional magnetic resonance imaging in psychiatry. *Journal of Magnetic Resonance Imaging* 23(6): 851–861.

Molina, V., Reig, S., Sarramea, F., Sanz, J., Artaloytia, J. F., Luque, R., Aragüés, M., Pascau, J., Benito, C., Palomo, T., and Desco, M. 2003. Anatomical and functional brain variables associated with clozapine response in treatment-resistant schizophrenia. *Psychiatry Research - Neuroimaging* 124(3): 153–161.

Moore, G. J., Bebchuk, J. M., Hasanat, K., Chen, G., Seraji-Bozorgzad, N., Wilds, I. B., Faulk, M. W., Koch, S., Glitz, D. A., Jolkovsky, L., and Manji, H. K. 2000. Lithium increases N-acetyl-aspartate in the human brain: In vivo evidence in support of bcl-2's neurotrophic effects? *Biological Psychiatry* 48(1): 1–8.

Mueller, S. G., Schuff, N., and Weiner, M. W. 2006. Evaluation of treatment effects in Alzheimer's and other neurodegenerative diseases by MRI and MRS. *NMR in Biomedicine* 19(6): 655–668.

Mueller, S. G., Weiner, M. W., Thal, L. J., Petersen, R. C., Jack, C. R., Jagust, W., Trojanowski, J. Q., Toga, A. W., and Beckett, L. 2005. Ways toward an early diagnosis in Alzheimer's disease: The Alzheimer's Disease Neuroimaging Initiative (ADNI). *Alzheimer's and Dementia* 1(1): 55–66.

Nordberg, A. 2004. PET imaging of amyloid in Alzheimer's disease. *Lancet Neurology* 3(9): 519–527.

Ogawa, M., Tsukada, H., Hatano, K., Ouchi, Y., Saji, H., and Magata, Y. 2009. Central in vivo nicotinic acetylcholine receptor imaging agents for positron emission tomography (PET) and single photon emission computed tomography (SPECT). *Biological and Pharmaceutical Bulletin* 32(3): 337–340.

Paterson, D., and Nordberg, A. 2000. Neuronal nicotinic receptors in the human brain. *Progress in Neurobiology* 61(1): 75–111.

Patil, S. T., Zhang, L., Martenyi, F., Lowe, S. L., Jackson, K. A., Andreev, B. V., Avedisova, A. S., Bardenstein, L. M., Gurovich, I. Y., Morozova, M. A., Mosolov, S. N., Neznanov, N. G., Reznik, A. M., Smulevich, A. B., Tochilov, V. A., Johnson, B. G., Monn, J. A., and Schoepp, D. D. 2007. Activation of mGlu2/3 receptors as a new approach to treat schizophrenia: a randomized Phase 2 clinical trial. *Nature Medicine* 13(9): 1102–1107.

Perret, E. 1974. Left frontal lobe of man and suppression of habitual responses in verbal categorical behavior. *Neuropsychologia* 12(3): 323–330.

Petersen, R. C., and Knopman, D. S. 2006. MCI is a clinically useful concept. *International Psychogeriatrics* 18(03): 394–402. doi:10.1017/S1041610206223925

Pilatus, U., Lais, C., Rochmont, A. d. M. d., Kratzsch, T., Frölich, L., Maurer, K., Zanella, F. E., Lanfermann, H., and Pantel, J. 2009. Conversion to dementia in mild cognitive impairment is associated with decline of N-actylaspartate and creatine as revealed by magnetic resonance spectroscopy. *Psychiatry Research - Neuroimaging* 173(1): 1–7.

Pimlott, S. L., and Ebmeier, K. P. 2007. SPECT imaging in dementia. *British Journal of Radiology* 80(SPEC. ISS. 2): S153–159.

Premkumar, P., Fannon, D., Kuipers, E., Peters, E. R., Anilkumar, A. P. P., Simmons, A., and Kumari, V. 2009. Structural magnetic resonance imaging predictors of responsiveness to cognitive behaviour therapy in psychosis. *Schizophrenia Research* 115(2–3): 146–155.

Rademakers, R., and Rovelet-Lecrux, A. 2009. Recent insights into the molecular genetics of dementia. *Trends in Neurosciences* 32(8): 451–461.

Ridha, B. H., Anderson, V. M., Barnes, J., Boyes, R. G., Price, S. L., Rossor, M. N., Whitwell, J. L., Jenkins, L., Black, R. S., Grundman, M., and Fox, N. C. 2008. Volumetric MRI and cognitive measures in Alzheimer disease: Comparison of markers of progression. *Journal of Neurology* 255(4): 567–574.

Risacher, S. L., Saykin, A. J., West, J. D., Shen, L., Firpi, H. A., and McDonald, B. C. 2009. Baseline MRI predictors of conversion from MCI to probable AD in the ADNI cohort. *Current Alzheimer Research* 6(4): 347–361.

Robinson, D. G., Woerner, M. G., Alvir, J. M., Geisler, S., Koreen, A., Sheitman, B., Chakos, M., Mayerhoff, D., Bilder, R., Goldman, R., and Lieberman, J. A. 1999. Predictors of treatment response from a first episode of schizophrenia or schizoaffective disorder. *The American Journal of Psychiatry* 156(4): 544–549.

Rusinek, H., De Santi, S., Frid, D., Tsui, W. H., Tarshish, C. Y., Convit, A., and De Leon, M. J. 2003. Regional Brain Atrophy Rate Predicts Future Cognitive Decline: 6-Year Longitudinal MR Imaging Study of Normal Aging. *Radiology* 229(3): 691–696.

Scherk, H., and Falkai, P. 2006. Effects of antipsychotics on brain structure. *Current Opinion in Psychiatry* 19(2): 145–150.

Schott, J. M., Crutch, S. J., Frost, C., Warrington, E. K., Rossor, M. N., and Fox, N. C. 2008. Neuropsychological correlates of whole brain atrophy in Alzheimer's disease. *Neuropsychologia* 46(6): 1732–1737.

Schuff, N., Woerner, N., Boreta, L., Kornfield, T., Shaw, L. M., Trojanowski, J. Q., Thompson, P. M., Jack, C. R., and Weiner, M. W. 2009. MRI of hippocampal volume loss in early Alzheimers disease in relation to ApoE genotype and biomarkers. *Brain* 132(4): 1067–1077.

Shaw, L. M., Korecka, M., Clark, C. M., Lee, V. M. Y., and Trojanowski, J. Q. 2007. Biomarkers of neurodegeneration for diagnosis and monitoring therapeutics. *Nature Reviews Drug Discovery* 6(4): 295–303.

Shaw, L. M., Vanderstichele, H., Knapik-Czajka, M., Clark, C. M., Aisen, P. S., Petersen, R. C., Blennow, K., Soares, H., Simon, A., Lewczuk, P., Dean, R., Siemers, E., Potter, W., Lee, V. M. Y., and Trojanowski, J. Q. 2009. Cerebrospinal fluid biomarker signature in alzheimer's disease neuroimaging initiative subjects. *Annals of Neurology* 65(4): 403–413.

Silverstone, P. H., Wu, R. H., O'Donnell, T., Ulrich, M., Asghar, S. J., and Hanstock, C. C. 2003. Chronic treatment with lithium, but not sodium valproate, increases cortical N-acetyl-aspartate concentrations in euthymic bipolar patients. *International Clinical Psychopharmacology* 18(2): 73–79.

Simonsen, A. H., McGuire, J., Hansson, O., Zetterberg, H., Podust, V. N., Davies, H. A., Waldemar, G., Minthon, L., and Blennow, K. 2007. Novel panel of cerebrospinal fluid biomarkers for the prediction of progression to Alzheimer dementia in patients with mild cognitive impairment. *Archives of Neurology* 64(3): 366–370.

Simpson, S. W., Jackson, A., Baldwin, R. C., and Burns, A. 1997. 1997 IPA/Bayer Research Awards in Psychogeriatrics. Subcortical hyperintensities in late-life depression: acute response to treatment and neuropsychological impairment. *International Psychogeriatrics* 9(3): 257–275.

Singh, I., and Rose, N. 2009. Biomarkers in psychiatry. *Nature* 460(7252): 202–207.

Smoller, J. W., and Finn, C. T. 2003. Family, Twin, and Adoption Studies of Bipolar Disorder. *American Journal of Medical Genetics - Seminars in Medical Genetics* 123 C(1): 48–58.

Stephan, K. E., Magnotta, V. A., White, T., Arndt, S., Flaum, M., O'Leary, D. S., and Andreasen, N. C. 2001. Effects of olanzapine on cerebellar functional connectivity in schizophrenia measured by fMRI during a simple motor task. *Psychological Medicine* 31(6): 1065–1078.

Stork, C., and Renshaw, P. F. 2005. Mitochondrial dysfunction in bipolar disorder: evidence from magnetic resonance spectroscopy research. *Molecular Psychiatry* 10(10): 900–919.

Stroop, J. R. 1935. Studies of interference in serial verbal reactions. *Journal of Experimental Psychology* 18: 643–662.

Strozyk, D., Blennow, K., White, L. R., and Launer, L. J. 2003. CSF Aβ 42 levels correlate with amyloid-neuropathology in a population-based autopsy study. *Neurology* 60(4): 652–656.

Taragano, F. E., Bagnatti, P., and Allegri, R. F. 2005. A double-blind, randomized clinical trial to assess the augmentation with nimodipine of antidepressant therapy in the treatment of "vascular depression". *International Psychogeriatrics* 17(3): 487–498.

Taylor, W. D., Hsu, E., Krishnan, K. R. R., and MacFall, J. R. 2004. Diffusion tensor imaging: Background, potential, and utility in psychiatric research. *Biological Psychiatry* 55(3): 201–207.

Taylor, W. D., Steffens, D. C., MacFall, J. R., McQuoid, D. R., Payne, M. E., Provenzale, J. M., and Krishnan, K. R. R. 2003. White matter hyperintensity progression and late-life depression outcomes. *Archives of General Psychiatry* 60(11): 1090–1096.

Thaker, G. K. 2007. Schizophrenia endophenotypes as treatment targets. *Expert Opinion on Therapeutic Targets* 11(9): 1189–1206.

Tost, H., Hakimi, S., and Meyer-Lindenberg, A. (2009). Magnetic resonance imaging biomarkers in schizophrenia research. In M. Ritsner (Ed.), *The handbook of neuropsychiatric biomarkers, endophenotypes, and genes* (Vol. II, pp. 123–144). New York: Springer.

Tsuchiya, K. J., Byrne, M., and Mortensen, P. B. 2003. Risk factors in relation to an emergence of bipolar disorder: A systematic review. *Bipolar Disorders* 5(4): 231–242.

Verbeek, M. M., De Jong, D., and Kremer, H. P. H. 2003. Brain-specific proteins in cerebrospinal fluid for the diagnosis of neurodegenerative diseases. *Annals of Clinical Biochemistry* 40(1): 25–40.

Videbech, P., and Ravnkilde, B. 2004. Hippocampal volume and depression: A meta-analysis of MRI studies. *American Journal of Psychiatry* 161(11): 1957–1966.

Wang, S. J., and Yang, T. T. 2005. Role of central glutamatergic neurotransmission in the pathogenesis of psychiatric and behavioral disorders. *Drug News and Perspectives* 18(9): 561–566.

White, T., Nelson, M., and Lim, K. O. 2008. Diffusion tensor imaging in psychiatric disorders. *Topics in Magnetic Resonance Imaging* 19(2): 97–106.

Wilkinson, D., and Halligan, P. 2004. The relevance of behavioural measures for functional-imaging studies of cognition. *Nature Reviews Neuroscience* 5(1): 67–73.

Winblad, B., Frisoni, G. B., Frolich, L., Johannsen, P., Johansson, G., Kehoe, P., Lovestone, S., Olde-Rikkert, M., Reynish, E., Vasser, P. J., and Vellas, B. 2008. EADC (European Alzheimer disease consortium) recommendations for future Alzheimer disease research in Europe. *Journal of Nutrition, Health and Aging* 12(10): 683–684.

Yildiz-Yesiloglu, A., and Ankerst, D. P. 2006. Review of 1H magnetic resonance spectroscopy findings in major depressive disorder: A meta-analysis. *Psychiatry Research - Neuroimaging* 147(1): 1–25.

Yildiz, A., Sachs, G. S., Dorer, D. J., and Renshaw, P. F. 2001. 31P nuclear magnetic resonance spectroscopy findings in bipolar illness: A meta-analysis. *Psychiatry Research - Neuroimaging* 106(3): 181–191.

Zellner, M., Veitinger, M., and Umlauf, E. 2009. The role of proteomics in dementia and Alzheimer's disease. *Acta Neuropathologica* 118(1): 181–195.

9
Autonomic Nervous System Markers for Psychophysiological, Anxiety, and Physical Disorders

Richard Gevirtz, PhD, BCIAC

The Autonomic Nervous System

Focus on the central nervous system in recent decades has often led to reduced interest in and research on the peripheral nervous systems, specifically the autonomic nervous system (ANS). But recent work on ANS physiology and psychophysiology has reemphasized the importance of ANS processes in understanding not only cardiac pathology (Armour 2008), but also human emotions, behavior and even cognition.

J. N. Langley (1921) and later, Walter Cannon (1939), brilliantly described the autonomic nervous system and pioneered the view that persists to this day that the sympathetic branch of the system (SNS), which they described as a "fight-or-flight" system, dominates and shapes our concept of stress. The parasympathetic branch (PNS) was relegated to the basic restoration tasks once sympathetic activity subsided. The more modern view, however, has challenged this conceptualization in several ways. First, the sympathetic nervous system (SNS) has been found to be a system that is "fractionated;" that is, it does not act as an orchestra or in "sympathy," but has a variety of specific autonomous functions. As Morrison (2001) stated:

> With advances in experimental techniques, the early views of the sympathetic nervous system as a monolithic effector activated globally in situations requiring a rapid and aggressive response to life threatening danger have been eclipsed by an organizational model featuring an extensive array of functionally specific output channels that can be simultaneously activated or inhibited in combinations that result in the patterns of autonomic activity supporting behavior and mediating homeostatic reflexes. With this perspective, the defense response is but one of many activational states of the central autonomic network. (p. 683)

He further elaborated that:

> compared with our ancestors, modern lifestyles have all but eliminated the danger of predatory aggression, and we rarely engage the central autonomic networks developed by evolutionary pressure to cope with the most stressful challenges to homeostasis. (p.683)

A second source of challenge has come from modern investigations into the PNS. Stephen Porges (1995a; 1995b; 2003) has been especially influential in shifting attention away from SNS-based concepts of stress, and toward a recognition of the importance of the PNS in day-to-day regulation. He noted that a branch of the vagus nerve emanating from the nucleus ambiguous modulates heart rate in an oscillatory manner, accounting for most of the variance in heart rate variability (HRV) (Porges 1995a,b). This branch of the vagal efferent system, which evolved most recently in higher mammals, seems to mediate social engagement in social species (Porges 2003). Thus, in modern daily life, the PNS is responsible for most of the regulation in the ANS and probably plays an important role in both producing and preventing stress-related disorders.

Recently, vagal afferents have been studied, and they seem to play an important role in homeostatic regulation. That the afferent pathways have more influence than the efferent pathways has stimulated research into the "little brain" in the heart (Armour 2007). Some authors speculate that diaphragmatic vagal afferent pathways may affect both limbic circuitry and pre-frontal executive functions (Brownet al. 1993; Brown and Gerbarg 2005). Thus, interactions between brain centers and peripheral processes are not easily separated. Integration of these systems offers hope of understanding the unique configurations that make up a large number of presenting symptoms with no obvious pathology, often labeled (misleadingly) "unexplained medical disorders" (Speckens et al. 1996), including somatoform disorders.

Heart Rate Variability: A Window into the ANS

The advent of inexpensive, high-speed signal processing equipment has led to an explosion of interest in the study of heart rate variability (HRV). Variations in inter-beat interval (IBI) over time have been shown to reflect autonomic influences on the heart and, by extension, other target organs in the viscera. The primary oscillator in this system is vagal braking triggered by exhalation. Reflexes in the medulla inhibit vagal breaking during inhalation and reengage during the expiratory phase of the breathing cycle. Thus, IBI shows a periodicity that reflects the respiratory cycle (0.15–0.5 Hz or 12–30 cycles per minute). The resulting IBI rhythm is called "respiratory sinus arrhythmia" (RSA) and has been shown to be mediated by parasympathetic pathways with the neurotransmitter acetylcholine (Berntson et al. 1997). Additionally, two other oscillators have an effect on total HRV. Blood pressure rhythms (at about 0.1Hz or 6 cycles per minute) are reflected in changes in aortic and carotid baroreceptor activity that mediate HR speeding or slowing through the vagus (for slowing) and the sympathetic nerve (for speeding). This baroreflex system is the frontline, short-term, blood pressure regulator. The resulting changes in IBI are another source of variation in HRV. Finally, a very slow (0.01Hz or 1–2 cycles per minute) oscillation appears to reflect vascular rhythms, again translated into IBI oscillations by the baroreceptor feedback loop (Seidel et al. 1997).

Taken as a whole, these influences create a pattern of variability that has been described as chaotic, but represents the sum of the various oscillatory influences. The standard deviation of a series of normal R wave IBIs (called SDNN) reflects the total variability in a time domain. Variability of about 10 percent (that is, an SDNN of 100 ms for an average IBI of 1000 ms or 60 beats per minute) is considered reflective of cardiovascular health, while values less than 50 ms are considered reflective of sub-optimal function. Other measures are commonly used in a similar way (root mean square of successive differences [RMSSD], or percent of sequential IBIs greater that 50 ms [pNN50]). These are considered "time domain" measures, in that they only give us a general index of how all oscillators are contributing to maintaining homeostasis in the organism.

Analyses that deconstruct the combined complex waveforms are used to specify parasympathetic/sympathetic influences. Thus, the respiratory frequencies (known as high frequency, or HF) reflect only parasympathetic influences on the sino-atrial (SA) node. Under stationary conditions, HF amplitudes reflected by area under the curve of a spectral analysis (FFT, DFT, or autoregressive spectral analysis) are thought to reflect the extent of vagal traffic between the brainstem and the SA node. This is sometimes labeled *vagal tone*, although caution is urged in making this assertion (Grossman et al. 1990; Grossman and Kollai 1993; Wilhelm et al. 2005). The oscillations emanating from the blood pressure rhythms are called low frequency (LF) and show up on a spectral analysis at 0.08–0.12 Hz. This wave is thought to reflect both sympathetic and parasympathetic influences, and is sometimes used the ratio LF/HF to reflect sympathovagal tone (Malliani et al. 1994a; Malliani et al. 1994b; Bernardi et al. 2001; Yeragani et al. 2004); however, some authors (Eckberg 1997) have been critical of this formulation. Despite its controversial nature, this ratio has often been used as a marker of autonomic balance (Martinez-Lavin et al. 1998; Cohen et al. 2000a; Cohen et al. 2000b; Sevarese-Levine 2007). The slowest rhythm, called very low frequency (VLF), has not been extensively studied, but is thought to reflect slow vascular rhythms that may be sympathetically mediated (Task 1996).

Most of the measurement standards have been formulated using 24-hour Holter or ambulatory monitoring. It should be noted, however, that shorter recording periods can reflect risk factors, and probably autonomic processes, accurately (Bigger et al. 1993). However, to do this, one must insure that the recording does not reflect changes in respirator, rate, tidal volume, or movement (Grossman and Kollai 1993; Wilhelm et al. 2004). A particularly egregious error is to interpret the LF/HF oscillatory bands of naturally slow breathers (or subjects breathing slowly during the study) as having very high sympathovagal tone. When one breathes down into the 0.08–0.14 Hz frequencies, HF waves disappear and the data look like that of a person with dominant LF activity, usually a biomarker with poor outcomes. Thus, a popular commercial program used for medical assessments based on the three bands described above posted a red alert for an experienced Indian yogi who happened to breathe at 8 breaths per minute (0.13 Hz), despite his remarkable good health status.

Despite these problems, HRV measurement offers a unique window into ANS functioning. Increasingly, the effects of the ANS are being considered as important players in health and optimum performance.

We recently analyzed a large HRV dataset provided by the Brain Resource International Database (BRID; www.BrainResource.com) which illustrates the potential use of HRV as a biomarker in a variety of settings.

Age Predicting Heart Rate and HRV

Age was found to be a significant predictor of HR (bpm) for the eyes-closed resting condition. This relationship explained similar amounts of variance for both females ($F = 292.7$, $p<.0001$, adj $R^2 = 0.24$; see Figure 9-1) and males ($F = 243.9$, $p <.0001$, adj $R^2 = 0.20$).

These findings show that age predicts an exponential decrease in heart rate in both females and males. This means that the rate of decrease is faster at a younger age. For children under 10, HRV expressed in peak-to-valley differences decreases between 1.5 and 1 beats per year. For adults between 30 and 40 years of age, HRV decreases between 0.28 and 0.21 beats per year.

Table 9-1 Summary of the Heart Rate Variability Measures Commonly Used

Method	*Name*	*Units*	*Description*
Time-Domain	SDNN	ms	Mean and SD of normal RR (NN) intervals
	SDANN	ms	Standard deviation of the averages of NN intervals in all 5-minute segments of the entire recording
	SDNN Index	ms	Mean of the standard deviation in all 5-minute segments of an entire recording
	RMSSD	ms	The square root of the mean of the sum of the squares of differences between adjacent NN intervals
	NN50	Count	Number of pairs of adjacent NN intervals differing by more than 50 ms in the entire recording
	pNN50	%	NN50 count divided by total number of NN intervals
	Total power	ms^2	The variance of NN intervals over the temporal segment (< 0.4 Hz).
	ULF	ms^2	Power in ultra-low-frequency range (< 0.003 Hz)
Frequency-Domain	VLF	ms^2	Power in ultra-low-frequency range (range >0.003 to <0.04Hz)
	LF	ms^2	Power in ultra-low-frequency range (range >0.04 to <0.15Hz)
	HF	ms^2	Power in ultra-low-frequency range (range >0.15 to <0.4Hz)
	LF/HF ratio		Ratio of LF to HF

Abbreviations:

SDNN: standard deviation of a series of normal R wave inter-beat intervals

RR:

NN: sequential IBIs

SDANN:

RMSSD,:root mean square of successive differences

NN50: sequential IBIs over 50 sec.

pNN50: percent of sequential IBIs over 50 sec.

ULF: ultra low frequency

VLF: very low frequency

LF: low frequency

HF: high frequency

Age was found to be a similarly significant predictor of HRV (using the RMSSD measure) for the eyes-closed resting condition. This relationship also explained similar amounts of variance for both females (F F = 333.7, $p < .0001$, Adj. $R^2 = 0.26$; see Figure 9-2) and males ($F = 364.5$, $p < .0001$, Adj. $R^2 = 0.27$). These findings indicate a decrease of 1.5 to 2 percent per year.

HRV and Self-Regulation Correlates

Self-regulation markers for negativity-positivity bias have been shown previously to correlate with HR (Gordon 2000) and to be related to self-report measures of depression, anxiety, and stress.

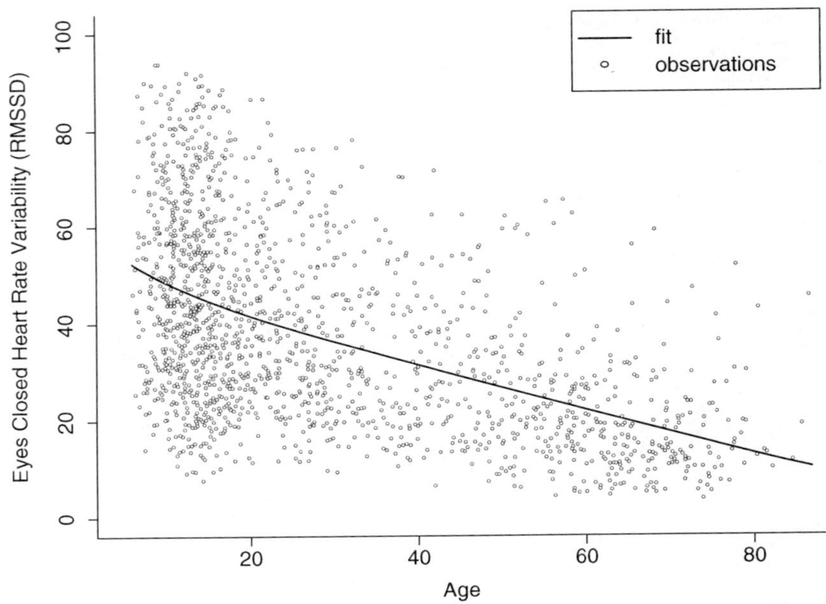

Figure 9-1 Heart rate variability (RMSSD) versus age (from 6 to 88 years). The fit for heart rate variability uses the natural log model ($F = 319.3$, R-Squared = 0.288, $p<0.0001$). Heart rate variability declines steadily with increasing age.

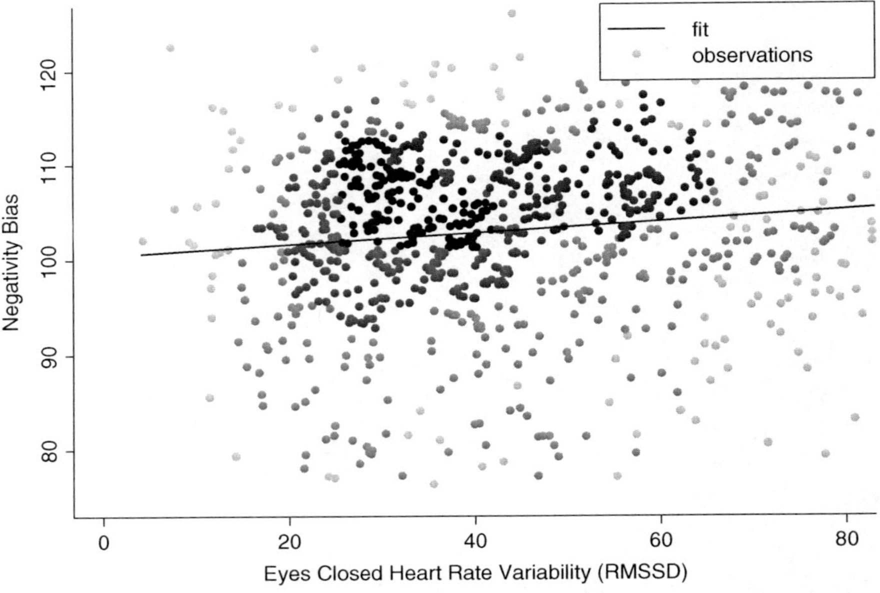

Figure 9-2 Heart rate variability (RMSSD) versus negativity bias. This analysis was limited to subjects between 6 and 27 years of age. The regression fit for heart rate variability uses the natural log model ($F = 14.07$, R-Squared = 0.016, $p<0.0001$). Negativity bias (lower scores are more negative) decreases with increasing heart rate variability.

HRV (RMSSD) was predictive of negativity bias only when age groups were considered in accord with the age-related distribution of HRV (<30 years, 30–55 years, and >50 years).

For the younger group (<30 years), HRV (RMSSD) showed a small predictive relationship with negativity bias, which followed a linear model, as illustrated in Figure 9-2 above ($F = 4.53$, $p = .039$, Adj. $R^2 = 0.005$). Correspondingly, HRV showed a small predictive relationship with negativity bias in those 60 and older, which followed a linear model ($F = 6.035$, $p = 0.015$, Adj. $R^2 = 0.033$).

HRV and EEG Brain Function

Regression analysis showed strongly predictive associations between HRV and EEG measures of relative power, including Alpha power, as shown in Figure 9-3 ($F = 50.43$, $p < .0001$, Adj. $R^2 = 0.027$).

HRV and Amygdala Volume

HRV predicted gray matter for the lateral prefrontal cortex (LFPC) volume bilaterally ($F = 12.57$, R-Squared $= 0.122$, $p < 0.0001$). Increasing HRV was associated with greater LPFC volume. HRV predicted gray matter for the amygdala bilaterally ($F = 4.2$, R-Squared $= 0.044$, $p < 0.05$). Increasing HRV was associated with greater amygdala volume (see Figure 9-4).

Resonant Frequency of HRV as a Marker

One characteristic of the ANS regulation of the cardiovascular system is *resonant frequency*. For some years, Paul Lehrer and I, with our colleagues and students, have been

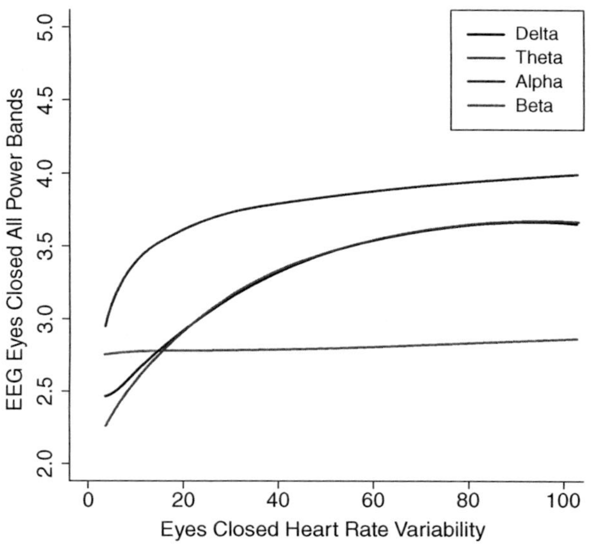

Figure 9-3 Regression analysis showed strongly predictive associations between HRV and EEG measures of relative power, including Alpha power ($F = 50.43$, $p < .0001$, Adj. $R2 = 0.027$).

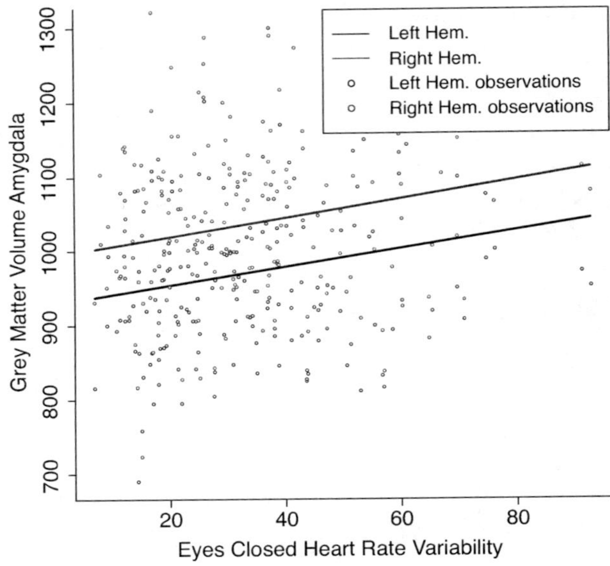

Figure 9-4 Heart rate variability (RMSSD) versus MRI amygdala volume for the left and right hemispheres. The regression fit uses the natural log model ($F = 4.2$, R-Squared $= 0.044$, $p<0.05$). Increasing heart rate variability is associated with greater amygdala volume in both hemispheres.

pioneering HRV biofeedback (Gevirtz 2000; Gevirtz and Lehrer 2003; Lehrer et al. 2003; Lehrer et al. 2006; Lehrer 2007). When subjects manipulate their breathing rate so as to breathe at rates of between five and seven breaths per minute (0.08–0.10 Hz), the RSA pattern (the corresponding heart rate to inspiration and expiration) greatly increases in amplitude. As Vashillo et al. (2002; 2006) have shown, each person has a specific frequency at which these peak-valley amplitudes are maximized. There appears to be resonance created by the juxtiposition of the blood pressure/baroreceptor oscillations and the respiratory oscillation at this specific frequency. This resonant frequency (RF) appears to be stable over time and dependent on the physical characteritics of the individual (Vaschillo et al. 2002; Vaschillo et al. 2006). In figure 9-5, the transfer function data from Vashillo et al. are shown for a representative five subjects.

When subjects are trained to breathe at this designated pace for ten to twenty minutes daily, improved homeostasis of the ANS, specifically the baroreflex, appears to be the result (Gevirtz 2000; Lehrer et al. 2000; Lehrer et al. 2003; Lehrer et al. 2004; Lehrer 2007).

Data now exist for benefits in: asthma (Lehrer et al. 2004), COPD (Giardino et al. 2004), fibromyalgia (FM) (Hassett et al. 2007), depression (Karavidas et al. 2007), hypertension (Schein et al. 2001; Reinke et al. 2007), congestive heart failure (CHF) (Swanson et al. 2006), irritable bowel syndrome (IBS) (Humphreys and Gevirtz 2000; Sowder et al. 2005), PTSD (Zucker and Gevirtz 2009), and other conditions.

HRV as a Biomarker for Various Disorders

In recent years, HRV measurements have been used as an assessment tool for a wide variety of disorders, usually with remarkably good sensitivity and specificity. Table 9-2 gives a sample of very recent findings.

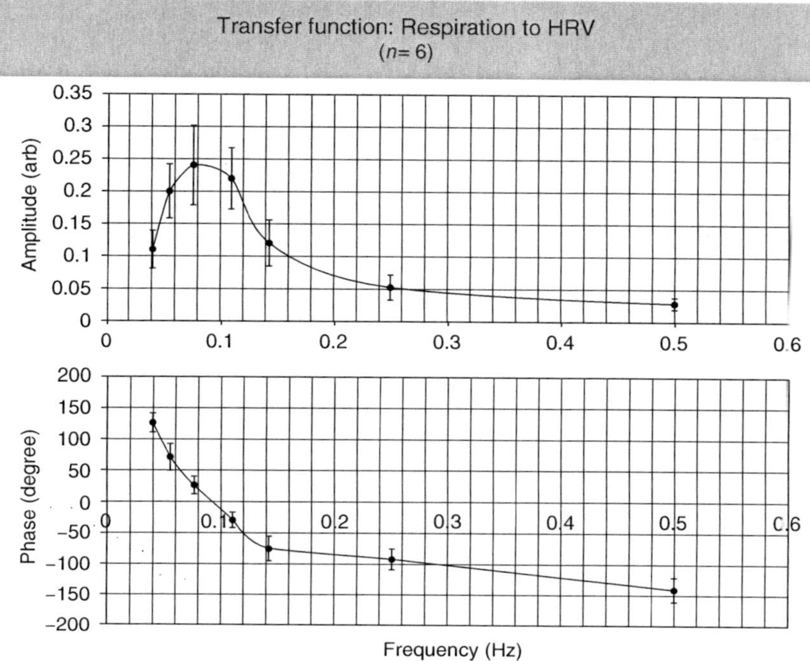

Figure 9-5 Transfer function analyses for 6 Ss. Breath pace was manipulated while HR was recorded and analyzed for RSA amplitude (Vaschillo et al. 2002; Vaschillo et al. 2006).

Personalized Medicine for Medically Unexplained Disorders or Somatization Disorders

Unexplained medical disorders, including somatoform disorders and somatization disorders, are given diagnotic labels that represent a heterogenous group of symptom clusters (Wessely et al. 1996; Barsky and Borus 1999; Hollifield et al. 1999; Tan et al. 2005). Typically included within these categories are clusters that have been labeled by their specialty groups such as IBS, myofascial pain syndrome (MPD; lumbar strain, cervical

Table 9-2 Markers for Diagnosis: Status and Future

Disorder or Symptom	*Sensitvity*	*Specificity*	*Reference*
Ventricular tacycardia	Na (but high r)	na	(Ozdemir et al. 2003)
Aneurysm formation	92	88	(Yildirim et al. 2007)
Atrial fibrillation	91.4	92.9	(Park et al. 2009)
Cardiomyopathy	95	83.3	(Valencia et al. 2009)
Trauma triage	80	75	(King et al. 2009)
Pregnancy-induced hypertension	na	na	(Pal et al. 2009)
Fall detection	70.4	80	(Nocua et al. 2009)
Sleep disordered breathing	99	96.7	(Al-Abed et al. 2009)
Children's externalizing and internalizing symptoms over time	na, but highly sig. predictor	na	(Hinnant and El-Sheikh 2009)

strain, etc.), Fibromyalgia (FM), chronic fatigue syndrome(CFS), multiple chemical sensitivity (MCS), and many more. These disorders are prevalent (Smith et al. 1986; Gureje et al. 1997; Gureje and Simon 1999), costly (deGruy et al. 1987a; deGruy et al. 1987b), and create a burden on patient and physician alike. While some authors have stressed the commonality among these disorders (Wessely et al. 1999), there may be a substantial advantage to treating them as somewhat distinct and applying autonomic or respiratory biomarkers to define treatment protocols. Our group has developed "mediational" models for many of these disorders to reduce stigma, personalize treatment, and increase efficacy (Gevirtz 2007). As shown in figure 9-6, a mediational model is developed for each disorder based on the best available psychophysiological evidence.

The model is constructed to focus first on symptom presentation from a somatic point of view. Physiological systems, such as those described above, are postulated to be mediating (or at least associated with) the specific symptom cluster. From there, emotional or psychological processes that are known or thought to drive the physiology are described. This psychophysiological path takes the place of the old "hysteria" pathway, which was puely descriptive of mysterious physical forces creating symptoms or report of symptoms. As seen, the symptoms themselves, especially when unexplained, often become the primary source of stress or psychological distress. In this model, early developmental factors and (at least for now) genetic factors are moved to the background. Social and cultural factors are always factored into both psychological processes and symptom presentation. We have developed models for several of the above-named clusters. Two examples follow.

Chronic Muscle Pain

Chronic muscle pain represents a substantial percentage of musculoskelatal complaints. Musculoskelatal complaints are second only to cardiovascular disease in terms of economic burden in North America and Europe. "Compared to other diseases MSD (musculoskeletal disorder) consumes a substantial proportion of healthcare resouces" (p. 2466). The particular set of symptoms that seems to elude effective medical treatment consists of local pain areas of muscular origin that have a characteristic referral pattern, are sensitive to local palpation, can be quite severe, and are aggravated by certain physical triggers, such as isometric contraction. Travell and Simons (1983) have labeled these disorders as myofascial pain syndrome (MPS) and posited that "trigger points" in the periphery account for most of

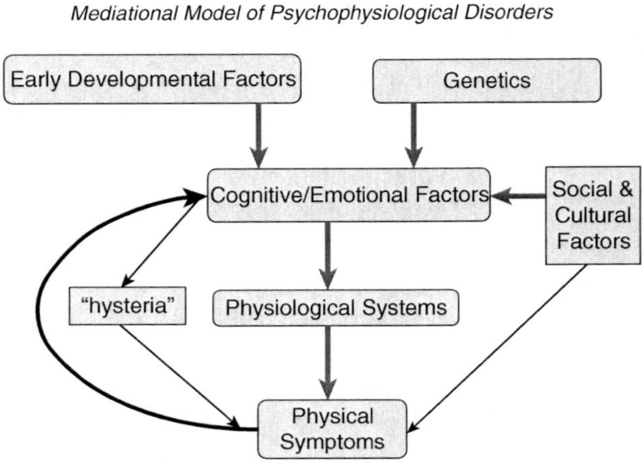

Figure 9-6 The generic mediational model for "unexplained medical disorders."

the variance in the pain and stiffness. Our group (Hubbard and Berkoff 1993; Gevirtz et al. 1996; Hubbard 1996; Hubbard 1998) has traced a psychophysiological pathway from limbic centers to the stellate ganglia to the muscle spindle that is alpha sympathetic in nature (Hubbard and Berkoff 1993; McNulty et al. 1994). This, together with studies looking at internalizing emotional styles, was the impetus for creating the mediational model for MPS. Figure 9-7 shows the variation of the mediational model for MPS. Using this as a heuristic model, we have designed personalized treatment protocols that use the particulars of the individual to create the model and then design a treatment program. For example, a typical patient whom we will call Mary was referred for neck and shoulder pain with "tension" headaches as the frequent result. She presented with a typical pain referral pattern after all diagnostic tests were negative. We were able to identify a series of painful trigger points in her trapezius and suboccipital musculature. A psychophysiological stress profile showed much symapthetic tonic activity, strong reactivity that was accompanied by low vagal tone (low resting levels of RSA). She described herself as a conscientious, perfectionistic, sensitive, unassertive person. Her friends always said she was "too nice for her own good." The model for Mary was thus a good fit for the model in figure 9-7, with her personality particulars substituted in the top boxes. The treatment plan followed from this model and was readily accepted by the patient without any perception of stigma. After five sessions of resonance frequency biofeedback with daily practice, Mary was ready to look at some cognitive interventions such as cognitive-behavioral therapy (CBT) or acceptance and commitment therapy (ACT) (Hayes, Strosahl, and Wilson 1999). Within six weeks she was pain free, and equipped with a series of tools to use in handling stress and pain.

As one example of outcome data based on this approach, we randomly assigned emergency department (ED), uncomplicated whiplash patients to one of two groups: treatment as usual (TAU) or a psychoeducational video which was twelve minutes in length; the patient was followed for six months (Oliveira et al. 2006). The results were quite striking, in that at the six-month follow-up, almost 37 percent of the TAU group was using narcotic medication for neck pain, compared to 3 percent of the video group. It appeared that by changing the explanatory model from one that included the possibility of severe tissue damage to a more psychophysiological model, we prevented chronicity. Since we develop a unique path model for each person, we personalize the treatment in such a way as to lower resistance to change and reduce the stigma often associated with mental health diagnoses.

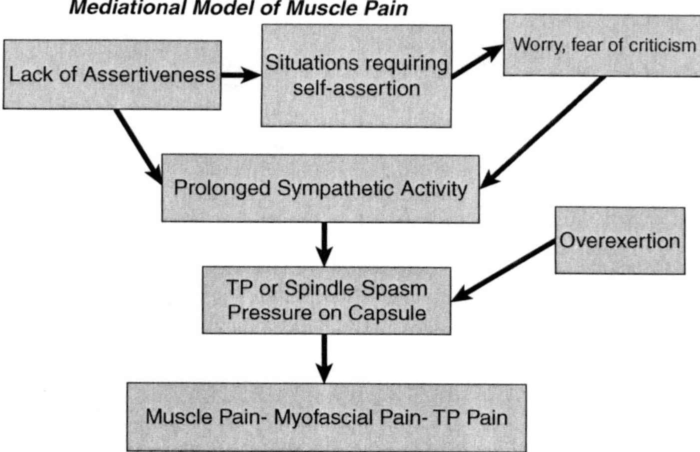

Figure 9-7 The mediational model for chronic muscle pain disorders.

Irritable Bowel Syndrome

Irritable Bowel Syndrome (IBS) is the most common diagnosis in gastroenterology practice, and is very common in primary care as well (Longstreth and Wolde-Tsadik 1993; Gralnek 1998). It is costly and often not effectively treated (Gralnek 1998). This disorder is poorly understood by specialty providers, since it does not fit into the diagnose-and-treat model to which they are accustomed. Over the last decade, several groups have pioneered research that shows the complexity of mind–gut interactions (Mayer et al. 1998; Mayer 1999; Mayer 2000; Mayer et al. 2000; Naliboff et al. 2000; Mayer et al. 2001; Mayer et al. 2001; Whitehead 2006). It is clear that IBS patients have less than optimal homeostatic control in their autonomic nervous system (Iovino et al. 1995; Gupta et al. 2002; Waring et al. 2004; Iovino et al. 2006). Based on these findings, we again have formulated a mediational model for IBS (shown in figure 9-8).

Here, the model differs a bit, because we postulate that too little restorative vagal or parasympathetic activity may be a major contributor to this condition. This follows from a formulation from Porges (1995) that emphasizes the role of the vagal efferent pathway in day-to-day stress responses. The proposed signature of minor stresses is a reduction in RSA that seems to accompany events as subtle as worrying or generalized anxiety. IBS patients have been shown to have higher levels of depression and anxiety than various comparison groups (Drossman et al. 1999) and especially high levels of worry (Hazlett-Stevens et al. 2003; Gros et al. 2009). They also have high levels of "conscientiousness" on the NEO Five-Factor Inventory™ (NEO-FFI) (Farnam et al. 2007; Farnam et al. 2008).

More specifically, IBS patients seem to be especially sensitive to visceral sensations which, in turn, lead to high levels of worry, anxiety, and sometimes depression (Gros et al. 2009). Thus, a signature that combines personality factors and autonomic markers can be seen. When treatment protocols are used that elucidate this model for the patient and supply them with self-regulatory tools that can restore homeostatic function, efficacy is greatly increased and obstacles to treatment are reduced. We (Sowder et al. 2005; Sowder et al. 2006) recently looked at autonomic mechanisms in children with recurrent abdominal pain (a disorder often considered as on the IBS spectrum). Using an ambulatory HR and respiratory monitor, we were able to obtain autonomic measures that reflect vagal tone during a typical school day, both before and after treatment. Treatment consisted of a psychoeducational session with children and parents and the teaching of resonance frequency

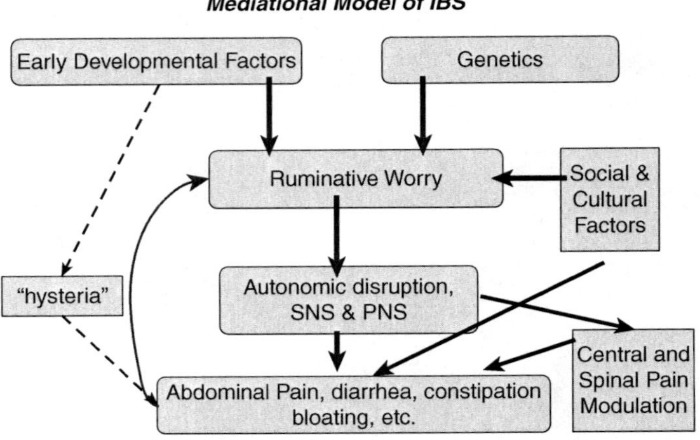

Figure 9-8 A mediational model for IBS.

biofeedback, as stated above. Remember, both Paul Lehrer's group and ours have documented that twenty minutes of daily, resonance frequency breathing practice produces substantial improvements in baroreflex, which we postulate represents improvements in autonomic homeostasis (Gevirtz and Lehrer 2003; Lehrer et al. 2003; Lehrer 2007). The training dramatically reduced symptoms (severe abdominal pain, diarrhea, constipation, nausea), but, more to the present point, did so by increasing resting vagal tone. Symptom reduction was correlated with increased vagal tone (r=0.62, p=.004). Thus, over 38 percent of the variance in symptom reduction was accounted for by vagal tone increases.

John (not his real name) is a typical patient in this symptom cluster. John is a 16-year-old high school junior who came to our clinic with complaints of severe abdominal pain, constipation, and bloating. He had missed almost a month of school, just when he needed to do well to get into a good college. He was typical in that he reported internalizing mental processes (worry, perfectionism, conscietiousness, sensitivity to the needs of others, etc). Also typical was his extreme sensitivity to visceral sensations (represented by the red line in the model shown in figure 9-8). We spent an hour with him explaining the mediational model after we had measured his physiological profile at rest, during stress, and at recovery. Once we were assured that he had "bought into" the model, we started the biofeedback training. With home practice, he became proficient within two weeks and immediately saw a reduction in symptoms. We then focused on psychological and coping factors that we hypothesized were driving the physiology (worry, especially about gut sensations, etc.). Within a few more weeks he had almost no symptoms, and, on those occasions that his gut began to cramp, he was able to "breathe his way out of it." We often add CBT or ACT to the treatment in the later stages. It is remarkable to observe how little resistence there is to these psychotherapies once they are embedded in the model.

Other Disorders

Models have been developed for other disorders as well, including non-cardiac chest pain (DeGuire et al. 1992; DeGuire et al. 1996), asthma (Lehrer et al. 2000), FM (Hassett et al. 2007), and anxiety/panic (Meuret et al. 2003; Meuret et al. 2004; Meuret et al. 2006; Meuret et al. 2008a; Meuret et al. 2008b). These models can serve not only as treatment templates, but as useful heuristics for future research. In each case, the challenge is to discern the physiological systems involved in the disorder, and then trace these mechanisms to psychological and emotional factors and, perhaps, to the central limbic structures that mediate them. Thayer and colleagues, for example, have postulated a "neurovisceral" hypothesis that attempts to describe a number of anxiety and attentional disorders (Thayer and Siegle 2002b) using both autonomic and central systems, namely, the central autonomic network (Friedman and Thayer 1998; Thayer and Friedman 2002a; Thayer et al. 2009).

Conclusions

In this chapter, I have attempted to expand the concept of biomarkers from genetic and brain markers, to include autonomic and respiratory markers, thus broadening the scope of what might be possible in personalizing medicine. Just as knowing how gene expression might help target a treatment, autonomic markers, when combined with psychological factors, should be able to help target disorders that have heretofore been treated less than optimally. Some examples from drawn from chronic pain and gastroenterology were presented, with the hope that practitioners and researchers could see the value of a simple marker (heart rate variability) that reflects imbalance between the body's primary excitatory and inhibitory systems. That context can hopefully be readily expanded to link to gene-brain personalized medicine markers underlying excitation-inhibitory imbalances in brain-based disorders.

References

Al-Abed, M. A., Manry, M., Burk, J. R., Lucas, E. A., & Behbehani, K. (2009). Sleep disordered breathing detection using heart rate variability and R-peak envelope spectrogram. *Conf Proc IEEE Eng Med Biol Soc, 1*, 7106–7109.

Armour, J. A. (2007). The little brain on the heart. *Cleveland Clinic Journal of Medicine, 74 Suppl 1*, S48–51.

Armour, J. A. (2008). Potential clinical relevance of the 'little brain' on the mammalian heart. *Experimental Physiology, 93*(2), 165–176.

Barsky, A. J., & Borus, J. F. (1999). Functional somatic syndromes. *Ann Intern Med, 130*(11), 910–921.

Bernardi, L., Passino, C., Wilmerding, V., Dallam, G. M., Parker, D. L., Robergs, R. A., et al. (2001). Breathing patterns and cardiovascular autonomic modulation during hypoxia induced by simulated altitude. *J Hypertens, 19*(5), 947–958.

Berntson, G. G., Bigger, J. T., Jr., Eckberg, D. L., Grossman, P., Kaufmann, P. G., Malik, M., et al. (1997). Heart rate variability: origins, methods, and interpretive caveats. *Psychophysiology, 34*(6), 623–648.

Bigger, J. T., Fleiss, J. L., Rolnitzky, L. M., & Steinman, R. C. (1993). The ability of several short-term measures of RR variability to predict mortality after myocardial infarction. *Circulation, 88*(3), 927–934.

Brown, R. P., & Gerbarg, P. L. (2005). Sudarshan Kriya yogic breathing in the treatment of stress, anxiety, and depression: part I-neurophysiologic model. *Journal of Alternative and Complementary Medicine, 11*(1), 189–201.

Brown, T. E., Beightol, L. A., Koh, J., & Eckberg, D. L. (1993). Important influence of respiration on human R-R interval power spectra is largely ignored. *J Appl Physiol, 75*(5), 2310–2317.

Cannon, W. B. (1939). *The Wisdom of the Body* (2nd ed.). New York: Norton.

Cohen, H., Benjamin, J., Geva, A. B., Matar, M. A., Kaplan, Z., & Kotler, M. (2000). Autonomic dysregulation in panic disorder and in post-traumatic stress disorder: application of power spectrum analysis of heart rate variability at rest and in response to recollection of trauma or panic attacks. *Psychiatry Res, 96*(1), 1–13.

Cohen, H., Neumann, L., Shore, M., Amir, M., Cassuto, Y., & Buskila, D. (2000). Autonomic dysfunction in patients with fibromyalgia: application of power spectral analysis of heart rate variability. *Semin Arthritis Rheum, 29*(4), 217–227.

deGruy, F., Columbia, L., & Dickinson, P. (1987). Somatization disorder in a family practice. *J Fam Pract, 25*(1), 45–51.

deGruy, F., Crider, J., Hashimi, D. K., Dickinson, P., Mullins, H. C., & Troncale, J. (1987). Somatization disorder in a university hospital. *J Fam Pract, 25*(6), 579–584.

DeGuire, S., Gevirtz, R., Hawkinson, D., & Dixon, K. (1996). Breathing retraining: a three-year follow-up study of treatment for hyperventilation syndrome and associated functional cardiac symptoms. *Biofeedback Self Regul, 21*(2), 191–198.

DeGuire, S., Gevirtz, R., Kawahara, Y., & Maguire, W. (1992). Hyperventilation syndrome and the assessment of treatment for functional cardiac symptoms. *Am J Cardiol, 70*(6), 673–677.

Drossman, D. A., Creed, F. H., Olden, K. W., Svedlund, J., Toner, B. B., & Whitehead, W. E. (1999). Psychosocial aspects of the functional gastrointestinal disorders. *Gut, 45 Suppl 2*, II25–30.

Eckberg, D. L. (1997). Sympathovagal balance: a critical appraisal. *Circulation, 96*(9), 3224–3232.

Farnam, A., Somi, M. H., Sarami, F., & Farhang, S. (2008). Five personality dimensions in patients with irritable bowel syndrome. *Neuropsychiatr Dis Treat, 4*(5), 959–962.

Farnam, A., Somi, M. H., Sarami, F., Farhang, S., & Yasrebinia, S. (2007). Personality factors and profiles in variants of irritable bowel syndrome. *World J Gastroenterol, 13*(47), 6414–6418.

Friedman, B. H., & Thayer, J. F. (1998). Autonomic balance revisited: panic anxiety and heart rate variability. *J Psychosom Res, 44*(1), 133–151.

Gevirtz, R. (2000). Resonant frequency training to restore homeostasis for treatment of psychophysiological disorders. *Biofeedback, 27*, 7–9.

Gevirtz, R., Hubbard, D., & Harpin, E. (1996). Psychophysiologic Treatment of Chronic Low Back Pain. *Professional Psychology: Research and Practice, 27*(6), 561–566.

Gevirtz, R., & Lehrer, P. (2003). Resonant Frequency Heart Rate Biofeedback. In M. S. F. Andrasik (Ed.), *Biofeedback: A Practitioners Guide* (3rd ed.). NY: Guilford.

Gevirtz, R. N. (2007). Psychophysiological Perspectives on Stress Related and Anxiety Disorders. In P. Lehrer & W. Sime (Eds.), *Stress Management: Principals and Practice* (2nd ed.). New York: Guilford Press.

Giardino, N. D., Chan, L., & Borson, S. (2004). Combined heart rate variability and pulse oximetry biofeedback for chronic obstructive pulmonary disease: preliminary findings. *Appl Psychophysiol Biofeedback, 29*(2), 121–133.

Gordon, E. (2009). Brain Resources Database. In P. communication (Ed.).

Gordon, E. e. a. (2000). get ref.

Gralnek, I. M. (1998). Health care utilization and economic issues in irritable bowel syndrome. *Eur J Surg Suppl*, (583), 73–76.

Gros, D. F., Antony, M. M., McCabe, R. E., & Swinson, R. P. (2009). Frequency and severity of the symptoms of irritable bowel syndrome across the anxiety disorders and depression. *J Anxiety Disord, 23*(2), 290–296.

Grossman, P., & Kollai, M. (1993). Respiratory sinus arrhythmia, cardiac vagal tone, and respiration: within- and between-individual relations. *Psychophysiology, 30*(5), 486–495.

Grossman, P., van Beek, J., & Wientjes, C. (1990). A comparison of three quantification methods for estimation of respiratory sinus arrhythmia. *Psychophysiology, 27*(6), 702–714.

Gupta, V., Sheffield, D., & Verne, G. N. (2002). Evidence for autonomic dysregulation in the irritable bowel syndrome. *Dig Dis Sci, 47*(8), 1716–1722.

Gureje, O., & Simon, G. E. (1999). The natural history of somatization in primary care. *Psychol Med, 29*(3), 669–676.

Gureje, O., Simon, G. E., Ustun, T. B., & Goldberg, D. P. (1997). Somatization in cross-cultural perspective: a World Health Organization study in primary care. *Am J Psychiatry, 154*(7), 989–995.

Hassett, A. L., Radvanski, D. C., Vaschillo, E. G., Vaschillo, B., Sigal, L. H., Karavidas, M. K., et al. (2007). A pilot study of the efficacy of heart rate variability (HRV) biofeedback in patients with fibromyalgia. *Appl Psychophysiol Biofeedback, 32*(1), 1–10.

Hayes, S. C., Strosahl, K.D., Wilson, K.G. (1999). *Acceptance and Commitment Therapy*. New York: Guilford Press.

Hazlett-Stevens, H., Craske, M. G., Mayer, E. A., Chang, L., & Naliboff, B. D. (2003). Prevalence of irritable bowel syndrome among university students: the roles of worry, neuroticism, anxiety sensitivity and visceral anxiety. *J Psychosom Res, 55*(6), 501–505.

Hinnant, J. B., & El-Sheikh, M. (2009). Children's externalizing and internalizing symptoms over time: the role of individual differences in patterns of RSA responding. *Journal of Abnormal Child Psychology, 37*(8), 1049–1061.

Hollifield, M., Tuttle, L., Paine, S., & Kellner, R. (1999). Hypochondriasis and somatization related to personality and attitudes toward self. *Psychosomatics, 40*(5), 387–395.

Hubbard, D. (1996). Chronic and Recurrent Muscle Pain: Pathophysiology and treatment, a review of pharmocologic studies. *J. of Musculoskelatal Pain, 4*(1/2), 123–143.

Hubbard, D. (1998). Persistent muscular pain: Approaches to relieving trigger points. *J. of Musculoskelatal Medicine, 15*(5), 16–26.

Hubbard, D. R., & Berkoff, G. M. (1993). Myofascial trigger points show spontaneous needle EMG activity. *Spine, 18*(13), 1803–1807.

Humphreys, P. A., & Gevirtz, R. N. (2000). Treatment of recurrent abdominal pain: components analysis of four treatment protocols. *J Pediatr Gastroenterol Nutr, 31*(1), 47–51.

Iovino, P., Azpiroz, F., Domingo, E., & Malagelada, J. R. (1995). The sympathetic nervous system modulates perception and reflex responses to gut distention in humans. *Gastroenterology, 108*(3), 680–686.

Iovino, P., Tremolaterra, F., Consalvo, D., Sabbatini, F., Mazzacca, G., & Ciacci, C. (2006). Perception of electrocutaneous stimuli in irritable bowel syndrome. *Am J Gastroenterol, 101*(3), 596–603.

Karavidas, M. K., Lehrer, P. M., Vaschillo, E., Vaschillo, B., Marin, H., Buyske, S., et al. (2007). Preliminary results of an open label study of heart rate variability biofeedback for the treatment of major depression. *Appl Psychophysiol Biofeedback, 32*(1), 19–30.

King, D. R., Ogilvie, M. P., Pereira, B. M., Chang, Y., Manning, R. J., Conner, J. A., et al. (2009). Heart rate variability as a triage tool in patients with trauma during prehospital helicopter transport. *Journal of Trauma, 67*(3), 436–440.

Langley, J. N. (1921). *The Autonomic Nervous System, Part I.* Cambridge: Heffer and Sons.

Lehrer, P. (2007). Biofeedback training to increase heart rate variability. In R. W. Paul Lehrer, Wes Sime (Ed.), *Principles and practiceof Stress Management* (3rd ed.). New York: Guilford Press.

Lehrer, P., Smetankin, A., & Potapova, T. (2000). Respiratory sinus arrhythmia biofeedback therapy for asthma: a report of 20 unmedicated pediatric cases using the Smetankin method. *Appl Psychophysiol Biofeedback, 25*(3), 193–200.

Lehrer, P., Vaschillo, E., Lu, S. E., Eckberg, D., Vaschillo, B., Scardella, A., et al. (2006). Heart rate variability biofeedback: effects of age on heart rate variability, baroreflex gain, and asthma. *Chest, 129*(2), 278–284.

Lehrer, P. M., Vaschillo, E., Vaschillo, B., Lu, S. E., Eckberg, D. L., Edelberg, R., et al. (2003). Heart rate variability biofeedback increases baroreflex gain and peak expiratory flow. *Psychosom Med, 65*(5), 796–805.

Lehrer, P. M., Vaschillo, E., Vaschillo, B., Lu, S. E., Scardella, A., Siddique, M., et al. (2004). Biofeedback treatment for asthma. *Chest, 126*(2), 352–361.

Longstreth, G. F., & Wolde-Tsadik, G. (1993). Irritable bowel-type symptoms in HMO examinees. Prevalence, demographics, and clinical correlates. *Dig Dis Sci, 38*(9), 1581–1589.

Malliani, A., Lombardi, F., Pagani, M., & Cerutti, S. (1994). Power spectral analysis of cardiovascular variability in patients at risk for sudden cardiac death. *J Cardiovasc Electrophysiol, 5*(3), 274–286.

Malliani, A., Pagani, M., & Lombardi, F. (1994). Physiology and clinical implications of variability of cardiovascular parameters with focus on heart rate and blood pressure. *Am J Cardiol, 73*(10), 3C-9C.

Martinez-Lavin, M., Hermosillo, A. G., Rosas, M., & Soto, M. E. (1998). Circadian studies of autonomic nervous balance in patients with fibromyalgia: a heart rate variability analysis. *Arthritis Rheum, 41*(11), 1966–1971.

Mayer, E. A. (1999). Emerging disease model for functional gastrointestinal disorders. *Am J Med, 107*(5A), 12S-19S.

Mayer, E. A. (2000). Spinal and supraspinal modulation of visceral sensation. *Gut, 47 Suppl 4*, iv69–72; discussion iv76.

Mayer, E. A., Chang, L., & Lembo, T. (1998). Brain-gut interactions: implications for newer therapy. *Eur J Surg Suppl, 582*, 50–55.

Mayer, E. A., Derbyshire, S., & Naliboff, B. D. (2000). Cerebral activation in irritable bowel syndrome. *Gastroenterology, 119*(5), 1418–1420.

Mayer, E. A., Naliboff, B. D., & Chang, L. (2001). Basic pathophysiologic mechanisms in irritable bowel syndrome. *Dig Dis, 19*(3), 212–218.

Mayer, E. A., Naliboff, B. D., Chang, L., & Coutinho, S. V. (2001). V. Stress and irritable bowel syndrome. *Am J Physiol Gastrointest Liver Physiol, 280*(4), G519–524.

McNulty, W. H., Gevirtz, R. N., Hubbard, D. R., & Berkoff, G. M. (1994). Needle electromyographic evaluation of trigger point response to a psychological stressor. *Psychophysiology, 31*(3), 313–316.

Meuret, A. E., Rosenfield, D., Hofmann, S. G., Suvak, M. K., & Roth, W. T. (2008). Changes in respiration mediate changes in fear of bodily sensations in panic disorder. *J Psychiatr Res.*

Meuret, A. E., White, K. S., Ritz, T., Roth, W. T., Hofmann, S. G., & Brown, T. A. (2006). Panic attack symptom dimensions and their relationship to illness characteristics in panic disorder. *J Psychiatr Res, 40*(6), 520–527.

Meuret, A. E., Wilhelm, F. H., Ritz, T., & Roth, W. T. (2003). Breathing training for treating panic disorder. Useful intervention or impediment? *Behav Modif, 27*(5), 731–754.

Meuret, A. E., Wilhelm, F. H., Ritz, T., & Roth, W. T. (2008). Feedback of end-tidal pCO2 as a therapeutic approach for panic disorder. *J Psychiatr Res, 42*(7), 560–568.

Meuret, A. E., Wilhelm, F. H., & Roth, W. T. (2004). Respiratory feedback for treating panic disorder. *J Clin Psychol, 60*(2), 197–207.

Morrison, S. F. (2001). Differential control of sympathetic outflow. *Am J Physiol Regul Integr Comp Physiol, 281*(3), R683–698.

Naliboff, B. D., Chang, L., Munakata, J., & Mayer, E. A. (2000). Towards an integrative model of irritable bowel syndrome. *Prog Brain Res, 122*, 413–423.

Nocua, R., Noury, N., Gehin, C., Dittmar, A., & McAdams, E. (2009). Evaluation of the autonomic nervous system for fall detection. *Conf Proc IEEE Eng Med Biol Soc, 1*, 3225–3228.

Oliveira, A., Gevirtz, R., & Hubbard, D. (2006). A psycho-educational video used in the emergency department provides effective treatment for whiplash injuries. *Spine (Phila Pa 1976), 31*(15), 1652–1657.

Ozdemir, O., Soylu, M., Demir, A. D., Topaloglu, S., Alyan, O., Geyik, B., et al. (2003). Increased sympathetic nervous system activity as cause of exercise-induced ventricular tachycardia in patients with normal coronary arteries. *Texas Heart Institute Journal, 30*(2), 100–104.

Pal, G. K., Shyma, P., Habeebullah, S., Shyjus, P., & Pal, P. (2009). Spectral analysis of heart rate variability for early prediction of pregnancy-induced hypertension. *Clinical and Experimental Hypertension, 31*(4), 330–341.

Park, J., Lee, S., & Jeon, M. (2009). Atrial fibrillation detection by heart rate variability in Poincare plot. *Biomed Eng Online, 8*, 38.

Porges, S. W. (1995a). Cardiac vagal tone: a physiological index of stress. *Neurosci Biobehav Rev, 19*(2), 225–233.

Porges, S. W. (1995b). Orienting in a defensive world: mammalian modifications of our evolutionary heritage. A Polyvagal Theory. *Psychophysiology, 32*(4), 301–318.

Porges, S. W. (2003). Social engagement and attachment: a phylogenetic perspective. *Ann N Y Acad Sci, 1008*, 31–47.

Reinke, A., Gevirtz, R., & Mussgay, L. (2007). Effects of heart rate variability feedback in reducing blood pressure. *Applied Psychophysiology and Biofeedback, 32*(abstract), 134.

Schein, M. H., Gavish, B., Herz, M., Rosner-Kahana, D., Naveh, P., Knishkowy, B., et al. (2001). Treating hypertension with a device that slows and regularises breathing: a randomised, double-blind controlled study. *J Hum Hypertens, 15*(4), 271–278.

Seidel, H., Herzel, H., & Eckberg, D. L. (1997). Phase dependencies of the human baroreceptor reflex. *American Journal of Physiology, 272*(4 Pt 2), H2040–2053.

Sevarese-Levine, J., Mianassian,A., Gevirtz, R.N., & Perry, W. (2007). Physiological differences between "Morning" and "Evening" individuals using heart rate variability as a measure of circadian rhythm. Paper presented at the Presented at the 62nd annual meeting of the Society of Biological Psychiatry, San Diego, CA.

Smith, G. R., Jr., Monson, R. A., & Ray, D. C. (1986). Patients with multiple unexplained symptoms. Their characteristics, functional health, and health care utilization. *Arch Intern Med, 146*(1), 69–72.

Sowder, E., Gevirtz, R. N., Kotay, A., & Shapiro, W. (2006). Autonomic function in patients with recurrent abdominal pain and normal controls, *American College of Gastroenterology*. Las Vegas, NV.

Sowder, E., Kotay, A., Cannon, C., Murphy, J., Gevirtz, R., & Shapiro, W. (2005). *Heart rate variability with recurrent abdominal pain: A multiple case history.* Paper presented at the ISARP, Hamburg, Germany.

Speckens, A. E., Van Hemert, A. M., Bolk, J. H., Rooijmans, H. G., & Hengeveld, M. W. (1996). Unexplained physical symptoms: outcome, utilization of medical care and associated factors. *Psychol Med, 26*(4), 745–752.

Swanson, K., Gevirtz, R., Spira, J., & Guarneri, E. (2006). Does biofeedback increase heart rate variability in patents with known New York Heart Association class I-III heart failure. *Applied Psychophysiology and Biofeedback, 2*, abstract.

Tan, G., Hammond, D. C., & Joseph, G. (2005). Hypnosis and irritable bowel syndrome: a review of efficacy and mechanism of action. *Am J Clin Hypn, 47*(3), 161–178.

Task. (1996). Heart rate variability. Standards of measurement, physiological interpretation, and clinical use. Task Force of the European Society of Cardiology and the North American Society of Pacing and Electrophysiology. *Eur Heart J, 17*(3), 354–381.

Thayer, J. F., & Friedman, B. H. (2002). Stop that! Inhibition, sensitization, and their neurovisceral concomitants. *Scand J Psychol, 43*(2), 123–130.

Thayer, J. F., & Siegle, G. J. (2002). Neurovisceral integration in cardiac and emotional regulation. *IEEE Eng Med Biol Mag, 21*(4), 24–29.

Thayer, J. F., Yamamoto, S. S., & Brosschot, J. F. (2009). The relationship of autonomic imbalance, heart rate variability and cardiovascular disease risk factors. *Int J Cardiol.*

Travell, J., & Simons, R. (1983). *Myofascial Pain Syndrome: The Trigger Popint Manual.* Baltimore: Williams and Wilkins.

Valencia, J. F., Vallverdu, M., Schroeder, R., Voss, A., Vazquez, R., Bayes de Luna, A., et al. (2009). Complexity of the short-term heart-rate variability. *IEEE Engineering in Medicine and Biology Magazine, 28*(6), 72–78.

Vaschillo, E., Lehrer, P., Rishe, N., & Konstantinov, M. (2002). Heart rate variability biofeedback as a method for assessing baroreflex function: a preliminary study of resonance in the cardiovascular system. *Appl Psychophysiol Biofeedback, 27*(1), 1–27.

Vaschillo, E. G., Vaschillo, B., & Lehrer, P. M. (2006). Characteristics of Resonance in Heart Rate Variability Stimulated by Biofeedback. *Appl Psychophysiol Biofeedback.*

Waring, W. S., Chui, M., Japp, A., Nicol, E. F., & Ford, M. J. (2004). Autonomic cardiovascular responses are impaired in women with irritable bowel syndrome. *J Clin Gastroenterol, 38*(8), 658–663.

Wessely, S., Chalder, T., Hirsch, S., Wallace, P., & Wright, D. (1996). Psychological symptoms, somatic symptoms, and psychiatric disorder in chronic fatigue and chronic fatigue syndrome: a prospective study in the primary care setting. *Am J Psychiatry, 153*(8), 1050–1059.

Wessely, S., Nimnuan, C., & Sharpe, M. (1999). Functional somatic syndromes: one or many? *Lancet, 354*(9182), 936–939.

Whitehead, W. E. (2006). Hypnosis for irritable bowel syndrome: the empirical evidence of therapeutic effects. *Int J Clin Exp Hypn, 54*(1), 7–20.

Wilhelm, F. H., Grossman, P., & Coyle, M. A. (2004). Improving estimation of cardiac vagal tone during spontaneous breathing using a paced breathing calibration. *Biomed Sci Instrum, 40*, 317–324.

Wilhelm, F. H., Grossman, P., & Roth, W. T. (2005). Assessment of heart rate variability during alterations in stress: complex demodulation vs. spectral analysis. *Biomed Sci Instrum, 41*, 346–351.

Yeragani, V. K., Pohl, R., & Balon, R. (2004). Complex demodulation of cardiac interbeat intervals: increased cardiac sympathovagal interaction during human sexual activity. *Arch Sex Behav, 33*(1), 65–69.

Yildirim, A., Soylu, O., Dagdeviren, B., Eksik, A., & Tezel, T. (2007). Sympathetic overactivity in patients with left ventricular aneurysm in early period after anterior myocardial infarction: does sympathetic activity predict aneurysm formation? *Angiology, 58*(3), 275–282.

Zucker, T. L., Samuelson, K. W., Muench, F., Greenberg, M. A., & Gevirtz, R. N. (2009). The effects of respiratory sinus arrhythmia biofeedback on heart rate variability and posttraumatic stress disorder symptoms: a pilot study. *Appl Psychophysiol Biofeedback, 34*(2), 135–143.

10 Personalized Medicine in Sleep Health

Ronald Grunstein, MD

Introduction

When thinking about personalized medicine in general, and how it applies to sleep disorders specifically, it is important to recognize that the basic conceptualization of this topic is not new. Hippocrates of Kos is alleged to have said, "It's far more important to know what person the disease has than what disease the person has," perhaps showing foresight in recognizing the role of inter-individual differences in health care.

The definition of personalized medicine (PM), and its uptake in the regulatory and media domains, have focused predominantly on the concept that managing a patient's health should be advanced by new methods of molecular analysis, in order to better manage a patient's disease or predisposition toward a disease. Indeed, the U. S. Congress defines personalized medicine as "the application of genomic and molecular data to better target the delivery of health care, facilitate the discovery and clinical testing of new products, and help determine a person's predisposition to a particular disease or condition" (United States Congress 2006). However, information dissemination on personalized medicine by some government organizations recognizes that human biological variation is complex (House of Lords Science and Technology Committee: Genomic Medicine Inquiry 2009).

Given the current limited state of knowledge of genomic and molecular influences on sleep and its disorders, this chapter will focus on potential rather than actual approaches to PM in sleep health, particularly those relevant to sleep loss and the most common sleep disorders. There have been some recent advances in genetic discoveries in sleep health that will be related to inter-individual differences. These findings allow some speculation on how a PM agenda in sleep health will operate in the future.

Current approaches to individualized management in the clinical sleep sciences have predominantly focused on phenotypes, such as anatomical structure, presence of sleepiness, or neurophysiological biomarkers. These will be briefly reviewed, including phenotypic markers of disease severity and treatment outcome. Importantly, unlike many other conditions considered in this book, treatment of one of the commonest sleep disorders, obstructive sleep apnea, involves a range of mechanical treatments. This will be discussed in the context of new potential applications of PM.

Summary of Relevant Background Information

Inter-Individual Differences in Sleep Biology and Symptomatology

Individuals differ markedly in their responses to extended work hours and shift-work schedules (Van Dongen 2006). There are obvious important environmental conditions that may explain this, including age, social factors, ability to sleep outside (and nap within) work periods, and ambient light levels. However, there are clearly endogenous neurobiological differences that will, in part, explain why under identical circumstances of work and sleep, one individual will be alert and responding to the environment, while another may be drowsy and in danger of causing some catastrophic event.

Increasing recognition of these potentially catastrophic consequences is driving this line of research. Disasters, such as the nuclear meltdowns at Three Mile Island and Chernobyl, were linked to impaired vigilance in the workforce managing these sites. The Institute of Medicine (IOM) report on "Sleep Disorders and Sleep Deprivation" (Colten and Altaveogt 2006) also highlighted the role of loss of alertness at work as being implicated in the space shuttle *Challenger* disaster, and the oil spill by the tanker the *Exxon Valdez*. Other research has shown how sleep loss impairs operations such as airport security (Basner 2008). Moreover, increasing awareness of sleep-related motor vehicle crashes means that there is a pressing agenda to understand sleepiness in all dimensions of the social environment and the workplace. In Australia, this has a widened legal dimension, as even executives of transport companies or their customers may be liable in truck crashes related to falling asleep. Detection of the "at risk" driver through biomarkers, or even risk prediction through genetic investigation, are attractive strategies to such executives. The IOM Committee on Optimizing Graduate Medical Trainee (Resident) Hours and Work Schedules to Improve Patient Safety concluded recently that working for more than sixteen consecutive hours without sleep is unsafe for both physicians in training (residents) and their patients (Ulmer, et al. 2008).

In the military setting, alertness and resistance to sleep are crucial factors in the sustained operations of modern warfare. This is even highlighted in the logistic requirements of responding to disasters. Future progress in aerospace will depend on better understanding the determinants of individual sleepiness. Estimates of productivity losses related to sleepiness in the workplace are huge and run into the billions of dollars. Therefore, a cogent, PM-based approach is required to ensure both vigilance in the workplace and the military, and minimization of conditions, such as shift-work sleep disorder (Czeisler et al. 2005) or shift-work-related insomnia.

Sleep–Wake Regulation: Circadian and Homeostatic Factors

Sleep is composed of two very different physiologic states: rapid eye movement (REM) sleep and non-rapid eye movement (NREM) sleep. NREM is further divided into four stages that are characterized by specific neurophysiological criteria. Stage 1 is the lightest stage of sleep. Stage 2 sleep has a higher arousal threshold and is the stage in which most time asleep is spent. Stages 3 and 4 are collectively referred to as "deep sleep," "delta sleep," or "slow wave sleep," (based upon their characteristic EEG profiles) and are associated with a high arousal threshold. More stage 3 and stage 4 sleep occurs in the first half of the night, while more REM sleep takes place in the last half (Krueger et al. 2008). This sleep pattern can be interrupted by awakenings that may be extremely brief or of prolonged duration. Sleep can be quantified in more detail in the research setting by quantitative EEG analysis, yielding more precise measures of activity of specific EEG waveforms that go beyond conventional clinical sleep "staging". (Dijk et al. 2009; Dijk et al. 1987).

The circadian pacemaker and sleep homeostat are both major determinants of the sleep–wake cycle. The sleep–wake cycle drives the sleep homeostat, which builds up a pressure for sleep across time spent awake and dissipates that pressure across time spent asleep. This also feeds back onto the circadian pacemaker, which is synchronized by the light–dark cycle (Franken and Dijk 2009).

It has been established that variation in the timing of sleep is associated with variation in the timing of rhythms driven by the suprachiasmatic nuclei (SCN). The core body temperature, cortisol, and melatonin rhythms of healthy early sleepers are set to an earlier phase compared to late sleepers. The differences in the timing of these circadian rhythms persist when the sleep–wake cycle, meal cycles, and the light–dark cycle are temporarily "removed" by keeping people awake for one to two days under constant, routine conditions. The timing of endogenous physiological rhythms associates with inter-individual variation in sleep timing in people without sleep complaints (Franken and Dijk 2009).

In effect, the homeostatic process produces increasing sleep pressure with increasing sleep loss, and the circadian process yields an opposing pressure for wakefulness during the day and withdraws that pressure at night (Borbely Borbély,1982; Saper, Scammell, and Lu 2005). These processes interact and produce dynamic changes in sleepiness and performance over time. Furthermore, they affect the timing and duration of sleep, and thereby exert indirect influence on sleepiness and performance, as well. For example, the typical shift worker is forced to work at times that are suboptimal for good performance, and to go to bed at times that are suboptimal for good sleep (Czeisler et al. 2005).

Given the involvement of two interacting processes, other neurobiological traits besides vulnerability to sleep loss are likely to contribute to individual differences in responses to extended work hours and shift work (Van Dongen 2006). These traits may include individual differences in sleep need, recovery rate during sleep following sleep deprivation, circadian timing (phase), circadian amplitude, and rate of circadian adjustment to schedule changes.

The individual variability in circadian phase position has been studied more than many of these other parameters. This encompasses the well-recognized trait of morningness–eveningness commonly described as "larks versus owls" and has been recognized to be involved in tolerance for shift work (Horne and Östberg 1977; Van Dongen 2006).

This normal sleep-wake process can be influenced by a wide variety of physiological, psychological, and environmental factors. Physiological studies of sleep have also documented marked changes in sleep structure in healthy older people (Adam et al. 2006; Duffy et al. 2009). Key changes include a striking age-related reduction in slow wave sleep (SWS) and computer-detected slow wave activity (SWA), as well as reductions in sleep continuity and sleep maintenance parameters. Reductions in REM sleep and total sleep time (TST) are also observed with aging, although these changes are less striking. The assumption has been that sleep propensity would increase in healthy aging, because of these changes in sleep structure. However, such an assumption is premature, as there is evidence that healthy aging is associated with a reduction in daytime sleep propensity, sleep continuity, and SWS. The age-related decline in SWS, and the corresponding reduction in daytime sleep propensity, may reflect a decrease in homeostatic sleep requirement. In fact, healthy older adults without sleep disorders can expect to be less sleepy during the daytime than young adults (Adam et al. 2006; Duffy et al. 2009).

Current Relevant Applications and Future Potential Applications

Sleep Disorders

The medical importance of sleep was recognized in the twentieth century because of the concomitant rapidly increasing knowledge about sleep, including the discovery of REM sleep and sleep apnea. This has resulted in the development of sleep medicine as an adult and pediatric specialty. By the 1970s in the United States, clinics and laboratories devoted to the study of sleep and sleep disorders had been founded, and practice standards had been developed (Colten and Altaveogt 2006).

Sleep disorders present with a wide variety of symptoms. In the not-too-distant past, patients presenting with excessive daytime sleepiness were often ignored, considered lazy, or thought to have some primary psychiatric problem. Now, we understand that this symptom is almost always a component of an identifiable and treatable sleep disorder in the absence of volitional sleep deprivation.

The most common sleep disorders in clinical practice include insomnia; sleep-breathing disorders such as sleep apnea; primary causes of excessive daytime sleepiness, such as narcolepsy and its variants; and restless leg syndrome (Kryger, Dement, and Roth 2009). Even within the broad spectrum of insomnia, many different phenotypes exist. These include sleep phase disorders in which there are extremes of circadian behavior, including the advanced sleep phase syndromes (ASPS) and delayed sleep phase syndromes (DSPS). A discussion of how PM may operate in common sleep disorders is at the end of this section.

Trait Differences in Vulnerability to Sleep Loss: Pathway for Fitness for Duty Selection?

There are substantial individual differences in performance impairment resulting from sleep deprivation, as well as significant evidence demonstrating that these individual differences constitute a trait (Leproult et al. 2003; Frey, Badia, and Wright 2004; Van Dongen et al. 2004; Van Dongen et al. 2005). Compelling evidence of this has only come from recent studies centered on sleep deprivation and sleep restriction protocols in laboratory environments with a high degree of experimental control. For example, in one landmark study, Van Dongen and colleagues (2004) subjected twenty-one healthy adults (ages 21–38; 12 men, 9 women) to 36 hours of total sleep deprivation on three separate occasions. In the week prior to each sleep deprivation session, the subjects either restricted their sleep to 6 hours per day (prior sleep restriction condition) or extended their time in bed for sleep to 12 hours per day (prior sleep satiation condition). The prior sleep restriction condition occurred only once, in randomized, counterbalanced order. During each laboratory sleep deprivation session, a battery of subjective sleepiness measures and cognitive performance tests was administered every 2 hours. This battery included a word detection test (WDT) and a psychomotor vigilance test (PVT). The PVT is a widely used, sustained attention reaction time task.

Data from this study are shown in Figure 10-1. Importantly, for every outcome measure, there was wide variability among individuals. However, within each subject, the responses to sleep deprivation were very similar for the two sleep deprivation sessions with prior sleep satiation. The percentage of variance explained by replicable individual differences ranged between 92.2 percent for the WDT, and 67.5 percent for the PVT. The individual differences persisted when baseline individual differences in sleepiness and performance

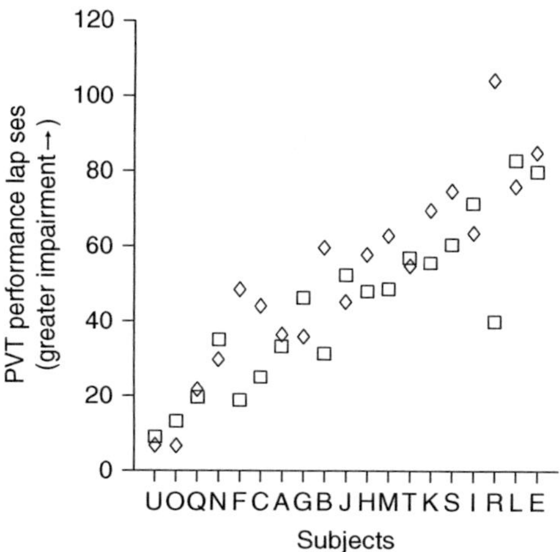

Figure 10-1 Inter-individual variation in psychomotor vigilance task performance in healthy volunteers with total sleep deprivation. Reproduced with permission from Van Dongen et al. 2004.

were taken into account. Moreover, in the sleep deprivation session with prior sleep restriction, the subjects' individual responses were also similar to those observed in the prior sleep satiation condition. This emphasizes that trait vulnerability to sleep loss is not related to habitual sleep, but rather reflects the ability of an individual to tolerate sleep loss.

Psychomotor vigilance task responses to total sleep deprivation, as determined by the averages over the previous twenty-four hours of total sleep deprivation, were measured at two timepoints. Data are shown for the psychomotor vigilance task (PVT). The abscissa of each panel shows the twenty-one individual subjects, labeled A through U, in arbitrary order. The subjects are ordered by the magnitude of their impairment (averaged over the two sleep deprivations), with the most resistant subjects on the left and the most vulnerable subjects on the right. Responses in the first exposure to sleep deprivation following seven days of sleep extension are marked by boxes; responses in the second exposure to sleep deprivation following seven days of sleep extension are marked by diamonds. The panels reveal that subjects differed substantially in their responses to sleep deprivation, while the responses were relatively stable within subjects between the two exposures to sleep deprivation.

These individual differences were substantial, both in absolute magnitude and relative to the group-average effects of sleep deprivation. Interestingly, while all the subjects studied had more lapses of attention when performing the PVT with sleep deprivation, some individuals averaged only ten lapses per 20-minute PVT test (about one lapse of attention every two minutes), whereas other individuals experienced nearly 100 lapses of attention per 20-minute PVT test (nearly five lapses of attention every minute). Therefore, while the performance of most people deteriorates significantly when wakefulness extends beyond approximately sixteen to eighteen hours, there are considerable inter-individual differences in the magnitude of this effect. There are likely also inter-individual differences in the effects of exposure to chronic partial sleep deprivation over the course of several weeks.

Explaining Inter-Individual Differences

As discussed above, when considering vulnerability to sleep loss, two distinct neurobiological processes play an important role in determining the temporal profile of performance responses to extended and shifted work times. One is the homeostatic process, which builds up a pressure for sleep across time spent awake, and dissipates that pressure across time spent asleep. The other is the circadian process, which yields an opposing pressure for wakefulness during the day and withdraws that pressure at night. Maximum vulnerability to falling asleep occurs when homeostatic pressure is high and circadian drive for alertness is withdrawn (Cohen 2010).

The implications of these findings are highlighted by very recent research showing marked performance deficits in individuals with even small amounts of chronic sleep "loss" when undertaking tasks during their biological night (Cohen 2010). When participants were preloaded with fatigue due to chronic sleep loss, the deterioration in performance during the biological night induced by acute sleep loss dwarfed the impairment induced by acute sleep loss alone in people who were not chronically sleep-restricted. Given that 16 percent of Americans routinely sleep six hours per night or less (29 Banks and Dinges 2007, this has major implications for occupational safety, particularly for those in safety-sensitive industries, such as long-haul transport.

Another aspect of "fitness for duty" and sleep loss has been the advent of new pharmacological approaches to wakefulness. Drugs, such as the wakefulness promoter modafinil (Czeisler et al. 2005), or others in pipeline development, can partly modify the performance decrements associated with sleep loss. Therefore, another dimension to trait vulnerability is inter-individual differences in the treatment response to these drugs (Bodenmann et al. 2009).

Obviously, there are compelling reasons to develop simple biomarkers of vulnerability to sleep loss that can be deployed easily, either clinically or in the workplace, and do not involve complex testing. Essentially, research into such biomarkers has involved EEG-based parameters or, more recently, genetic markers.

Waking and Sleep EEG

The sleep homeostat is an "hourglass" oscillator (Dijk, Beersma, and Daan 1987); it tracks the history of sleep and wakefulness, and thereby tracks sleep debt. Established markers of the sleep homeostat are SWA in the EEG during NREM sleep and theta EEG activity during wakefulness (Dijk and Archer 2010). It has been suggested that changes in these markers are related to some of the biochemical consequences of sleep and wakefulness, such as variation in extracellular adenosine concentration and other sleep regulatory substances (see below), or related to variation in connectivity, i.e., synaptic strength, in neuronal networks (Tononi and Cirelli 2006; Landolt 2008; Krueger et al. 2008). The circadian oscillator, located in the SCN of the hypothalamus, is a self-sustained oscillator that determines the preferred timing of sleep and wakefulness (Deboer et al. 2003; Schmidt et al. 2009). Established markers of the circadian process include plasma melatonin, cortisol, and core body temperature. There is now a wealth of data supporting the essential features of this model (Saper, Scammell, Lu 2005; Franken and Dijk 2009). Moreover, the neuroanatomical basis of the circadian regulation of sleep in particular has been elucidated in some detail and has generated a range of mathematical models (Achermann and Borbely 2003; Dijk and Von Schantz 2005; Phillips and Robinson 2009).

With the EEG, brain electrical activity can be measured repeatedly with high temporal resolution. Distinct EEG patterns in healthy individuals, such as resting activity in the alpha range (8–13 Hz), are genetically determined with heritability estimates of up to 90 percent (van Beijsterveldt et al. 1996). Not only the waking EEG, but also the sleep EEG, show distinct trait-like, inter-individual differences (Buckelmuller et al. 2006). The magnitude of these differences is even larger than the pronounced effects of prolonged wakefulness (Tucker, Dinges, and Van Dongen 2007). Sleep deprivation increases sleep pressure and gives rise to reliable, state-specific EEG changes in waking and sleep (Borbely and Achermann 1999). Independent of sleep pressure, heritability in the 8–16 Hz EEG band in NREM sleep may be as high as 96 percent (De Gennaro et al. 2008). This finding corroborates the notion that the sleep EEG is one of the most heritable traits in humans. Individual differences in EEG alpha activity may even be related to inter-individual differences in cognitive functions (Klimesch 1999).

Despite our increasing knowledge related to EEG markers, these parameters are rarely employed clinically, due to lack of standardized approaches to data collection. In addition, there are barriers to its use, including defects in the research translation pathway and lack of ease of deployment in the clinical or workplace setting. For many in sleep medicine, genetic markers seem to be attractive alternatives.

Adenosine and Sleep: Influence of Polymorphisms

Adenosine's importance in sleep–wake control is supported by many animal studies of adenosine receptor agonists and antagonists (Porkka-Heiskanen et al. 1997; Landolt 2008). Local adenosine levels rise in certain brain areas in rats and cats while awake, and decline during sleep. Local release of adenosine in the basal forebrain (BF) has been proposed as a signal for the homeostatic regulation of NREM sleep (Porkka-Heiskanen et al. 1997; Landolt 2008). Pharmacological studies revealed that inhibition of adenosine metabolism, carried out by blocking adenosine kinase, prolongs sleep and increases SWA in the EEG (Landolt 2008). Genes that contribute to the regulation of extracellular adenosine levels, such as adenosine deaminase (ADA) and S-adenosyl-homocysteine hydrolase (SAHH), modify the rate at which NREM sleep need accumulates during wakefulness (Bodenmann et al. 2007; Landolt 2008).

Strong support for a role for adenosine receptors in sleep–wake processes comes from the potent stimulant action of caffeine (Wyatt et al. 2004; Landolt 2008). In doses typically contained in coffee, tea, energy drinks, certain foods, and pharmaceutical formulations, caffeine acts as an adenosine receptor antagonist. The average daily caffeine consumption by adults in Western societies leads to low micromolar blood plasma levels. Disturbed sleep after caffeine is among the primary reasons why people voluntarily abstain from the stimulant. It is widely accepted that at micromolar concentrations, the stimulant acts primarily as a competitive antagonist at A1 and A2A receptors. However, recent data from knockout mice (this will be understood by readers) suggest that the A2A receptor is the main target for caffeine-induced wakefulness (Landolt 2008).

Low doses of caffeine (100 mg), when administered immediately prior to sleep, prolong sleep latency, reduce SWS in the first NREM/REM sleep cycle, and impair sleep efficiency (Landolt et al. 2004). The EEG spectral power shows changes consistent with increased arousal (enhanced beta activity), with the low delta range decreased and power in the spindle frequency range increased. Similar nocturnal effects are noted after a higher morning dose of caffeine (200 mg), although caffeine concentration in saliva is approaching zero.

These caffeine-induced EEG changes are robust and occur in both rested and sleep-deprived subjects, irrespective of prior history of caffeine intake or short- or long-term caffeine abstinence. These changes in sleep and the sleep EEG are similar to those seen in patients with primary insomnia (i.e., insomnia not related to another sleep, medical, or psychiatric disorder). Frontal parts of the cortex reflect the homeostatic process of sleep–wake regulation more sensitively than other cortical regions. During sleep deprivation, power in the theta band of the waking EEG increases more in an anterior derivation than in a posterior derivation, specifically in those reporting sleep-caffeine-sensitivity (Landolt 2008); this effect is attenuated by caffeine.

ADA 22G→A polymorphism

It has recently been observed in humans that a functional genetic variation in adenosine deaminase ADA has a profound impact on sleep and the sleep EEG (Rétey et al. 2005). The most frequent variant allele (ADA*2) of ADA that is asymptomatic in heterozygous carriers is caused by a G to A transition at nucleotide 22 of the ADA gene (*ADA* 22G→A polymorphism). This genotype is present in roughly 10 percent of healthy Caucasians, whereas the homozygous A/A genotype occurs in less than 1 percent of the population (Landolt 2008). Individuals with the *ADA* 22G→A polymorphism exhibit 20 to 30 percent lower enzymatic activity of ADA in blood cells than individuals with the G/G genotype. An 8-hour nocturnal sleep episode in subjects with the G/A genotype is characterized by 30 minutes more SWS than in subjects with the G/G genotype (Rétey et al. 2005). Moreover, not only is the duration of SWS longer—its intensity, as estimated from SWA, is higher in the G/A than in the G/G genotype.

These findings on increased "intensity" of sleep with this polymorphism are consistent with the functional reduction in ADA activity resulting in higher sleep "pressure" with the concomitant SWA findings. It has been suggested that the effects of this ADA polymorphism are similar in magnitude to the consequences of one night without sleep, and provide direct evidence of a role for adenosine in NREM sleep homeostasis in humans. Attempts to pharmacologically elevate the extracellular adenosine concentration by inhibiting ADA could, in principle, provide a promising treatment of sleep maintenance insomnia, although there may be unwanted side effects because of the ubiquity of adenosine receptors (Landolt 2008). It is also unknown whether the ADA 22G→A polymorphism plays a role in the prevalence of insomnia, or influences the development of secondary insomnia in patients with disorders, such as depression or chronic pain.

ADORA2A

Studies in knockout mice have shown that the wakefulness-promoting effect of caffeine requires functional A2A receptors. Adenosine A2A receptors are important for mediating central nervous effects of sleep- and wake-promoting substances, such as adenosine, prostaglandin D2, gamma-aminobutyric acid (GABA), histamine, and caffeine (Landolt 2008). Clear individual differences in the subjective response to caffeine exist, and this may relate to genetic variation (Rétey et al. 2006). For example, the c.1083T4C polymorphism in the adenosine A2A receptor gene (*ADORA2A*) modulates individual differences in symptoms of anxiety after caffeine, suggesting that this genetic variability has functional relevance (Landolt 2008).

Recent studies in mice provided strong evidence that caffeine promotes wakefulness by blocking adenosine A2A receptors. Large differences in these effects exist in distinct mouse strains that exhibit genetically determined differences in A2A receptor function. Moreover, as previously noted, certain frequencies in the EEG power spectrum in sleep and wakefulness are among the most heritable traits in humans and show large inter-individual variability and high intra-individual stability. These include the theta and alpha bands. Work from Rétey and colleagues in Zurich (2007) has demonstrated that power in the ≈7.5–10 Hz EEG (theta-low alpha) range is higher in subjects with the 1976C/C genotype of the A2A receptor gene than in subjects with the T/T genotype. This difference was also observed in NREM and REM sleep. The lack of wake and sleep differences suggests that this A2A receptor genotype plays a role in EEG-generating mechanisms, rather than in sleep-wake regulation. These findings indicated that certain aspects of the high inter-individual variability in human sleep and EEG variables reflecting sleep need are associated with polymorphisms in the adenosinergic system. It has long been recognized that the sleep effects of caffeine are highly variable among individuals (Landolt 2008). Prolonged wakefulness impairs optimal PVT performance more in self-rated caffeine-sensitive men than in caffeine-insensitive men (Landolt 2008), and this variance is attenuated by caffeine. Although caffeine does not necessarily counter the cognitive consequences of sleep loss, it has distinct effects on EEG topography.

Subsequent work by the Zurich group in young men has shown that the *ADORA2A* gene regulates individual sensitivity to subjective and objective effects of caffeine on sleep. The *ADORA2A* C/C is more common in caffeine-sensitive individuals compared with the T/T genotype, making them particularly susceptible to disturbed sleep after ingestion of caffeine (Rétey et al. 2007).

Given that the heritability of insomnia symptoms may be as high as 50 percent, these findings regarding the *ADORA2A* gene highlight the possibilities for a PM approach in insomnia. If further findings confirm the role of the c.1083T4C polymorphism modulating individual susceptibility to cortical hyperarousal induced by caffeine, there will be direct relevance to large numbers of caffeine consumers. Moreover, this polymorphism may have wider applicability in understanding and modulating vulnerability to insomnia.

Unpublished data have suggested that the *ADORA2A* haplotype may be associated with resistance to the effects of sleep loss on performance (Bodenmann et al. 2007). These preliminary data would indicate potential identification of those at risk of performance deterioration when they are sleep deprived.

Catechol-O-methyltransferase (COMT)

Catechol-O-methyltransferase (COMT) is a methylation enzyme that metabolizes released dopamine (Dauvilliers 2001; Dickinson and Elvevåg 2009). The *COMT* gene contains an evolutionarily recent G to A missense mutation that translates into a substitution of Met for Val at codon 108/158 (Val108/158 Met). The enzyme containing Met is unstable at 37 degrees Celsius and has one-quarter of the activity of the enzyme containing Val (36). The alleles are codominant, as heterozygous individuals have enzyme activity that is midway between that of homozygote individuals (Dauvilliers 2001; Bodenmann 2009a). COMT plays an important role in metabolizing cortical dopamine (Dickinson and Elvevåg 2009). In subjects with the high activity VAL variant (COMT-H), dopamine is catabolized at up to four times the rate of the MET variant (COMT-L), resulting in a significant reduction of synaptic dopamine following neurotransmitter release, ultimately reducing dopaminergic stimulation of the post-synaptic neuron.

The *COMT-H* polymorphism is associated with altered cortical dopamine D1 receptor availability in corticolimbic regions, and gray matter volume in the hippocampus and dorsolateral prefrontal cortex (Dickinson and Elvevåg 2009). As a consequence, neurons with COMT-H show higher levels of activation during certain cognitive tasks, as they require higher levels of neuron firing to achieve the same level of post-synaptic stimulation. Healthy carriers of *COMT-H* were previously found to less efficiently avoid perseverative errors than carriers of the *COMT-L* allele on tasks of executive functioning (Dickinson and Elvevåg 2009). Moreover, *COMT* polymorphisms have been studied extensively in neurological and psychiatric conditions involving catecholaminergic systems (Tunbridge, Harrison, and

Weinberger 2006; Dickinson and Elvevåg 2009). There is strong evidence to suggest that COMT-H affects the prefrontal cortex and associated areas, modulating executive functioning, working memory, and measures of attention (Tunbridge, Harrison, and Weinberger 2006; Dickinson and Elvevåg 2009).

The frequencies of the low-activity allele (*COMT-L*) vary significantly from 0.01 to 0.62, depending on the population. Europeans have nearly equal frequencies of the two alleles, while the high activity *COMT-H* allele is much more common in populations in all other parts of the world.

COMT and Sleep

A decrease in the activation and functioning of the prefrontal cortex may be critical for the waking consequences of sleep loss (Dang-Vu 2007), and this may, in part, relate to *COMT* polymorphisms. Certainly, *COMT* polymorphisms have been linked to inter-individual differences in a range of sleep-related parameters. A normal variation of the EEG is the low-voltage alpha (LVA) trait characterized by absence (or clear reduction) of a pronounced alpha rhythm in wakefulness. COMT-H is associated with an LVA phenotype (Enoch et al. 2003). Alpha EEG peak frequency in the waking EEG is 1.4 Hz lower in the *COMT-H* genotype than in the *COMT-L* genotype (Bodenmann et al. 2009a). In wakefulness, REM sleep and NREM sleep, *COMT-H* subjects have less EEG power in the eleven to thirteen Hz band than individuals with the *COMT-L* genotype. This difference is stable with repeat recordings one week apart. The differences persist despite concomitant sleep deprivation and the wakefulness promoter modafinil (Bodenmann et al. 2008a). *COMT-H* subjects also had decrements in executive function associated with these EEG changes. Genetic variation of *COMT* predicts robust inter-individual differences in functional aspects of the EEG. They suggest that mechanisms involving dopamine contribute to brain alpha oscillations in wakefulness and sleep. It is possible that the *COMT* polymorphism may predict characteristics in patients with sleep disorders—for example, contributing to impaired executive function related to sleep loss.

One important functional role of the *COMT* polymorphism has been collection of data related to modafinil pharmacogenomics. Modafinil may operate by promoting primarily dopaminergic neurotransmission. In men in whom the *COMT* polymorphism was characterized, modafinil potently improved vigor and well-being, and maintained baseline performance with respect to executive functioning and vigilant attention throughout forty hours of sleep deprivation in *COMT-H* subjects. Modafinil was hardly effective in subjects with the *COMT-L* genotype (Bodenmann 2009b; see Figure 10-2). OK Neither modafinil nor the *COMT-H* polymorphism affected distinct EEG markers of sleep homeostasis in recovery sleep.

Modafinil maintains cognitive performance during sleep deprivation in Val/Val (*COMT-H*) subjects only. Cognitive testing, including a ten-minute random-number generation (RNG) task preceded by a ten-minute psychomotor vigilance task (PVT), was administered every three hours across a forty-hour waking period beginning thirty minutes after awakening. The tick marks on x-axes are rounded up to the nearest hour. The time course of RNG redundancy (zero-order response stereotypy), PVT lapses (reaction times [RTs] > 500 ms, transformed by $\sqrt{x} + \sqrt{1 + x}$), slowest 10th percentile of reaction times (expressed as speed, 1/RT), and the 90th–10th interpercentile range of speed are illustrated. Broken vertical lines indicate 100 mg modafinil or placebo administration.

Importantly, these findings show that mechanisms involving prefrontal cortex dopamine contribute to specific aspects of impaired subjective state and cognitive performance after sleep deprivation. It may explain frequent anecdotal reports of either sensitivity or

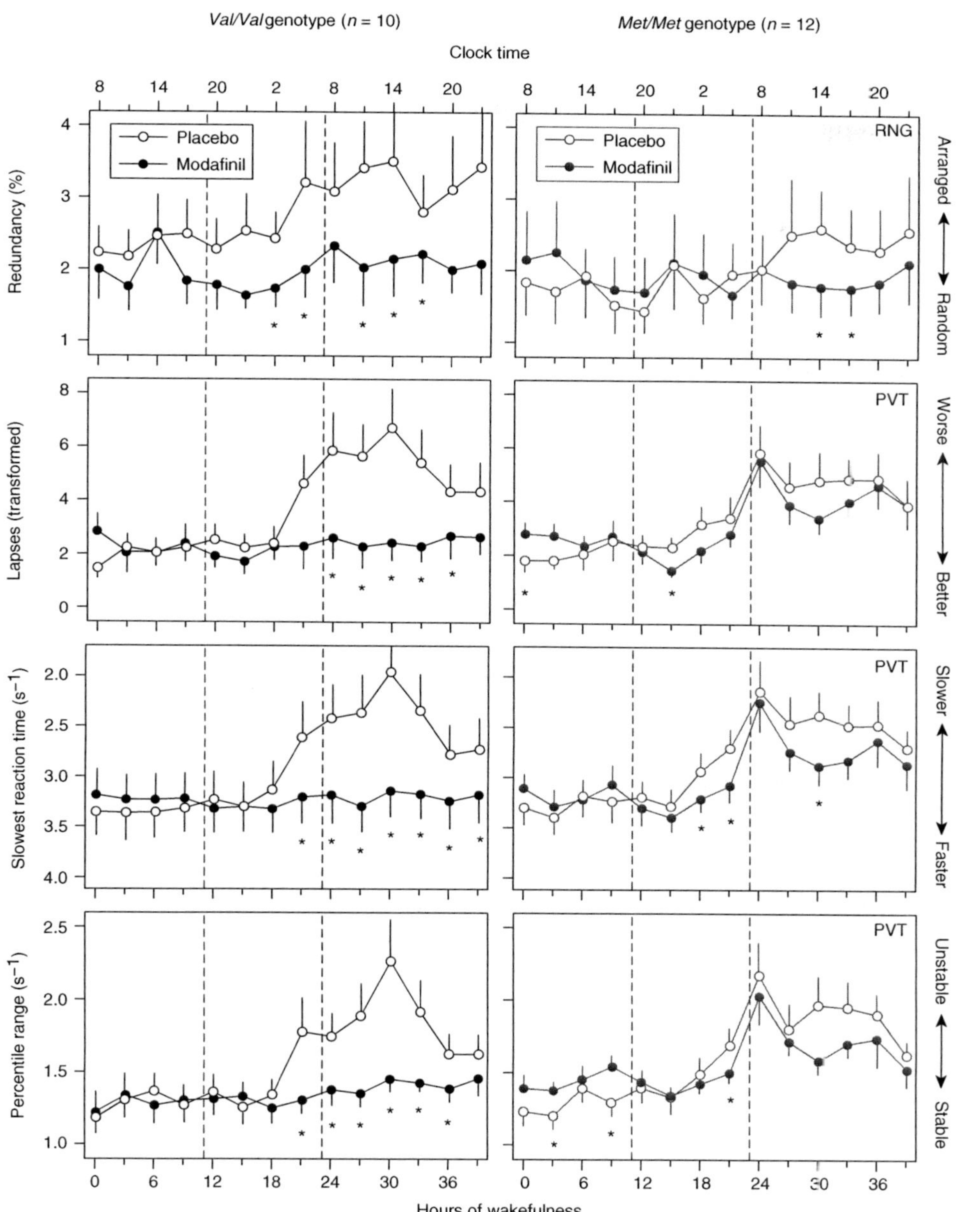

Figure 10-2 Pharmacogenomics for modafinil, a wakefulness promoter related to *COMT* polymorphism. **Black symbols:** *COMT-H* genotype (open circles: placebo condition; closed circles: modafinil condition). **Gray symbols:** *COMT-L* genotype (open circles: placebo condition; closed circles: modafinil condition). **Top panels:** modafinil reduced redundancy after sleep deprivation but the effect was higher in *COMT-H* than in *COMT-L* subjects. **Lower panels:** during sleep deprivation, modafinil maintained all measures of sustained vigilant attention (PVT lapses, slowest reaction times, interpercentile range) at baseline levels in *COMT-H* subjects, but was virtually ineffective in *COMT-L* subjects. Asterisks denote significant differences compared to the respective placebo values ($P \leq 0.05$, paired, two-tailed t-tests). Reproduced with permission from Bodenmann 2009b.

tolerance to the effects of modafinil. This is clearly a highly important future line of investigation. In this context, it is interesting to note that there is an apparent sexual dimorphism in *COMT* activity in human narcolepsy (Dauvilliers 2001). Women narcoleptics with high *COMT* activity fall asleep twice as quickly as those with low *COMT* activity during the multiple sleep latency test (MSLT), while the opposite was true for men (Dauvilliers 2001). *COMT* genotype also strongly affected the presence of sleep paralysis and the number of REM sleep onsets during the MSLT. Importantly, the *COMT* genotype distribution between men and women narcoleptics is associated with response to modafinil. The optimal daily dose of modafinil is approximately 100 mg lower in women narcoleptics, and lower in all narcoleptics with the low activity *COMT* genotype. It suggests that a sexual dimorphism in COMT activity affects the response to modafinil and probably to other dopaminergic stimulants. In ADHD, children with *COMT-H* had poorer sleep continuity on one week of sleep activity monitoring compared with children with the *COMT-L* genotype while receiving a placebo and while receiving methylphenidate (Gruber et al. 2006). However, in the case of Parkinson disease, there does not appear to be an explanatory effect of *COMT* polymorphisms for sleepiness and sleep attacks (Rissling et al. 2006).

PER3

Variations in genes involved in the generation of circadian rhythmicity have been recognized and have been related to some individual differences in sleep and circadian phenotypes (Takahashi et al. 2008; Dijk and Archer 2010). *PER3* is a member of the family of PERIOD genes. It has been identified widely in hypothalamic areas related to circadian rhythmicity in animal species and also expressed in other brain areas, including the gyrus dentatus, medial amygdaloid nucleus, cingulate cortex, hippocampal pyramidal cells, cerebral cortex and nucleus tractus solitarius. *PER3* is also expressed in the ventro-lateral preoptic nucleus (VLPO), which is a key area in sleep–wake regulation (Dijk and Archer 2010). In fact, expression of *PER3* in the VLPO appears to be in anti-phase with expression in the SCN.

In humans and other primates, the *PER3* gene contains a variable number tandem repeat (VNTR) polymorphism. In humans, a 54-nucleotide coding-region segment is repeated four or five times. This polymorphism is not present in non-primate mammals. Population studies have shown that in most populations, around 10 percent of individuals are homozygous for the 5-repeat ($PER3^{5/5}$), whereas approximately 50 percent are homozygous for the 4-repeat ($PER3^{4/4}$). In some populations in Papua New Guinea, the prevalence of the various genotypes appears to be reversed (Dijk and Archer 2010) though reasons for this are unknown.

The frequency of people homozygous for the 5-repeat is higher in the morning-preference types, or "larks," than in evening types, or "night owls." Moreover, in delayed sleep phase syndrome (DSPS) the prevalence of the 5-repeat allele is very low (Dijk and Archer 2010). This association may be age-dependent.

$PER3^{5/5}$ individuals displayed many of the sleep characteristics previously observed in "morning types." They have shorter sleep latency, more SWS, and more SWA, in particular, in the first part of the nocturnal sleep episode. Quantitative EEG analyses revealed frequency-specific changes in the EEG during NREM sleep (marked increase in alpha activity in both wakefulness and REM sleep, as well as the increase in delta activity in NREM sleep) that were very similar to the changes induced by sleep deprivation, and consistent with the concept that $PER3^{5/5}$ individuals live under a higher sleep pressure. Further evidence in support of this hypothesis was derived from the analysis of a recovery sleep episode in which the differences between the two genotypes persisted (Viola et al. 2007).

Analysis of the waking EEG during the sleep deprivation period demonstrated a more rapid increase of theta/alpha activity in *PER3*$^{5/5}$ individuals, similar to the more rapid increase of theta activity observed in morning types. *PER3*$^{5/5}$ individuals also showed increased slow eye movements during sleep deprivation. Alpha activity in REM sleep is increased and REM sleep is suppressed after sleep deprivation in those with *PER3*$^{5/5}$, suggesting that REM sleep pressure is lower, rather than higher, in this genotype. This can be interpreted within the context of the homeostatic regulation of SWS: more SWS leads to a suppression of REM sleep (Dijk and Archer 2010).

During sleep deprivation, and, in particular, during the morning hours approximately two to six hours after the melatonin peak, performance deteriorates faster in *PER3*$^{5/5}$ individuals, compared with *PER3*$^{4/4}$ individuals. Tasks probing executive function were most affected by the sleep deprivation in the vulnerable genotype (Groeger et al. 2008). This is important because executive functions have been considered to be most sensitive to the effects of sleep deprivation in general.

Using functional Magnetic Resonance Imaging (fMRI), brain responses to an executive task decline in *PER3*$^{5/5}$ individuals in a frontal area implicated in executive control (Vandewalle et al. 2009). This decline was not observed in *PER3*$^{4/4}$ individuals. In *PER3*$^{4/4}$ individuals, sleep loss was associated with an increase in blood-oxygen-level dependent (BOLD) responses in many frontal and parietal areas. By contrast, in *PER3*$^{5/5}$ individuals, sleep deprivation led to a reduction in BOLD responses in many frontal and parietal areas. Thus, these fMRI data confirm the differential response to sleep loss in the two genotypes consistent with previous observations.

During chronic partial sleep deprivation (PSD), *PER3*$^{5/5}$ subjects also exhibited slightly, but reliably, higher SWA than *PER3*$^{4/4}$ subjects. However, unlike with total sleep deprivation (above), *PER3* polymorphism variants did not differ on baseline sleep measures or in their physiological sleepiness, or cognitive executive functioning, or subjective responses to chronic PSD. Thus, the *PER3* VNTR polymorphism is not necessarily a genetic marker of differential vulnerability to the cumulative neurobehavioral effects of chronic PSD, unlike total SD. More work is required to confirm whether the *PER3* VNTR is a genuine, practical genetic marker for individual differences in the vulnerability to sleep loss (Dijk and Archer 2010).

Predicting Individual Differences in Vulnerability to Sleep Loss

There could be considerable benefit to accurately predicting who is resistant or vulnerable to the adverse effects of sleep loss. A relatively small percentage of individuals may account for most of the risk posed by occupational fatigue and, therefore, they are a priority for identification. Sleep disorders and other clinical conditions can put individuals at elevated risk of performance impairment during sleep deprivation, but variability among individuals in the degree of impairment due to sleep loss may nevertheless be substantial. However, identifying these individuals requires precision that would endure legal scrutiny. Examination of demographics, age, baseline cognitive functioning, sleep traits and habits, circadian rhythm profiles, standard clinical outcomes of blood and urine chemistry, and psychological traits such as personality, has yielded no viable candidate predictors of vulnerability.

Low baseline levels of global brain activation, as measured with fMRI, make an individual more vulnerable to the effects of sleep deprivation on a working memory task, but this type of examination is impractical and not fully verified (Dang-Vu et al. 2007). Therefore, the use of genetic predictors (see above) of responsiveness to sleep deprivation is attractive. It is still unknown how much of the inter-individual variance in responses to sleep loss in the general population can be explained by these genetic predictors, nor do we know what their role would be in a PM approach to treatment. Another option for the prediction of performance impairment due to sleep loss at the level of individuals involves the use of biomathematical models of fatigue and performance.

Key Sleep Disorders: A Personalized Medicine Agenda

Obstructive Sleep Apnea

Sleep breathing disorders are categorized across a continuum. This spectrum includes simple snoring, more habitual forms of snoring, mild fragmentation of sleep due to complete or partial airway obstruction, more severe forms of sleep apnea with longer and more frequent breathing pauses and, finally, respiratory failure during sleep, often resulting over time in respiratory failure during wakefulness. The most commonly presented form of sleep breathing disorder in clinical practice is obstructive sleep apnea. This occurs in a symptomatic form in 5 percent of the adult population; sub-clinical forms exist in over 25 percent of the population (Bearpark et al. 2005). There is a strong association between sleep apnea, cardiovascular disorders such as hypertension, and cardiovascular risk parameters (Marshall et al. 2008). This association may be primarily mediated by confounding due to coexisting central obesity, but continuous positive airway pressure (CPAP) intervention studies have highlighted that obstructive sleep apnea alone may confer the cardiovascular risk (Marin et al. 2005). The most common symptom, apart from snoring, is excessive daytime sleepiness due to sleep fragmentation from arousals caused by upper airway obstruction, as well as a contribution from exposure to hypoxemia during sleep. There are a great variety of symptomatic presentations. One of the important characteristics of an obstructive sleep apnea (OSA) is the absence of reliable pharmacotherapy (Marshall et al. 2008), in contrast to many other disorders, such as hypertension, arthritis and lipid disorders, for which there are medications available.

The dominant forms of therapy for obstructive sleep apnea are mechanical or surgical (Cistulli and Grunstein 2005). CPAP machines and mandibular advancement dental splints are the most commonly prescribed forms of mechanical therapy. Surgery is used for mild forms of sleep apnea and heavy snoring, but is less commonly deployed in patients with more severe forms of the disease. Given the relatively invasive nature of mechanical treatments, such as CPAP, it is not surprising that adherence problems exist for these therapies. It is important to note that although constant improvements are occurring in the technology of mechanical treatments, adherence is largely determined by behavioral factors (Weaver and Grunstein 2008).

The severity of the sleep breathing disorder is usually measured objectively by counting abnormal breathing and/or hypoxemic events during sleep investigation or overnight respiratory monitoring. This paradigm is changing, as simpler methods of "triage" diagnosis (Rofail et al. 2010), or even biometric methods of screening, will be deployed (Lee et al. 2009). Excessive daytime sleepiness is measured by subjective report using common instruments, such as the Epworth Sleepiness Scale, or less commonly objective testing of sleepiness by quite laborious methods, such as the multiple sleep latency or maintenance of wakefulness tests (Sullivan and Kushida 2008). Importantly, the relationship between frequency of abnormal breathing events and sleepiness is not robust (Patel et al. 2004; Dempsey et al. 2010). Many patients with severe forms of sleep apnea do not feel sleepy and report reasonable productivity in the workplace. Others, with milder forms of the disease, complain on investigation of marked decrement in daytime alertness. Differences between individuals with OSA may be directly related to the nature of their sleep disordered breathing events. Some individuals with OSA have marked sleep fragmentation with little, if any, intermittent hypoxia, while others may show marked intermittent hypoxia with lesser degrees of fragmentation. Within an individual, the magnitude of hypoxia and arousals may vary across the sleep period (Dempsey et al. 2010).

The discrepancy between severity of the sleep breathing disorder on neurophysiological testing in the sleep laboratory and symptoms is a great challenge to therapy and provides an opportunity for a PM approach. In this context, the presence of sleepiness is an important determinant of outcomes of therapy, such as CPAP. Patients who have significant sleep apnea but do not have sleepiness comply poorly with CPAP (Weaver and Grunstein 2008). Some researchers have argued that even some of the vascular consequences of sleep apnea are different in patients who have daytime sleepiness and those without this complaint (Robinson et al. 2008). We know that many individuals with OSA do not have excessive sleepiness, and only about 50 percent of OSA subjects develop hypertension. See above now The question is, why do some subjects with OSA develop hypertension and cardiovascular disease while others do not? One possibility is that individuals who develop hypertension and cardiovascular disease have increased pathogenic responses or decreased adaptive responses for equivalent degrees of OSA. Such heterogeneity in response might be related to differences in protective mechanisms. For example, individuals vary in their antioxidant ability, which is complex, as it involves dietary factors and multiple intrinsic antioxidant systems. Candidates for measurement are numerous, including antioxidant enzymes (such as superoxide dismutases and glutathione peroxidases), non-enzymatic antioxidants (such as glutathione, bilirubin and vitamins C and E), and melatonin, which has a role in the regulation of antioxidant enzyme activity and expression. Polymorphisms that may relate to these candidates are also of interest.

Adaptive responses that affect sympathetic activation are also potentially heterogeneous between individuals with OSA. Likely candidates include differential downregulation of adrenergic receptors and differences in norepinephrine clearance rates, both of which act to diminish the stimulatory response and the increased sympathetic activity in OSA (Dempsey et al. 2010).

Differences between those with OSA with comorbidities and those without could be genetic in origin. There is now a large literature on genes conferring risk for hypertension, insulin resistance, and cardiovascular disease. Polymorphisms in genes affecting oxidative stress, inflammation, and sympathetic activity are also of high interest. There are a limited but growing number of candidate gene studies addressing the question of gene variants and their pathophysiological effect in the OSA population. Proteomic and microarray approaches are being initiated (Arnardottir et al. 2009).

Given the association between vascular disorders and sleep apnea, it has been argued that patients with moderate to severe forms of sleep apnea without sleepiness should still be prescribed CPAP treatment on a prognostic rationale. This is naturally difficult, when the main therapy is mechanical and intrusive, rather than once-per-day pharmacotherapy. This conundrum may be improved by PM approaches that characterize the degree of cardiovascular risk in a particular patient with OSA and the importance of treatment.

A similar rationale could be developed for the neurobiological consequences of OSA. Recurrent sleep fragmentation and hypoxemia may cause neurobiological damage in the absence of symptoms. Prediction of this effect would be helpful in OSA. The corollary is that if patients are at reduced risk because of specific phenotypes and genotypes, treatment may not be required. The challenge is to find these simpler biomarkers that would allow the better characterization of patient risk and potential outcome. The advancing research on traits for sleepiness may provide future understanding of inter-individual differences in the neurobiological effects of sleep apnea.

Personalized Medicine in OSA: Biomarkers, Pharmacogenomics, and Therapyxogenomics

It is unlikely that a new dominant form of pharmacotherapy will be available in the immediate future specifically for OSA. Anti-obesity medications have an impact in most cases of sleep apnea, but there is only a partial response to these therapies (Yee et al. 2007). Large inter-individual differences occur. It is possible that there may be particular patient phenotypes, such as adiposity distribution in the face and upper airway, that may be more responsive to weight loss, and that may have an impact on sleep apnea severity. For example, a drug that has a specific effect on central fat stores or selective reduction of airway or neck fat would be useful in this context. Similarly, it is now postulated that certain forms of breathing regulatory phenotypes may be important in pathogenesis of OSA (Dempsey et al. 2010). These involve sensitivity of the respiratory system to perturbations from arousal and asphyxia, labeled as the "loop gain" of the system (Wellman et al. 2008). Certain drugs or agents may modify loop gain through central or peripheral actions. Developing simple markers to detect loop gain differences between patients could inform treatment approaches, particularly in those not compliant with mechanical treatments.

Nevertheless, mechanical treatments are the mainstay of treatment in the short and medium term. Although CPAP is the most efficacious treatment and is most likely to reduce OSA severity, dental splints may be just as effective, as patient compliance is higher (Cistulli and Grunstein 2005). What are the determinants of treatment effectiveness? For example, a number of airway measurements may predict efficacy of mandibular advancement splints. Determinants of airway anatomy and a similar predictive approach have been used before surgery (Chan and Cistulli 2010; Chan et al. 2010). Moreover, as discussed above, the absence of sleepiness is certainly a phenotype that is associated with poor and often futile compliance with CPAP. As cognitive-behavioral therapy (CBT) approaches dramatically improve CPAP compliance, are there biomarkers that could predict response to CBT?

With pharmacological treatments, pharmacogenomic approaches have stimulated the PM agenda. There is no apparent recognition of potential for similar advances for mechanical treatments. We have coined the term *therapyxogenomics* (from the ancient Greek *pyxion*, for "box") to describe future potential approaches that will combine genetic information with response and outcomes of mechanical therapies—particularly, but not exclusively, for sleep apnea. Therapyxogenomics would involve such diverse methods as linking genes that have an impact on cognition, respiratory control, and airway anatomy with treatment prediction and outcome. Some genes influencing respiratory control have been identified, so this may not be entirely speculative (Borday et al. 2003).

An outline of how a PM agenda could occur in sleep apnea is described in Figure 10-3. First, it is important to establish whether the disease has an immediate impact or not, recognizing that there may be a potential impact "down the line" in some patients with increased cardiovascular disease or neurobiological deficits. This approach requires evidence that this latter issue is a real problem. Perhaps for these patients a strategy for neuro- or cardioprotection is the best alternative. Recognizing the presence of immediate disease impact requires accurate phenotyping. Selection of treatment again will require biomarker selection. Given the problems with delivering effective treatment (poor compliance with

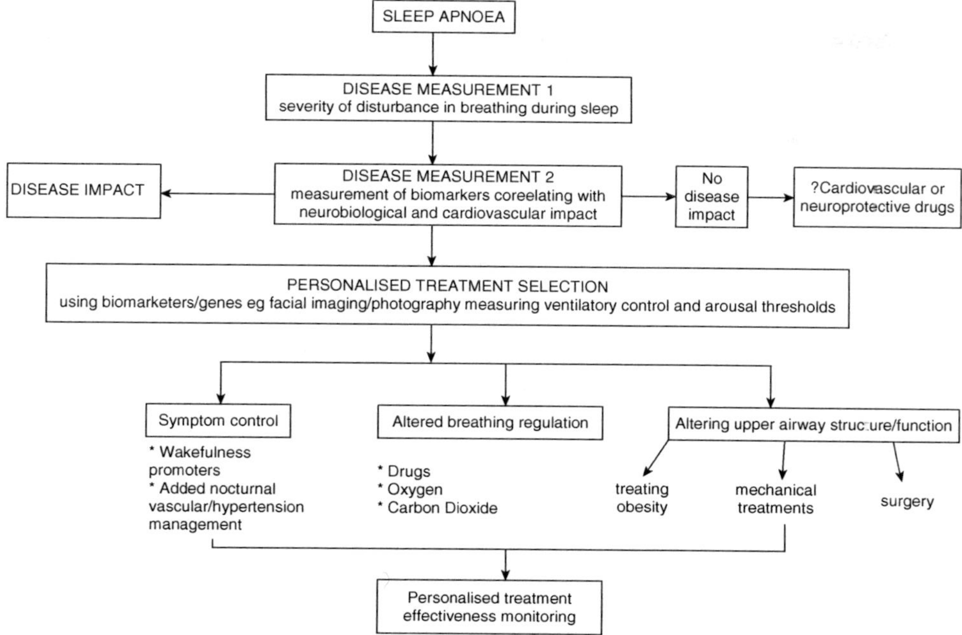

Figure 10-3 Potential personalized medicine management approach to sleep apnea.

mechanical treatments and limited efficacy of alternatives), an approach that is personalized based on biomarkers that can predict treatment response is attractive.

Personalized Medicine in Insomnia

Insomnia is the most common sleep disorder. Nearly 30 percent of adults have insomnia-type complaints, and approximately 5 percent of adults complain of chronic insomnia with impaired daytime function (Mai and Buysse 2008; Ohayon and Reynolds 2009). Insomnia is more common in women and the elderly and is also a common feature of depression and chronic pain states. However, obvious inter-individual differences exist (Buysse 2000; Mai and Buysse 2008). For example, not all patients with depression suffer from insomnia, and some suffer from excessive daytime sleepiness (Buysse 2000). The biological explanation for this variation is unknown. Similarly, patients may suffer similar degrees of pain during wakefulness, but only a subgroup of will develop problems with insomnia. It would be reasonable so speculate that genetic factors may cause some of this variability (see section on adenosine, above). In one small group of patients with insomnia—children with autism spectrum disorder (ASD)—presence of polymorphisms related to melatonin synthesis may prove relevant to management by indicating which patients may benefit from melatonin treatment for insomnia (Melke et al. 2008; Wasdell et al. 2008).

There is a range of treatment options available for insomnia. Insomnia has been a major target for pharmaceutical companies, which have developed a range of hypnotics. However, drug development recently has been quite problematic because of the realization that not all hypnotics are effective in different insomnia phenotypes. There also is emerging evidence for the success of behavioral treatments, such as CBT (Morin 2009).

Selection of treatment options for an individual patient are often based on referral patterns and geography of resources, rather than any evidence-based, individualized approach. For example, in Australia, patients presenting to primary care are prescribed hypnotics, preferentially, rather than CBT-I (CBT for insomnia). Use of biomarkers for phenotyping insomnia is rare, although recently it has been suggested that specific EEG markers may predict of efficacy of CBT-I (Krystal and Edinger 2010). At this stage, biomarkers or genetic analyses have not been used to predict treatment outcome in insomnia in larger-scale studies.

One insomnia-producing disorder, restless legs syndrome, has a strong genetic basis (Trenkwalder, Högl, Winkelmann 2009), but how these genetic discoveries influence diagnosis and treatment is not known. Interestingly, given nascent associations between insomnia and cardiovascular risk, increased EEG beta-rhythm activity (a marker of brain activation during sleep) has been linked to higher nocturnal blood pressure in patients with insomnia (Lanfranchi et al. 2009). This type of finding may well play a role in individualized patient selection for treatment. However, selecting patients on the basis of specific EEG in markers, such as high beta-rhythm activity, has not been done in clinical trials.

A potential schema of the role of PM in insomnia is included in Figure 10-4. This figure illustrates the potential for treatment of sleep disorders as an exemplar of personalized medicine, with genomic brain EEG and body sympathetic arousal markers providing much information for consideration of personalized treatment options that include sleep

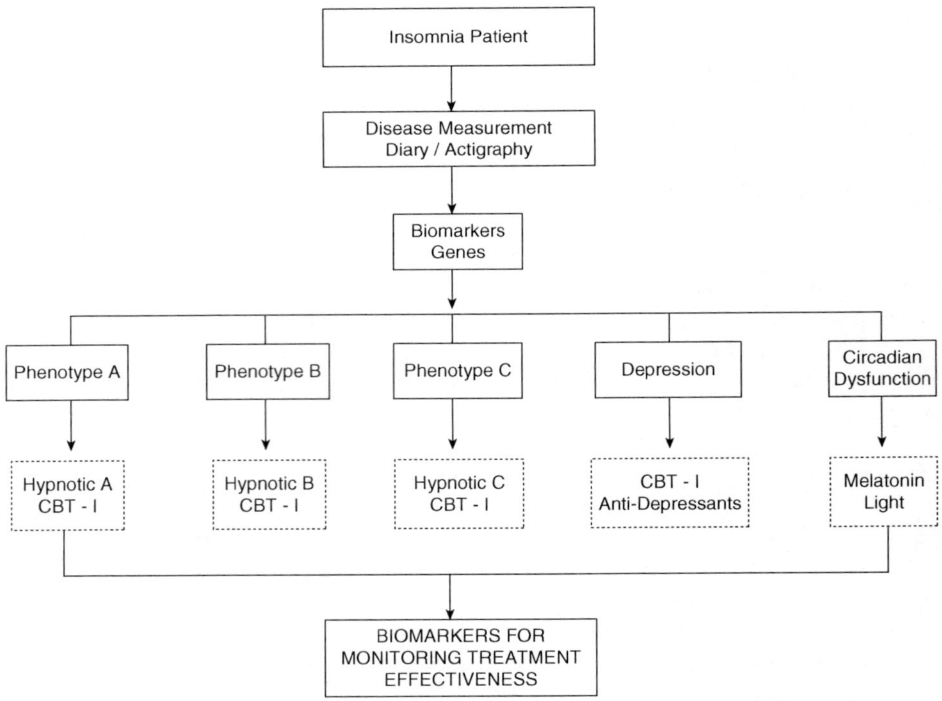

Figure 10-4 Putative personalized medicine approach in insomnia. Insomnia is a heterogeneous disorder. Many researchers are working on uncovering biomarkers for these various phenotypes. Treatment effectiveness is important in developing a PM agenda for this condition (CBT-I = cognitive behavioral therapy for insomnia).

hygiene, CBT-I, mechanical intervention (CPAP and mandibular splints), weight loss, and hypnotics.

Insomnia is often treated as a homogeneous clinical problem, whn, in fact, there are many phenotypes related to the symptom complex, including presence of shifts in the circadian rhythm for melatonin/temperature (circadian dysfunction) and comorbid depression or anxiety. Even polymorphisms related to adenosine or clock genes may be relevant. Development of biomarkers for these phenotypes would add to the management pathway, as well as aid monitoring of treatment effectiveness.

Limitations in the Field and Speculations About the Future of Personalized Medicine and Health Care

Clinical integration of a PM agenda barely exists in sleep medicine. Recognition of basic phenotypes (e.g., facial dysmorphia in OSA or absence of sleepiness in a patient with moderate sleep apnea) influence decision-making in terms of prescribing therapies, but an integrative interdisciplinary approach has not been implemented. One of the major limitations in sleep medicine is the lack of large-scale clinical trials to inform decision-making using phenotype, or even genotype, as a basis for the decision. In the U. S., there is a reasonably well-resourced professional organization for sleep medicine clinicians, but its focus is essentially on promotion of the medical discipline and the economic issues of its members. There is a lack of a driving force toward larger-scales studies, standardization of phenotype measurement, and integration of genetic data collection.

Certainly, one important way forward for the development of a PM approach in sleep health is better standardization and networking of clinical centers. Literally millions of sleep studies have been performed worldwide, but there is not integration of these data in a standardized format with basic phenotype collection. The risk now is that sleep centers may contract with the advent of simplified diagnosis of sleep apnea. In fact, this is a strong rationale for maintenance of such centers in sleep disorder treatment pathways, given the interdisciplinary growth of the field and the recognition of individualized medicine in care. Their survival depends on a "hub and spoke" model of chronic care in sleep medicine that could be driven by PM, using evidence-based data on biomarkers related to "burden of disease" and treatment effectiveness.

At this point, there is an emerging effort to establish databases of large patient datasets which have been obtained using standardized measures. In Europe, the European Sleep Apnea Data base (ESADA), funded by the European Union, is an attempt to do this. In the United States and Australasia, sleep-center consortia are being formed to drive standardization of databases. This is mirrored by efforts to develop international clinical trial networks. Given that most academic centers still use polygraphy for diagnosis of sleep problems, data can be reduced to various common electronic formats. Inclusion of sleep measures in other disease databases would also serve to advance an integrative healthcare agenda.

Glossary

Electroencephalogram (EEG): *Measure of electrical potentials at the surface of the head.*

Wakefulness: *Vigilance state characterized by a low amplitude, high frequency, mixed EEG pattern.*

NREM sleep:	*Vigilance state characterized by high amplitude, low frequency oscillations, dominated by the slow/delta and spindle oscillations (see below) and a relaxed muscle tone.*
REM sleep:	*Vigilance state characterized by an EEG resembling wake or stage-1 sleep in humans, in association with muscle atonia.*
Slow wave sleep (SWS):	*The deepest stages of NREM sleep (stages 3 and 4 in humans) during which slow/ delta waves are especially prevalent and arousal thresholds highest.*
Slow/delta waves:	*EEG oscillations within the 0.75–4.5 Hz frequency range. High amplitude slow/ delta waves are a defining feature of NREM sleep stages-3 and 4, i.e., SWS.*
Slow wave activity (SWA) (or delta power):	*Mathematical variable extracted from the EEG that quantifies the amplitude and prevalence of slow/delta waves. The relative value of SWA at the onset of NREM sleep can be accurately predicted on the basis of prior wake duration. Because of this, SWA is widely acknowledged to reflect sleep need and is thought to reflect a sleep homeostatic process.*
Sleep propensity:	*Probability or tendency to be in, or to transition into, a sleep state. At the global level, sleep propensity can be assessed behaviorally. At the local level, it can be assessed using electrophysiological markers of the sleep state.*
Sleep spindles:	*Transient (0.5–2 s), spindle-shaped 10–15 Hz EEG features that prevail during early sleep (stage 2 in humans), and also herald NREM-REM sleep transitions. Sigma activity refers to the prevalence and amplitude of spindle oscillations in the 10–15 Hz range.*
Theta oscillations:	*EEG oscillations in the 5–10 Hz range.*
Synaptic strength (or weight):	*Input–output relationship of a synapse.*

References

Acherman, P., and A. A. Borbély. 2003. Mathematical models of sleep regulation. *Front Biosci* 8: S683–S693.

Adam, M, J. V. Rétey, R. Khatami, and H. P. Landolt. 2006. Age-related changes in the time course of vigilant attention during 40 hours without sleep in men. *Sleep* 29: 55–57.

Arnardottir, E. S., M. Mackiewicz, T. Gislason, K. L. Teff, and A. I. Pacl. 2009. Molecular signatures of obstructive sleep apnea in adults: a review and perspective. *Sleep* 32(4): 447–70.

Banks, S, Dinges D. F. 2007. Behavioral and physiological consequences of sleep restriction. *J Clin Sleep Med* 3: 519–28.

Basner, M., J. Rubinstein, K. M. Fomberstein, M. C. Coble, A. Ecker, D. Avinash, and D. F. Dinges. 2008. Effects of night work, sleep loss and time on task on simulated threat detection performance. *Sleep* 31(9): 1251–59.

Bearpark, H., L. Elliott, R. Grunstein, S. Cullen, H. Schneider, W. Althaus, and C. Sullivan. 1995. Snoring and sleep apnea. A population study in Australian men. *Am J Respir Crit Care Med.* 151(5): 1459–65.

Bodenmann, S., C. Hohoff, C. Grietag, J. Deckert, J. Rétey, and H. P. Landolt. 2007. Genetic variation in the adenosine A2A receptor gene modulates performance on the psychomotor vigilance task. *Sleep and Biological Rhythms* 5: A47.

Bodenmann, S., T. Rusterholz, R. Dürr, C. Stoll, V. Bachmann, E. Geissler, K. Jaggi-Schwarz, and H. P. Landolt. 2009a. The functional Val158Met polymorphism of COMT predicts interindividual differences in brain alpha oscillations in young men. *J Neurosci.* 29: 10855–62.

Bodenmann, S., S. Xu, U. F. Luhmann, M. Arand, W. Berger, H. H. Jung, and H. P. Landolt. 2009b. Pharmacogenetics of modafinil after sleep loss: Catechol-O-methyltransferase genotype modulates waking functions but not recovery sleep. *Clin Pharmacol Ther.* 85(3): 296–304.

Borbély, A. A. 1982. A two process model of sleep regulation. Hum Neurobiol. 1: 195–204.

Borbély AA, Achermann P. 1999. Sleep homeostasis and models of sleep regulation. *J Biol Rhythms.* 14: 557–68.

Borday, C., V. Abadie, F. Chatonnet, M. Thoby-Brisson, J. Champagnat, and G. Fortin. 2003. Developmental molecular switches regulating breathing patterns in CNS. *Respir Physiol Neurobiol.* 135: 121–32.

Buckelmüller, J., H. P. Landolt, H. H. Stassen, and P. Achermann. 2006. Trait-like individual differences in the human sleep electroencephalogram. *Neuroscience* 138: 351–56.

Buysse, D. J. 2000. Rational pharmacotherapy for insomnia: Time for a new paradigm. *Sleep Med Rev.* 4: 521–27.

Chan, A. S., and P. A. Cistulli. 2010 (in press). Oral appliance treatment of obstructive sleep apnea: An update. *Curr Opin Pulm Med.*

Chan, A. S., R. W. Lee, V. K. Srinivasan, M. A. Darendeliler, R. R. Grunstein, and P. A. Cistulli. 2010 (in press). Nasopharyngoscopic evaluation of oral appliance therapy for obstructive sleep apnoea. *Eur Respir J.*

Chuah, Y. M., V. Venkatraman, D. F. Dinges, and M. W. Chee. 2006. The neural basis of interindividual variability in inhibitory efficiency after sleep deprivation. *J Neurosci.* 26: 7156–62.

Cistulli, P. A., and R. R. Grunstein. 2005. Medical devices for the diagnosis and treatment of obstructive sleep apnea. *Expert Rev Med Devices* 2(6): 749–63.

Cohen, D. A., W. Wang, J. K. Wyatt, R. E. Kronauer, D. J. Dijk, C. A. Czeisler, and E. B. Klerman. 2010. Uncovering residual effects of chronic sleep loss on human performance. *Sci. Transl. Med.* 2: 14ra3.

Colten, H. R., and B. M. Alteveogt, eds. 2006. *Sleep disorders and sleep deprivation: An unmet public health problem.* Washington, D.C.: National Academies Press.

Cornelis, M. C., A. El Sohemy, and H. Campos. 2007. Genetic polymorphism of the adenosine A2A receptor is associated with habitual caffeine consumption. *Am J Clin Nutr.* 86(1): 240–44.

Czeisler, C. A., J. K. Walsh, T. Roth, R. J. Hughes, K. P. Wright, Jr., L. Kingsbury, et al. 2005. Modafinil for excessive sleepiness associated with shift work sleep disorder. *N Engl J Med.* 353: 476–86.

Dang-Vu, T. T., M. Desseilles, D. Petit, S. Mazza, J. Montplaisir, and P. Maquet. 2007. Neuroimaging in sleep medicine. *Sleep Med.* 8: 349–72.

Dauvilliers, Y., E. Neidhart, M. Lecendreux, M. Billiard, and M. Tafti. 2001. MAO-A and COMT polymorphisms and gene effects in narcolepsy. *Mol Psychiatry.* 6: 367–72.

De Gennaro, L, C. Marzano, F. Fratello, F. Moroni, M. C. Pellicciari, F. Ferlazzo, S. Costa, A. Couyoumdjian, G. Curcio, E. Sforza, A. Malafosse, L. A. Finelli, P. Pasqualetti, M. Ferrara, M. Bertini, and P. M. Rossini. 2008. The electroencephalographic fingerprint of sleep is genetically determined: A twin study. *Ann Neurol.* 64(4): 455–60.

Deboer, T., M. J. VanSteensel, L. Detari, and J. H. Meijer. 2003. Sleep states alter activity of suprachiasmatic nucleus neurons, *Nat Neurosci.* 6: 1086–90.

Dempsey, J. A., S. C. Veasey, B. J. Morgan, and C. P. O'Donnell. 2010. Pathophysiology of sleep apnea. *Physiol Rev.* 90: 47–112.

Dickinson, D., and B. Elvevåg. 2009. Genes, cognition and brain through a COMT lens. *Neuroscience.* 164: 72–87.

Dijk, D. J., and S. N. Archer. 2010 (in press). PERIOD3, circadian phenotypes, and sleep homeostasis. *Sleep Med Rev.*

Dijk, D. J., D. G. M. Beersma, and S. Daan. 1987. EEG power density during nap sleep: Reflection of an hourglass measuring the duration of prior wakefulness. *J Biol Rhythms* 2: 207–19.

Dijk, D. J., and M. Von Schantz. 2005. Timing and consolidation of human sleep, wakefulness, and performance by a symphony of oscillators. *J Biol Rhythms* 20: 279–90.

Duffy, J. F., H. J. Willson, W. Wang, and C. A. Czeisler. 2009. Healthy older adults better tolerate sleep deprivation than young adults. *J Am Geriatr Soc.* 57: 1245–51.

Enoch, M. A., K. Xu, E. Ferro, C. R. Harris, and D. Goldman. 2003. Genetic origins of anxiety in women: A role for a functional catechol-O-methyltransferase polymorphism. *Psychiatr Genet.* 13: 33–41.

Franken, P., and D. J. Dijk. 2009. Circadian clock genes and sleep homeostasis, *Eur J Neurosci.* 29: 1820–29.

Frey, D. J., P. Badia, and K. P. Wright, Jr. 2004. Inter- and intra-individual variability in performance near the circadian nadir during sleep deprivation. *J Sleep Res.* 13(4): 305–15.

Groeger, J. A., A. U. Viola, J. C. Lo, S. M. von Archer, and D. J. Dijk. 2008. Early morning executive functioning during sleep deprivation is compromised by a PERIOD3 polymorphism. *Sleep* 31: 1159–67.

Gruber, R., N. Grizenko, G. Schwartz, L. Ben Amor, J. Gauthier, R. de Guzman, and R. Joober. 2006. Sleep and COMT polymorphism in ADHD children: Preliminary actigraphic data. *J Am Acad Child Adolesc Psychiatry* 45: 982–89.

Horne, J. A., and O. Östberg. 1977. Individual differences in human circadian rhythms, *Biol Psychol.* 5: 179–90.

Klimesch, W. 1999. EEG alpha and theta oscillations reflect cognitive and memory performance: A review and analysis. *Brain Res Brain Res Rev.* 29: 169–95.

Krueger, J. M., D. M. Rector, S. Roy, H. P. Van Dongen, G. Belenky, and J. Panksepp. 2008. Sleep as a fundamental property of neuronal assemblies. *Nat Rev Neurosci.* 9: 910–19.

Kryger, M. H., W. C. Dement, and T. Roth. 2009. *Principles and practice of sleep medicine*, 5th ed. Philadelphia: W. B. Saunders.

Krystal, A. D., and J. D. Edinger. 2010 (in press). Sleep EEG predictors and correlates of the response to cognitive behavioral therapy for insomnia. *Sleep.*

Landolt, H. P. 2008. Sleep homeostasis: A role for adenosine in humans? *Biochem Pharmacol.* 75: 2070–79.

Landolt, H. P., J. V. Rétey, K. Tönz, J. M. Gottselig, R. Khatami, I. Buckelmüller, and P. Achermann. 2004. Caffeine attenuates waking and sleep electroencephalographic markers of sleep homeostasis in humans. *Neuropsychopharmacology* 29(10): 1933–39.

Lee, R. W., P. Petocz, T. Prvan, A. S. Chan, R. R. Grunstein, and P. A. Cistulli. 2009. Prediction of obstructive sleep apnea with craniofacial photographic analysis. *Sleep* 32(1): 46–52.

Leproult, R., E. F. Colecchia, A. M. Berardi, R. Stickgold, S. M. Kosslyn, and E. Van Cauter. 2003. Individual differences in subjective and objective alertness during sleep deprivation are stable and unrelated. *Am J Physiol Regul Integr Comp Physiol.* 284: R280–90.

Mai, E., and D. J. Buysse. 2008. Insomnia: Prevalence, impact, pathogenesis, differential diagnosis, and evaluation. *Sleep Med Clin.* 3: 167–74.

Marin, J. M., S. J. Carrizo, E. Vicente, and A. G. Agusti. 2005. Long-term cardiovascular outcomes in men with obstructive sleep apnoea-hypopnoea with or without treatment with continuous positive airway pressure: An observational study. *Lancet* 365: 1046–53.

Marshall, N. S., K. K. Wong, P. Y. Liu, S. R. Cullen, M. W. Knuiman, and R. R. Grunstein. 2008. Sleep apnea as an independent risk factor for all-cause mortality: The Busselton Health Study. *Sleep* 31(8): 1079–85.

Marshall, N. S., B. J. Yee, A. V. Desai, P. R. Buchanan, K. K. Wong, R. Crompton, K. L. Melehan, N. Zack, S. G. Rao, R. M. Gendreau, J. Kranzler, and R. R. Grunstein. 2008b. Two randomized placebo-controlled trials to evaluate the efficacy and tolerability of mirtazapine for the treatment of obstructive sleep apnea. *Sleep* 31: 824–31.

Melke, J., H. Goubran Botros, P. Chaste, C. Betancur, G. Nygren, H. Anckarsäter, M. Rastam, O. Ståhlberg, I. C. Gillberg, R. Delorme, N. Chabane, M. C. Mouren-Simeoni, F. Fauchereau, C. M. Durand, F. Chevalier, X. Drouot, C. Collet, J. M. Launay, M. Leboyer, C. Gillberg, and T. Bourgeron. 2008. Abnormal melatonin synthesis in autism spectrum disorders. *Mol Psychiatry* 13: 90–98.

Morin, C. M., A. Vallières, B. Guay, H. Ivers, J. Savard, C. Mérette, C. Bastien, and L. Baillargeon. 2009. Cognitive behavioral therapy, singly and combined with medication, for persistent insomnia: A randomized controlled trial. *JAMA* 301: 2005–15.

Ohayon, M. M., and C. F. Reynolds, III. 2009. Epidemiological and clinical relevance of insomnia diagnosis algorithms according to the DSM-IV and the International Classification of Sleep Disorders ICSD. *Sleep Med.*10: 952–60.

Patel, S. R., D. P. White, A. Malhotra, M. L. Stanchina, and N. T. Ayas. 2003. Continuous positive airway pressure therapy for treating sleepiness in a diverse population with obstructive sleep apnea: Results of a meta-analysis. *Arch Intern Med.* 163: 565–71.

Phillips, A. J., and P. A. Robinson. 2007. A quantitative model of sleep–wake dynamics based on the physiology of the brainstem ascending arousal system. *J Biol Rhythms* 22: 167–79.

Porkka-Heiskanen, T., R. E. Strecker, M. Thakkar, A. A. Bj rkum, R. W. Greene, and R. W. McCarley. 1997. Adenosine: A mediator of the sleep-inducing effects of prolonged wakefulness. *Science* 276: 1265–68.

Rétey, J. V., M. Adam, J. M. Gottselig, R. Khatami, R. Dürr, P. Achermann, and H. P. Landolt. 2006. Adenosinergic mechanisms contribute to individual differences in sleep deprivation-induced changes in neurobehavioral function and brain rhythmic activity. *J Neurosci.* 26(41): 10472–79.

Rétey, J. V., M. Adam, E. Honegger, R. Khatami, U. F. Luhmann, H. H. Jung, W. Berger, and H. P. Landolt. 2005. A functional genetic variation of adenosine deaminase affects the duration and intensity of deep sleep in humans. *Proc Natl Acad Sci U S A.* 25(102): 15676–81.

Rétey, J. V., M. Adam, R. Khatami, U. F. Luhmann, H. H. Jung, W. Berger, et al. 2007. A genetic variation in the adenosine A2A receptor gene ADORA2A contributes to individual sensitivity to caffeine effects on sleep. *Clin Pharmacol Ther.* 81(5): 692–98.

Rissling, I., B. Frauscher, F. Kronenberg, M. Tafti, K. Stiasny-Kolster, A. C. Robyr, Y. Körner, W. H. Oertel, W. Poewe, B. Högl, and J. C. Möller. 2006. Daytime sleepiness and the COMT val158met polymorphism in patients with Parkinson disease. *Sleep* 29: 108–11.

Robinson, G. V., B. A. Langford, D. M. Smith, and J. R. Stradling. 2008. Predictors of blood pressure fall with continuous positive airway pressure CPAP treatment of obstructive sleep apnoea OSA. *Thorax* 63(10): 855–59.

Rofail, L. M., K. K. H. Wong, G. Unger, G. B. Marks, and R. R. Grunstein. 2010 (in press). The utility of single channel nasal airflow pressure transducer in the diagnosis of OSA at home. *Sleep.*

Saper, C. B., T. E. Scammell, and J. Lu. 2005. Hypothalamic regulation of sleep and circadian rhythms. *Nature* 437: 1257–63.

Schmidt, C., F. Collette, Y. Leclercq, V. Sterpenich, G. Vandewalle, P. Berthomier, et al. 2009. Homeostatic sleep pressure and responses to sustained attention in the suprachiasmatic area. *Science* 324: 516–19.

Sullivan, S. S., and C. A. Kushida. 2008. Multiple sleep latency test and maintenance of wakefulness test. *Chest* 13: 854–61.

Takahashi, J. S., H. Hong, C. H. Ko, and E. L. McDearmon. 2008. The genetics of mammalian circadian order and disorder: Implications for physiology and disease. *Nat Rev Genet.* 9: 764–75.

Tononi, G., and C. Cirelli. 2006. Sleep function and synaptic homeostasis. *Sleep Med Rev.* 10: 49–62.

Trenkwalder, C., B. Högl, and J. Winkelmann. 2009. Recent advances in the diagnosis, genetics and treatment of restless legs syndrome. *J Neurol.* 256: 539–53.

Tucker, A. M., D. F. Dinges, and H. P. Van Dongen. 2007. Trait interindividual differences in the sleep physiology of healthy young adults. *J Sleep Res.* 16(2): 170–80.

Tunbridge, E. M., P. J. Harrison, and D. R. Weinberger. 2006. Catechol-o-methyltransferase, cognition, and psychosis: Val158Met and beyond. *Biol Psychiatry* 60: 141–51.

Ulmer, C., D. M. Wolman, and N. M. E. Johns. eds. 2008. *Resident duty hours: Enhancing sleep, supervision, and safety*. Washington, D.C.: The National Academies Press.

U. S. Congress. A bill to secure the promise of personalized medicine for all Americans by expanding and accelerating genomics research and initiatives to improve the accuracy of disease diagnosis, increase the safety of drugs, and identify novel treatments. 2006. S.976.

United Kingdom. Parliament. House of Lords Science and Technology Committee. *Genomic Medicine Inquiry*. 2009 www.parliament.uk/documents/upload/stGMHumanGeneticsCommission.pdf. (accessed June 10, 2010).

Van Dongen, H. P. A. 2006. Shift work and inter-individual differences in sleep and sleepiness. *Chronobiol Int.* 23(6): 1139–47.

Van Dongen, H. P. A., M. D. Baynard, G. Maislin, and D. F. Dinges. 2004. Systematic interindividual differences in neurobehavioral impairment from sleep loss: Evidence of trait-like differential vulnerability. *Sleep* 27: 423–33.

Van Dongen, H. P. A., K. M. Vitellaro, and D. F. Dinges. 2005. Individual differences in adult human sleep and wakefulness: leitmotif for a research agenda. *Sleep.* 28(4): 479–96.

Vandewalle, G., S. N. Archer, C. Wuillaume, E. Balteau, C. Degueldre, A. Luxen, et al. 2009. Functional MRI assessed brain responses during an executive task depend on interaction of sleep homeostasis, circadian phase, and PER3 genotype, *J Neurosci.* 29: 7948–56.

Viola, A. U., S. N. Archer, L. M. James, J. A. Groeger, J. C. Lo, D. J. Skene, et al. 2007. *PER3* polymorphism predicts sleep structure and waking performance. *Curr Biol.* 17: 613–18.

Wasdell, M. B., J. E. Jan, M. M. Bomben, R. D. Freeman, W. J. Rietveld, J. Tai, D. Hamilton, and M. D. Weiss. 2008. A randomized, placebo-controlled trial of controlled release melatonin treatment of delayed sleep phase syndrome and impaired sleep maintenance in children with neurodevelopmental disabilities. *J Pineal Res.* 44: 57–64.

Weaver, T. E., and R. R. Grunstein. 2008. Adherence to continuous positive airway pressure therapy: The challenge to effective treatment. *Proc Am Thorac Soc.* 5: 173–78.

Wellman, A., A. Malhotra, A. S. Jordan, K. E. Stevenson, S. Gautam, and D. P. White. 2008. Effect of oxygen in obstructive sleep apnea: Role of loop gain. *Respir Physiol Neurobiol.* 31(162): 144–51.

Wyatt, J. K., C. Cajochen, A. Ritz-De Cecco, C. A. Czeisler, and D. J. Dijk. 2004. Low-dose repeated caffeine administration for circadian-phase-dependent performance degradation during extended wakefulness. *Sleep* 27(3): 374–81.

Yee, B. J., C. L. Phillips, D. Banerjee, I. Caterson, J. A. Hedner, and R. R. Grunstein. 2007. The effect of sibutramine-assisted weight loss in men with obstructive sleep apnoea. *Int J Obes.* 31(1): 161–68.

11 Theranostics and Translation toward Personalized Medicine for Multiple Sclerosis

Ariel Miller, MD, PhD, Tamar Paperna, PhD,
Opher Caspi, MD, PhD, Izabella Lejbkowicz, PhD,
Elsebeth Staun-Ram, PhD, and Nili Avidan, PhD

Introduction

Multiple sclerosis (MS) is the most common chronic neurological disease affecting young adults in the Western world, with onset usually at twenty to forty years of age. Women are affected three to four times more frequently than men. The disease usually is characterized by episodes of neurological dysfunction, followed by periods of stabilization or partial-to-complete remission of symptoms. Patients may complain of double or blurred vision, numbness, weakness in one or two extremities, instability in walking, tremors, and problems with bladder control, sexual dysfunction, heat intolerance, and fatigue.

These symptoms can appear over a few hours or days (acute exacerbation) or can gradually worsen over weeks and months (progressive course). Depending on the course and subtype of the disease, these symptoms will either persist or slowly resolve over weeks or months, and may even culminate in complete remissions. The relapsing-remitting pattern (RRMS) is the most common form and is characteristic for this disease. The other disease phenotypes are primary-progressive (PPMS), which is more common in patients with disease onset at later ages, and the combined relapsing-progressive (RPMS) and secondary-progressive (SPMS). Additional subtypes also considered for different therapeutic approaches include those based on extreme disease activity (benign versus malignant MS), the age of onset, childhood MS versus classical MS, and familial versus sporadic MS. The etiology of MS, like many other autoimmune and immune-mediated diseases, is complex and multifactorial, with underlying genetic and environmental factors relevant to its onset and progression.

Clinical disability results from destruction of the Central Nervous System (CNS) myelin, a product of the oligodendrocytes, due to several core biological processes:

1. Inflammation leading to demyelination: immune cells (T cells, B cells, macrophages) with aberrant activity invade the brain and spinal cord and cause destruction of CNS myelin. This inflammatory process may also cause secondary neurodegeneration.

2. Primary neurodegeneration (axonal and neuronal loss) without prominent inflammation.

The inflammatory and neurodegenerative processes are followed by an attempt of the CNS to repair itself. However, this process is often partial. The acute MS attack (presented as paralysis, visual loss, etc.) is considered to result from an aberrant acute immune activation and inflammatory process in the CNS. The residual neurological deficits and chronic accumulating disability are considered to be from the neurodegenerative process and incomplete repair. Repair processes are mainly noted after the acute attack, and recovery of function can be spontaneous. However, in attacks associated with significant disability, there is often need for steroid treatment.

The cause of MS is still unknown. However, intense research in recent decades has led to a better understanding of the immunological cascade of the disease. As a result, several disease-modifying drugs (DMDs), which have been tested and approved in clinical trials over the last ten to fifteen years, are being used as long-term prophylactics to reduce relapse rate. The new immunotherapies include glatiramer acetate, interferon beta (IFN-β), and natalizumab. These medications can attenuate the disease (that is, reduce the number of relapses per year, and improve quality of life), but do not cure the patient. Also, these DMDs are beneficial in only approximately 40 percent of the RRMS patients. Notably, currently there are no treatments for patients with the progressive disease who have gradually increasing disability (Compston et al. 1998; Trapp and Nave 2008).

The goal of personalized medicine is that the patient will get the right dose of the right medicine at the right time. We are in a process of shifting paradigms, as we are moving from treating the disease to treating the patient, and we are seeing health care tailored according to the patient's specific needs and characteristics (Grossman and Miller 2010; Kirstein-Grossman et al. 2002; Miller et al. 2008).

Biomarkers for Clinical Disease Prognosis

Intensive efforts are being made to detect reliable biomarkers for early diagnosis of MS, disease subtypes, clinical prognosis, disease progression, and response to therapy. In general, a biomarker is an objectively measured and evaluated indicator of physiologic or pathogenic processes, or of responses to a therapeutic intervention (Bielekova and Martin 2004).For biomarkers to be of significant clinical utility, they must be sensitive, specific, and reproducible, as well as easily detected in accessible bodily fluids—such as blood, urine, or saliva—or as a brain-based or cognition marker. MS lesions are typically located in the white matter of the brain and spinal cord, strongly limiting the accessibility of the tissue to repeated measurements. Cerebrospinal fluid (CSF), sharing a close anatomical relationship with the brain and spinal cord, is regarded as an important source for measuring cellular and biochemical changes related to MS pathology. However, the invasive procedure associated with CSF collection, the lumbar puncture (LP), greatly restricts the availability of repeated measurements.

Additionally, only limited correlation has been found between CSF parameters and disease activity, disability, MS subtypes, and response to treatment (Tumani, Hartung et al. 2009). Blood and urine samples are easily collected, but blood samples show a diurnal variation in many soluble markers, depending, among other parameters, on the known circadian pattern of immune regulators, such as cortisol (Kanabrocki et al. 2007; Glass-Marmor et al. 2009). In addition, biomarker levels in the serum might be affected by liver degradation or kidney excretion, while protein levels in both urine and serum might be altered by systemic or bladder infections, fluid intake, or other factors (Bielekova and Martin 2004; Tumani et al. 2009a).

Biomarkers to Support Clinical MS Diagnosis and Treatment

Specific, sensitive, and validated biomarkers are highly valuable tools which support accurate clinical diagnosis and decision-making on early treatment initiation for the prevention of disease progression and accumulation of disability. At present, the only validated biomarker, in addition to results drawn from magnetic resonance imaging (MRI) of the CNS, is CSF oligoclonal IgG bands (OCGBs) (McDonald et al. 2001; Polman et al. 2005; Alvarez-Cermeno et al. 2008). However, many other markers have been proposed and are currently under further evaluation. Considerable efforts have been focused on the detection of MS-related laboratory biomarkers, including cytokines, chemokines, complement-related molecules, antibodies, adhesion molecules, apoptosis-related molecules, markers of blood-brain barrier disruption and antigen presentation, and markers of demyelination and axonal damage (Bielekova and Martin 2004). Table 11-1 summarizes these biomarkers.

Oligoclonal IgG/IgM Bands

The only validated biomarker for MS diagnosis in clinical practice, OCGB, can be detected in the CSF (McDonald et al. 2001; Polman et al. 2005; Alvarez-Cermeno et al. 2008).These oligoclonal bands are highly specific for MS and are detected in more than 95 percent of MS patients, although they may be non-detectable at early stages and may become evident only in more advanced stages of the disease (Andersson et al. 1994; Joseph et al. 2009; Masjuan et al. 2006; Villar et al. 2008).

Oligoclonal IgM bands (OCMB) can also be detected in the CSF of MS patients, and it has been suggested that they are associated with an aggressive disease course with a high relapse rate and disability progression. This is yet to be confirmed, however, before it can be approved as a biomarker in MS diagnosis or for response to therapy.

Neuromyelitis Optica (NMO) Antibody

The importance of a diagnostic biomarker has been emphasized following the discovery of a specific antibody distinguishing Neuromyelitis Optica (NMO), also known as Devic's disease, from MS. NMO was formerly regarded as a MS disease variant, but today it is recognized as a clinically distinct disease, characterized clinically and pathologically by optic neuritis and myelopathy, rapid progression to disability, distinctive features in MRI of the brain and spinal cord, and usually absence of OCB in the CSF. Distinction of NMO from MS is important because the two disorders respond differently to immunomodulatory therapies (Cross 2007). Recently, a specific serum immunoglobulin marker, the NMO antibody, directed against the astrocytic water channel aquaporin 4, was approved as a biomarker for distinguishing between NMO and MS (Lennon et al. 2004; Smith et al. 2009).

Potential Diagnostic Biomarkers

Several other biomarkers are currently being evaluated with promising preliminary results. In the future, they may be incorporated into MS diagnosis and follow-up treatment.

Viral infection has been suggested as a trigger for autoimmune response in individuals who are genetically predisposed to MS. However, not all of the viruses potentially involved in triggering MS onset may be biomarkers of the disease. For example, antibodies against

Table 11-1 Biomarkers for Clinical Disease Prognosis and MS Subtypes

Biomarker	*Bodily fluid*	*Primary findings*	*Comments*	*Reference*
OCGB	CSF	Present in more than 95% of MS patients. Highly associated with disease and disability progression and relapse rate. Specificity >86%, sensitivity >95%. Can predict conversion from CIS to MS.	The only presently recognized biomarker supporting clinical diagnosis of MS.	Alvarez-Cermeno et al. 2008; Joseph et al. 2009; Masjuan et al. 2006; McDonald et al. 2001; Polman et al. 2005
OCMB	CSF	Associated with aggressive disease course, fast conversion from CIS to MS and development of SPMS. Not found in PPMS.	A possible supporting biomarker for definition of MS subtype.	Villar et al. 2002; 2003; 2009
NMO Ab	serum	A specific immunoglobulin biomarker, directed against the astrocytic water channel aquaporin 4 (AQP4), distinguishing between MS and the clinically distinct disease NMO. Specificity >99%, sensitivity >60%.	A specific biomarker for diagnosis of NMO.	Jarius et al. 2007; Lennon et al. 2004; Wingerchuk et al. 2006
OPN	CSF, plasma	Higher in patients with active disease, no correlation with disability.	Not specific for MS.	Braitch et al. 2008; Chowdhury et al. 2008; Vogt et al. 2004
S100	CSF	Increased in MS, particularly during relapse. PPMS< SPMS < RRMS. Possible biomarker for disease progression.	Associated with extensive destruction of astrocytes.	Lamers et al. 1995; Massaro et al. 1985; Petzold et al. 2002
GFAP	CSF	Elevated in MS patients, correlating with disease and disability scales. Possible biomarker for disease progression.	Associated with extensive destruction of astrocytes.	Malmestrom et al. 2003; Petzold et al. 2002; Rosengren et al. 1995
MBP	CSF	Increased levels of MBP in MS patients during relapse.	A marker of myelin sheath injury.	Lamers et al. 1998; Lamers et al. 1995; Massaro et al. 1985
NF-L, NF-H	CSF	NF-L increased in MS patients highest in RRMS, correlating with relapse rate and degree of disability. Elevated levels seen already in CIS patients, correlating with later conversion to MS. NF-H : SPMS> RRMS. Correlation with disability scale, relapse rate and EDSS. May predict a poorer diagnosis.	Reflect the neurodegenerative process and axonal damage.	Brettschneider et al. 2006; Ehling et al. 2004; Lycke et al. 1998; Malmestrom et al. 2003; Petzold et al. 2005; Tumani et al. 2009a

Tau	CSF	Increased in MS patients, correlating with Gd-enhancing lesions and relapse activity. Together with NF-H predict conversion from CIS to RRMS. Result not confirmed by Valis 2008 and Teunissen 2009	Reflect axonal damage. Controversy reports on altered Tau level in MS patients.	Brettschneider et al. 2005; Brettschneider et al. 2006; Teunissen et al. 2009; Valis et al. 2008
NAA	CSF	Decreased levels in SPMS, in correlation with severity of disease and MRI parameters. SPMS<RRMS. Possible biomarker for subtype definition.	Reflect the degree of axonal dysfunction, damage or loss.	Jasperse et al. 2007; Teunissen et al. 2009
Ab against MOG or MBP	Serum	No associations between the presence of anti-MOG and anti-MBP IgM and IgG antibodies and progression from CIS to clinically definite multiple sclerosis.	Not potential biomarkers.	Kuhle et al. 2007
proteomics	CSF	Vitamin D-binding protein decreased in MS patients.	Vitamin D deficiency is suggested as a risk factor for MS.	Qin et al. 2009
proteomics	CSF	IgG free kappa light chain protein increased in CIS and RRMS patients. Apolipoprotein E isoform increased in RRMS.		Chiasserini et al. 2008
proteomics	CSF	Serin peptidase inhibitor upregulated and eight proteins down-regulated alpha-1-B glycoprotein, fetuin-A, apolipoprotein A4, haptoglobin, human zinc-alpha-2-glycoprotein ZAG, retinol-binding protein, superoxide dismutase 1, transferrin in CIS patients converted to RRMS.		Tumani et al. 2009a

Summary of studied biomarkers for clinical prognosis of Multiple Sclerosis and MS subtypes. The major outcomes from the presented articles are shortly described.

OCGB- oligoclonal IgG bands; OCMB - oligoclonal IgM bands; OPN- Osteopontin; GFAP- glial fibrillary acidic protein; MBP - Myelin basic Protein; NF-L,NF-H- Neurofilaments light and heavy; NAA- N-acetyl aspartic acid; MOG- myelin oligodendrocyte glycoprotein.

the Epstein-Barr virus (EBV) are found in 99 percent of MS patients, but they also are found in 90 percent of the general population, and, therefore, are not specific to MS (Bagert 2009). A combined intrathecal humoral immune response against neurotropic viruses, including measles (M), rubella (R), and varicella zoster virus (Z), called the MRZ reaction (MRZ-R), has been associated with autoimmune-mediated diseases of the CNS, including MS and CNS vasculitis (Felgenhauer et al. 1985; Reiber 1998; Robinson-Agramonte et al. 2007).The MRZ-R is present in 90 percent of MS patients, while absent in patients with other autoimmune CNS diseases, such as NMO (Jarius et al. 2008) and paraneoplastic neurological disorders (PND) (Jarius et al. 2009). Taking into account the rarity of CNS vasculitis, MRZ-R may be a specific biomarker for MS, but further investigations of MRZ-R in rarer autoimmune conditions are still required (Jarius et al. 2009).,The pathophysiological relevance of this MRZ reaction in MS remains unclear.

Biomarkers of MS Subtypes and Inflammatory Disease Activity

As indicated above, MS is a heterogeneous disease with several clinical disease subtypes. The development of biomarkers for those clinical subsets is an important goal in MS research, since different disease subsets may benefit from different treatment strategies. Such biomarkers may therefore be valuable as part of the decision-making process for optimal, patient-specific therapy.

Immunological markers are considered relevant to the acute exacerbation and relapsing phase of the disease, while markers of neurodegeneration are considered relevant to the progressive phase and disability accumulation.

Anti-Glycan Antibodies

Glycans are the predominant surface markers of immune cells. Serum anti-glycans have been implicated in several autoimmune diseases, and are suspected of interfering with the normal function of the immune system. Studies of the anti-glycan profile in MS patients have recently demonstrated elevated serum levels of antiGlc(a1.4) Glc(a) (GAGA4) antibodies in MS patients, compared to patients with other neurological diseases (Schwarz et al. 2008). In addition, higher levels of IgM antibodies to GAGA4 seem to assist in the prediction of clinically isolated syndrome (CIS) patients who are prone to an imminent relapse (Freedman et al. 2009). This biomarker may be valuable in determining which CIS patients are at high risk for development of definite MS, as well as rapid progression of disability, and who would benefit from early disease-modifying treatment. The efficacy of this possible biomarker is currently under evaluation in major retrospective and prospective studies.

Matrix Metalloproteinases (MMPs) and Tissue Inhibitors of MMPs (TIMPs)

MMPs are involved in the pathogenic cascade of MS, and have been implicated in blood-brain barrier (BBB) disruption, immune cell migration, and CNS demyelination. The profile of MMP expression and the ratio of MMP to TIMP are considered as potential serum biomarkers for disease type, disease activity and response to therapy. Elevated MMP-2, MMP-7, and MMP-9 expression have been found in CNS lesions and in peripheral blood lymphocytes (PBLs) from MS patients (Cuzner et al. 1996; Anthony et al. 1997; Lichtinghagen et al. 1999; Ozenci et al. 1999; Galboiz et al. 2001). Increased serum MMP-9 levels precede appearance of new gadolinium-enhancing lesions on MRI in RRMS patients

(Waubant et al. 1999). MMP-9/TIMP-1 serum level ratio predicts the appearance of new gadolinium-enhancing lesions on MRI in SPMS patients (Waubant et al. 2003). A significant reduction of PBL MMP-7 and MMP-9 mRNA levels was observed in the RRMS group, but not in the SPMS group, correlating with clinical improvement, as measured by relapse rate and Expanded Disability Status Scale (EDSS) score (Galboiz et al. 2001). These results suggest that the two clinical subtypes respond differentially to IFN-β treatment.

Markers of Neurodegeneration

Biomarkers of neurodegenerative processes in MS reflect neuronal and axonal damage, and are considered as hallmarks of disease progression and disability accumulation (Bjartmar and Trapp 2001; Tumani et al. 2009a). Although much research has been dedicated in recent years to elucidation of the neurodegenerative processes underlying MS, only a few and relative small studies on specific markers for neurodegeneration have been reported, and follow-up studies to clarify the relevance of these indicators are still required. During MS relapse, both S100 protein and glial fibrillary acidic protein (GFAP), two astrocyte-specific markers, have been found to be elevated in CSF (Massaro et al. 1985; Lamers et al. 1995; Rosengren et al. 1995; Petzold et al. 2002; Malmestrom et al. 2003). In addition, elevated myelin basic protein (MBP) in the CSF fluid of MS patients is a marker of myelin sheath injury (Massaro et al. 1985; Lamers et al. 1995; Lamers et al. 1998). Additional biomarkers for neuropathology and axonal damage include tau protein, 14-3-3 protein, neurofilament light chain (NF-L), neurofilament heavy chain (NF-H), and N-acetyl aspartic acid (NAA) protein (Ehling et al. 2004; Teunissen et al. 2005; Tumani, Hartung et al. 2009a).At present, NF-L seems to be the most promising neurodegenerative marker, as it was found to be specific to MS, and correlates with both relapse rate and degree of disability (Lycke et al. 1998; Malmestrom et al. 2003; Ehling et al. 2004; Teunissen et al. 2009). In addition, both NF-H and NAA might be good biomarkers for progressive MS subtype, as NF-H is especially elevated in progressive MS patients (Petzold et al. 2005). However NAA, a neuron-specific marker, is decreased in SPMS patients compared to those with RRMS (Jasperse et al. 2007; Teunissen et al. 2009) both correlating with severity of disease and MRI parameters (number of T2 lesions, gadolinium-enhancing lesions, normalized brain volume, black hole lesion load).

The potential of biomarkers for neurodegenerative processes would be of utmost importance for inclusion as part of the therapeutic arsenal for MS, once the efficacy of neuroprotective medications are assessed in clinical trials.

Protein Signatures in CSF and Serum

In recent years, novel technologies, such as proteomics and antigen microarrays, have been used in attempts to identify novel protein markers that could be predictors of health status, disease subtypes, or response to treatment. Although none of the results has yet been incorporated into clinical use, the methods and their possible applications have attracted a great deal of attention.

In several proteomics studies, the CSF protein expression profiles of healthy and MS patients were compared (Hammack et al. 2004; Lehmensiek et al. 2007; Chiasserini et al. 2008; Stoop et al. 2008; Qin et al. 2009b; Tumani et al. 2009; Kuhle et al. 2007). These studies may provide a protein signature that, in future, could be used to identify CIS patients at risk of progressing to MS. Interestingly, decreased expression of vitamin D-binding protein in RRMS patients was found in one of the studies (Qin et al. 2009), supporting the hypothesis that altered supplies of vitamin D might be a risk factor for developing MS. Other differentially expressed CSF proteins include the IgG free kappa light chain protein and an apolipoprotein E (ApoE) isoform (Chiasserini et al. 2008).

Antigen microarray analyses have been aimed at detecting unique serum autoantibody signatures that distinguish RRMS, SPMS, and PPMS patients from both healthy controls and those with other neurological or autoimmune diseases (Quintana et al. 2008). RRMS was found to be characterized by autoantibodies to heat shock proteins (HSPs) that were not observed in PPMS or SPMS. RRMS, SPMS and PPMS were each characterized by unique patterns of reactivity to CNS antigens and to specific antibody patterns to lipids and CNS-derived peptide (Quintana et al. 2008; Weiner 2009). However, the importance and relevance of these candidate biomarkers have yet to be confirmed (Hammack et al. 2004; Lehmensiek et al. 2007; Chiasserini et al. 2008; Stoop et al. 2008; Qin et al. 2009; Tumani et al. 2009).

Biomarkers Monitoring Disease Progress and Response to Therapy

Among the major challenges in MS research is the development of biomarkers to support clinical decisions associated with predicting and monitoring response to therapy. So far, there are no validated biological or imaging markers to predict and guide initial treatment selection in MS, and the choice of therapy is based mainly on clinical judgment, expert opinion, and experience, generally carried out by trial and error (Rudick and Polman 2009). The relapsing course of MS is highly unpredictable; however, the autoimmune response results in irreversible axonal loss. Therefore, identification of subjects with an unsatisfactory response to treatment is a crucial point for optimizing therapy in a theranostics-based personalized medicine approach—theranostics being a form of therapeutics guided by diagnostics (Pozzilli and Prosperini 2008).

The drugs currently approved for MS are all immunomodulatory agents, and include three forms of IFN-β (Avonex, Betaseron and Rebif), glatiramer acetate (GA, Copaxone), mitoxantrone (Novatrone), and natalizumab (Tysabri) (Rudick and Polman 2009). In addition, several new drugs are under development, including drugs designed to have a neuroprotective mode of action. It is estimated that up to two-thirds of RRMS patients do not respond to therapy, emphasizing the value of biomarkers for predicting response (Rudick and Polman 2009). Hence, there is intensive ongoing research for detection of biological markers of treatment response in MS.

Biomarkers for Response of IFN-β Treatment

Several candidate molecules have been investigated as biomarkers for IFN-β treatment efficacy, or biomarkers of adverse effects. These include neutralizing antibodies, cytokines, enzymes, and cell-surface molecules, measured in the serum or CSF (see Table 11-2). A few potential biomarkers for response to glatiramer acetate and natalizumab have been suggested (see Table 11-3), but at present, no recognized biomarkers are available.

Table 11-2 Biomarkers for Response to IFN-β

Biomarker	*Patient number*	*Bodily fluid*	*Primary findings*	*Comments*	*Reference*
Biomarkers of IFN-β bioactivity					
NAbs	78 MS	serum	NAbs reduce the clinical efficacy of IFN-β relapse rate and EDSS.	Routine testing for NAbs is necessary.	Malucchi et al. 2004; Rot et al. 2008
	2506 MS	serum	High NAbs levels not associated with worsening of symptoms. Poor response to IFN-β can also occur in patients with low NAbs level. Sensitivity > 97%.		
Nabs and MxA	106 MS	PBL	MxA protein levels increased in IFN-β treated patients, relapse < remission, responders > non-responders.	Routine analysis of both MxA and NAbs may serve as biomarker for bioactivity of IFN-β.	Kracke et al. 2000; Malucchi et al. 2008; Vallittu et al. 2002
	137 MS	PBMCs	Lower relapse-free survival RFS rates in MxA-negative than in MxA-positive patients. Lower RFS rate in NAbs-positive than NAbs-negative patients.		
	20 RRMS	PBLs	Development of NAbs correlated with reduction in MxA. Sensitivity >96%, specificity >81%.		
Cytokines					
IL-10, ratio IL-10/ IL-12	27 RRMS	serum	Responders: increased IL-10 level and ratio IL-10/IL-12. Non-responders: no change in IL-10, IL-12.	IL-10/IL-12 ratio as a possible serum marker predictive of favorable clinical response.	Carrieri et al. 2008; Rudick et al. 1998; Wiesemann et al. 2008
	107 MS	CSF	Increased IL-10 after 2 years of treatment, correlating with a favorable therapeutic response.		
	120 RRMS	serum	IL-10 and IL-5 up-regulated during IFN-β treatment, only IL-10 correlating with response to treatment.		

(*continued*)

Table 11-2 (continued)

Biomarker	*Patient number*	*Bodily fluid*	*Primary findings*	*Comments*	*Reference*
IL-12	26 RRMS	blood	Baseline levels of IL-12p35 mRNA lower in "good responders" than in "non-responders," correlating with and predicting the clinical outcome.		van Boxel-Dezaire et al. 2000
IFNγ IL-4, IL-6, IL10	29 RRMS	PBMC	No significant change in cytokines related to response to treatment.	Not supportive of biomarkers.	Sega et al. 2008
TRAIL	62 RRMS	PBMC	TRAIL mRNA: Responders > non-responders. Increased serum TRAIL before treatment predicted treatment response in the first year.	A possible prognostic marker of short treatment response only.	Wandinger et al. 2003; Buttmann et al. 2007
	30 RRMS	plasma	TRAIL did not predict long-term 1-2 years response to treatment.		
Adhesion molecules					
VCAM1, VLA-4	19 MS	serum	Increased VCAM1 levels correlated with treatment response and to decreased MRI activity. Stable or reduced VCAM1 levels in non-responders.	VCAM may be a possible biomarker for early response to IFN-β.	Rieckmann et al. 1998; Soilu-Hanninen et al. 2005
	24 MS		VLA-4 and VCAM1 increased in IFN-β-treated patients, VCAM1 correlating with response to treatment.		

ICAM-1	35 RRMS	serum	Increased ICAM-1 levels during first year only. Lower levels in Nab-positive patients.		Trojano et al. 1999
MMP-2,-7,-9 MT1-MMP TIMP-1,-2,	36 SPMS	serum	Decreased MMP-9 levels during first year of treatment, in correlation with reduced relapse rate, MRI activity and EDSS	MMP/TIMP ratio possible biomarker for response to therapy in RRMS.	Trojano et al. 1999; Waubant et al. 2003; Galboiz et al. 2001
	33 RRMS	serum	Ratio MMP-9/TIMP-1 predicted new lesions. IFN-β reduced ratio MMP-9/TIMP-1		
	6 RRMS, 6 SPMS	PBL	MMP-2,-7, -9 mRNA reduced and TIMP-2 increased in RRMS, while only MMP-7 reduced in SPMS, in response to treatment.		

Summary of studies of possible biomarkers for response to IFN-β treatment.

Nabs: neutralizing antibodies MxA: myxovirus-resistance-protein A IL: Interleukin TRAIL: TNF-related apoptosis-inducing ligand VCAM1- vascular cell adhesion molecule 1; VLA-4- very late antigen-4; ICAM-1- Intercellular adhesion molecule 1; MMPs- matrix metalloproteinases; TI MP – tissue inhibitor of MMP.

Development of Neutralizing Antibodies

IFN-β treatment leads to development of neutralizing antibodies (NAbs) in up to 35 percent of patients, usually appearing from six to twenty-four months after initial treatment (Rudick and Polman 2009). Patients who do not develop NAbs within twenty-four months are unlikely to develop NAbs later (Sorensen et al. 2005). Since NAbs reduce or abolish IFN-β bioactivity, reduce drug efficacy, and increase relapses and MRI lesions, they are an important factor for determining response to IFN-β therapy (Sorensen et al. 2003; Gilli et al. 2004; Hesse et al. 2007; Scagnolari et al. 2007; Rot et al. 2008). However, since some patients with low NAb levels still respond poorly to IFN-β, NAbs status is not a sufficient predictor for IFN-β response (Rot et al. 2008). About 6 percent of patients treated with natalizumab develop antibodies against the drug; this is associated with reduced clinical efficacy (Calabresi et al. 2007). Patients treated with glatiramer acetate also develop antibodies against the drug, but the clinical significance of these antibodies remains unclear (Salama et al. 2003; Farina et al. 2005; Fox et al. 2007).

Myxovirus-Resistance Protein A (MxA)

Induction of myxovirus-resistance protein A (MxA), an antiviral protein exclusively induced by type 1 IFNs, is a hallmark of the biologic function of IFN-β (Malucchi et al. 2008). In MS patients, lymphocyte MxA protein expression level clearly increases in patients receiving IFN-β (Kracke et al. 2000; Vallittu et al. 2002) and correlates with disease activity (Kracke et al. 2000). Abolition of MxA stimulation can indicate reduced biologic activity and efficacy of IFN-β treatment, because of the development of NAbs (Kracke et al. 2000; Vallittu et al. 2002). As a few patients with positive NAbs show no change in MxA protein levels (Vallittu et al. 2002), analysis of efficacy of IFN-β should include measuring levels of both NAbs and MxA, in addition to the clinical evaluation. MxA and mRNA quantification after one year of treatment with IFN-β has also been found to have a strong prognostic value for the risk of new relapses (Vallittu et al. 2002; Malucchi et al. 2008), and lack of MxA in vivo response in MS patients with NAbs is a reliable marker of a reduced biologic response to IFN-β, with no indication of residual bioactivity (Hesse et al. 2009). Hence, MxA mRNA levels are increasingly used in daily clinical practice (Hesse et al. 2009).

Matrix Metalloproteinases (MMPs) and Tissue Inhibitors of MMPs (TIMPs)

MMPs and their inhibitors, TIMPS, are involved in MS pathogenesis and have been proposed as biomarkers of response to therapy. IFN-β treatment decreases MMP-9 levels in serum of RRMS patients, correlating with both decreased relapse rates and improvement in disease activity as measured by the presence of MRI T_1 lesion (Trojano et al. 1999). In a two-year study of IFN-β-treated RRMS patients, TIMP-1 serum levels were increased in "good responders" only, suggesting that TIMP-1 could be a biomarker of positive response to treatment (Comabella, et al. 2009), whereas the decreased level of MMP-9 was not sustained over time and did not significantly correlate with treatment response.

Cytokines and Adhesion Molecules

Several cytokines, chemokines, and adhesion molecules have been suggested as biomarkers for prediction of response to IFN-β treatment; however, at present, none has been validated. One of the biomarkers that emerges from different studies is increased CSF level of

Table 11-3 Biomarkers for Response to Glatiramer Acetate or Natalizumab Treatment

Biomarker	*Patient number*	*Bodily fluid*	*Primary findings*	*Comments*	*Reference*
Response to Glatiramer Acetate					
proliferative response, IFNγ and IL-4	14 MS	PBMC	Responders: 86% met at least two out of the three criteria in response to GA: 1 reduced proliferative response to GA, 2 strong activation of IFN-γ-producing T cells, 3 activation of IL-4–producing T cells. Non-responders: only 22% met two criteria.	The immunological response assay possible biomarker for response to GA.	Farina et al. 2002
IL-5, IL-13	25 MS	serum	IL-13 and IL-5 increased in 80 % of GA responders, and in 0 % of non-responders. IL-13 not detectable in controls, untreated MS and non-responders. Increased IL-13 and IL-5 levels correlated with positive clinical response to GA-treatment.	IL-13 and IL-5 are potential biomarkers for positive response to GA.	Wiesemann et al. 2003
proliferative response, IFNγ, IL-5, IL-10	13 RRMS	PBMC	No change in cytokine levels. Reduced GA-specific proliferation response in GA- treated patients. Fluctuation of GA-specific proliferative responses was significantly lower in GA responders than in untreated patients and in non-responders.	The proliferation assay can be a potential marker for response to GA.	Weder et al. 2005
GA-Ab on IL-4, IL-10, IL-12	42 RRMS	T-cell lines	The normal obtained increase in IL-10, IL-4 and decrease in IL-12 and TNF-αlevels from GA treatment was reversed in the presence of GA antibodies.	GA antibodies may block the regulatory effects of GA on T cells.	Salama et al. 2003
BDNF, IFNγ, IL-4, IL-5, IL-10	20 RRMS	PBMC	BDNF, IL-4, IL-5 and IL-10 increased, while IFNγ decreased in response to treatment. No correlation found between BDNF secretion and cytokine response, lesional load, and measures of atrophy.		Sarchielli et al. 2007

(continued)

Table 11-3 (continued)

Biomarker	*Patient number*	*Bodily fluid*	*Primary findings*	*Comments*	*Reference*
BDNF, IFNγ IL-2	17 RRMS	PBMC	In responders: Decreased percentage of INFγ- and IL-2-producing CD4+ and CD8+ T cells, increased percentage of CD3+, CD4+ and CD4+/CD45RA+ T cells, and persistently increased BDNF production. In non-responders: no significant changes in these parameters.	Induction of BDNF may be part of GA mechanism of action.	Blanco et al. 2006
Response to Natalizumab					
MMP-9, OPN,TNFα, IFNγ, IL-23, IL-10	22 RRMS	CSF, PBMC	Natalizumab reduced CSF cell counts, IL-23, IFNγ, OPN and MMP-9 and increased IL-10. In contrast, in PBMCs, TNF-α and IFN-γ were increased. Patients being in remission at baseline showed the same deviations of mediators as those in relapse after treatment.	The affected mediators may be possible biomarkers for natalizumab therapy.	Khademi et al., Eur J Neurol 2009, 16 (4): 528–36
IFNγ, TNF, IL-17	28 MS	PBL	Frequency of CD4+ T cells producing IFNγ, TNF and IL-17 upon anti-CD3 stimulation increased 6 months after initiation of natalizumab treatment. Frequency of IL-2 and IL-17 producing CD8+ T cells was also increased following treatment.		Kivisakk et al. 2009

Summary of studied possible biomarkers for response to GA or natalizumab treatment.

IL: interleukin

Ab: antibody

BDNF: Brain-derived neurotrophic factor

MMPs: matrix metalloproteinases.

OPN: Osteopontin

interleukin 10 (IL-10), which in IFN-β-treated patients was shown to correlate with a favorable therapeutic response (Rudick et al. 1998; Carrieri et al. 2008; Wiesemann et al. 2008). The baseline mRNA level of another interleukin, IL-12p35, was shown to be lower in responders compared with nonresponders, and to correctly predict the clinical outcome in 81 percent of the patients (Van Boxel-Dezaire et al. 2000). Several studies have focused on tumor necrosis factor (TNF)–related, apoptosis-inducing ligand (TRAIL), which is upregulated by IFN-β (Chawla-Sarkar et al. 2001). Serum levels of TRAIL before IFN-β treatment were shown to predict treatment response in the first year (Wandinger et al. 2003); however, TRAIL levels do not correlate with long-term response or adverse effects (Buttmann et al. 2007).

Studies have also focused on adhesion molecules. Increased vascular cell adhesion protein 1 (VCAM1) serum levels in IFN-β treated patients were associated with decreased MRI activity after one year, whereas stable or reduced VCAM1 levels occurred more often in nonresponders (Rieckmann et al. 1998). The serum levels of intercellular adhesion molecule 1 (sICAM-1) were also found to be increased in MS patients treated with IFN-β, and tended to be lower in patients with NAbs (Trojano et al. 1999). In another study, during the first three to six months of IFN-β therapy, leukocyte integrin alpha-4 subunit (VLA-4) was downregulated and its counter-receptor, VCAM1, was upregulated in the serum of patients with favorable, long-term IFN-β treatment response (Soilu-Hanninen et al. 2005). In contrast, in a different study, no correlations between changes in levels of IFN-γ, IL-4, IL-6, IL-10, VLA-4, and ICAM-1 and response to treatment were found (Sega et al. 2008). Hence, the value and the relevance of these markers still remains to be determined. The inconsistencies among different studies reflect the lack of consensus of response definition, the short follow-up period, and the relatively small numbers of patients studied.

Biomarkers for Response to Glatiramer Acetate and Natalizumab

Biomarkers for response to glatiramer acetate (GA) treatment have only been carried out in a handful of studies on small patient groups, and results have not yet been replicated. In one study, GA responders had increased IL-4 and IFN-γ levels, as well as a reduced proliferative peripheral blood mononuclear cell (PBMC) response (Farina et al., 2002; Farina et al., 2005). In a different study, a correlation between reduced GA-specific cell proliferation and clinical response was found (Weder et al. 2005). IL-13 and IL-5 serum levels have been shown to be increased in GA-responders, correlating with positive clinical response, whereas IL-13 was undetectable in either healthy controls, MS patients not treated with GA, or GA-nonresponders (Wiesemann et al. 2003).

No studies have been published on biomarkers for response to natalizumab treatment, and only a few studies describe the immunomodulatory effects of this drug; however, these reports may provide some hints for future biomarkers. Natalizumab was shown to reduce expression of IFN-γ, IL-23, osteopontin, and MMP-9, while increasing IL-10 in CSF cells, and concurrently increasing TNF and IFN-γ mRNA in PBMCs in treated patients (Khademi et al. 2009). Another study also found that natalizumab increases the percentage of activated leukocytes producing proinflammatory cytokines in blood (Kivisakk et al. 2009). Further research on the mode of action of natalizumab and natalizumab treatment response might provide additional biomarkers, which could be assessed in appropriate studies.

Future Insights: Potential Biomarkers for Novel Neuroprotective and Neurorepair Therapy

Massive effort has been invested in promoting and developing neuroprotective and neuroregenerative drugs for MS treatment. Biomarkers reflecting axonal damage or repair are needed to evaluate and monitor the effects of these drugs. Several markers for remyelination have been studied, including neuronal cell adhesion molecule (Nr-CAM), ciliary neurotrophic factor (CNTF), brain-derived neurotrophic factor (BDNF), nerve growth factor (NGF), and neurotrophin-3 (NT-3) (Otten and Gadient 1995; Massaro et al. 1997; Massaro 2002; Sarchielli et al. 2002; Blanco et al. 2005). The suggested functions of NGF and other neurotrophins include preventing neural death, and favouring the recovery process, neural regeneration, and remyelination. Therefore, use of these neurotrophins has been proposed as a therapeutic strategy in neurodegenerative and inflammatory diseases of the CNS, including MS (Sarchielli et al. 2002).

One of the most studied molecules in this context is BDNF, which is produced by both neuronal and immune cells. BDNF receptors are located within the CNS. During relapse and recovery, significantly higher levels of BDNF were produced by RRMS patients' PBMCs than during remission, possibly in an attempt to repair damage and induce remyelination. Significantly lower BDNF values were found in patients with SPMS compared to control subjects, suggesting that lack of BDNF might contribute to the rapid demyelination and axonal loss in this progressive form of MS (Sarchielli et al., 2002; Sarchielli et al. 2007). Glatiramer acetate-specific T cells were shown to produce BDNF in vitro, and GA-treated patients show a gradual increase in BDNF levels in the first three months of treatment; thereafter, levels remain constant (Sarchielli et al. 2007). In addition, the GA-induced increase in BDNF levels was shown to be specific to GA responders (Blanco et al. 2006). Together, these findings suggest that GA therapy might have a neuroprotective effect, in addition to its known immunomodulatory effect, in MS patients (Hohlfeld 2004; Farina et al. 2005; Yong 2009).

The future will provide additional information on candidate biomarkers of neurorepair and neuroprotection, obtained from the current, ongoing clinical trials with new therapeutic agents. Intensive research efforts should be focused on evaluation of such biomarkers to improve clinical diagnostics and treatment management in MS.

Personalized Therapy Using Gene Expression Profiles

Experimental Approach to Gene Expression Profiling

Gene expression profiling allows the quantification of all known transcripts in one experiment, and the study of the effects of treatments, diseases, and developmental stages on the transcriptome. Gene expression classification is widely used to determine the molecular profile of different tumors, providing crucial information regarding prognosis and therapeutic strategies. In the last few years, commercially available diagnostic kits have been developed to measure the likelihood of cancer-specific recurrence, and to predict prognosis following chemotherapy (Paik 2006; Cheang, van de Rijn, and Nielsen 2008; Marchionni et al. 2008). Another example is the use of targeted drugs, such as Herceptin (trastuzumab) and Tykerb (lapatinib), that are effective in tumors expressing high levels of human epidermal growth factor receptor 2 (HER2), which constitute 20 to 30 percent of all breast tumors (Osako et al. 2008). As more targeted drugs are developed, individualized medications for patients having specific tumor profiles are expected to become more common, providing better treatments and lowering overall medical costs for cancer treatment. At present, only cancer therapy has effectively adapted expression profiling as a diagnostic and prognostic tool. However, it is expected that additional diseases might benefit from this approach in the future.

An analysis of longitudinal patterns of gene expression from IFN-β–treated patients is expected to aid in predicting and explaining the clinical response to the applied drug. Fortunately, the array-based approach has become a standard methodology. This approach is expected to identify gene expression profile signatures associated with patient drug response without previous hypotheses about their biological function. Proof of concept of high-throughput, genome-wide gene expression technology and analysis has been demonstrated, and finding new IFN-β-related biological pathways is now on the priority list of several investigators (De Jager et al. 2009; Baranzini, Galwey et al. 2009).

Highlights of Gene Expression Profiling Results

More than nineteen studies have been completed with highly variable results, as summarized in Table 11-4 (Achiron et al. 2004; Baranzini et al. 2005; Fernald et al. 2007; Iglesias et al. 2004; Koike et al. 2003; Lindberg et al. 2008; Malucchi et al. 2008; Petereit et al. 2002; Santos et al. 2006; Satoh et al. 2005; Sellebjerg et al. 2008; Sellebjerg et al. 2009; Singh et al. 2007; Sturzebecher et al. 2003; Van Baarsen et al. 2008; Weinstock-Guttman et al. 2003; Weinstock-Guttman et al. 2008; Weinstock-Guttman et al. 2007; Yamaguchi et al. 2008). The most common functional categories for drug-induced genes were cell differentiation and proliferation, migration and matrix degradation, antigen processing and presentation, apoptosis, and cytokine regulation. There is accumulating evidence suggesting that there are no or few differences in biological activity and routes of administration between different formulations of IFN-β. No significant differences have been found in the biological response levels of patients between weekly Avonex treatment and treatment with high-frequency injected formulations, such as Rebif and Betaferon, (Van Baarsen et al. 2008). Despite the wide variation in the genes included in different study designs, as well as differences in laboratory methodologies and gene selection algorithms, some overlap in the detected genes could be ascertained.

One of the most comprehensive studies was done by Weinstock-Guttman et al. (2008), in which the gene expression signatures of both single dose treatment and several months' worth of treatment with IFN-beta were correlated with a five-year clinical outcome in MS patients(Weinstock-Guttman et al. 2008). Their results highlight the heterogeneity and complexity of gene expression responses to IFN-β therapy. As anticipated, functional differences were found between the genes expressed after the first dose and after chronic dosing. Interestingly, the Partial Responder group seemed to have lower responses of numerous mRNAs than the Good Responder group after the first injection. Additionally, despite a change of treatment in Partial Responder patients, there was a poor clinical outcome during the following four years in this group, suggesting that the Partial Responder group might represent a more aggressive disease that responds poorly to immunomodulatory drugs.

Myxovirus Resistance Protein A (MXa) as a Marker for Treatment Response to IFN-β

The most prominent interferon-induced genes and cytokines, including *MX1* (gene for Myxovirus resistance protein 1), *IFI27* (gene for interferon alpha-inducible protein 27 or p27), *STAT1* (gene for signal transducer and activator of transcription 1-alpha/beta),

Table 11-4 Summary of Gene Expression Studies for MS Published through September 2009

Study	*Drug*	*Study size* **(Good responders %) (Poor responders %)**	*Response definition: Main measures & follow-up time*	*Study design*	*Results*
Petereit et al. 2002	IFN-β	33 patients **(16 48.5%)** **(17 51.5%)**	Relapse no. 2 years	IFN-gamma production measured before treatment	Low production of IFN-gamma predicted no exacerbations (69%), high production- relapses (71%)
Koike et al. 2003	IFN-β	13 patients	Undefined[1]	Pre-treatment samples, and after 3 and 6 months. CD3+ and CD3- fractions from PBMC, cDNA microarray of 1,263 sequences	21 DEGs
Sturzebecher et al. 2003	IFN-β	10 patients **(6 60%)** **(4 40%)**	MRI Nab to IFN-β clinical examinations	*Ex-vivo* and *in-vitro*, Mini-Lymphochip (MLC) cDNA microarrays containing 6,432 or 12,672 sequences	IL-8 significantly down-regulated, only in responders.
Weinstock-Guttman et al. 2003	IFN-β	8 patients	Undefined[1]	Monocyte depleted PBMC, multiple time points from 0 up to 1 week, over 4,000 genes	
Achiron et al. 2004	IFN-β, GA, IVIg	26 patients	Undefined[1]	Non-treated patients compared with patients treated with either of IFN-β, GA, or IVIg. 12,000 genes Affymetrix array	535 DEGs
Iglesias et al. 2004	IFN-β	17 patients, 12 controls[2]	Undefined[1]	PBMC expression of E2F pathway elements in patients and controls. 6,800 genes on Affymetrix array	Enhanced transcription of E2F pathway, validation in EAE.

Baranzini et al. 2005	IFN-β	52 patients **(33 63%)** **(19 37%)**	Relapse no. EDSS 2 years	Expression analysis of 70 candidate genes by real-time PCR, pre-treatment and additional time points	Mx1 predicts beta-IFN bioactivity; 9 gene triplets predicted response with high accuracy
Santos et al. 2006	IFN-β	52 patients	Undefined[1]	Pre-treatment blood samples, 4 and 8 hours after, real-time PCR; pre-treatment serum analyzed for NAbs-	NAb-positive patients had strongly attenuated gene expression response to IFN-induced mRNAs
Satoh et al. 2005	IFN-β	72 patients, 22 controls[2]	Relapse no. and severity EDSS MRI Patient satisfaction. 2 years	T cells analysis using 1,258 genes array, 286 DEGs used as discriminators	4 subgroups of MS identified; responders clustered in 2 of the 4.
Fernald et al. 2007	IFN-β	2 patients, 4 controls[2]	Undefined[1]	Multiple time points from 0 up to 1 week, over 22,000 genes, network analysis	
Singh et al. 2007	IFN-β	5 patients	Undefined[1]	PBMC pre-treatment sample, 24 hours and 6 months. 10,000 genes array	
Weinstock-Guttman et al. 2007	IFN-β	22 patients	Relapse no. EDSS MRI	Genotyping of 2 promoter SNPs in MxA; blood samples taken before and after drug dosing, mRNA levels of MxA and 5 more genes measured by real-time PCR	The 2 SNPs were not associated with altered MxA gene expression.
Malucchi et al. 2008	IFN-β	137 patients	Relapse no. EDSS 3 years	Pre-treatment blood samples and 12 months later, MxA mRNA, NAb and BAb binding-Ab) measured	Measurement of MxA and NAb predicts the risk of new relapses.

(*continued*)

Table 11-4 (continued)

Study	*Drug*	*Study size* **(Good responders** %) **(Poor responders** %)	*Response definition: Main measures & follow-up time*	*Study design*	*Results*
Weinstock-Guttman et al. 2008	IFN-β	22 patients	Relapse no. EDSS 2 years	Pre-treatment blood samples and at 1, 2, 4, 8, 24, 48, 120, 168 h after the first IFN-beta dose and at 1, 6 and 12 months after chronic dosing	JAK-STAT, TNFRSF10B, IL6, TGF-β, retinoic acid and CDC42 pathways were differentially modulated.
Van Baarsen et al. 2008	IFN-β	16 patients initial 30 for validation	Undefined[1]	Pre-treatment blood samples and 1 month later	15 IFN response genes
Sellebjerg et al. 2008	IFN-β	10 patients	Undefined[1]	PBMC samples collected before and during treatment	74 DEGs
Sellebjerg et al. 2009	IFN-β	25pateints	Neutralizing antibodies	Pre-treatment blood samples12h, 48h and 6 months later	Measurement of MxA, *CCL2*, *CXCL10* and *IFI27* correlate with NAb titer.
Yamaguchi et al. 2008	IFN-β	16 patients, 6 controls[2]	Undefined[1]	Naïve and IFN-β-treated patients, samples taken pre-treatment and at additional 3 time points	Enrichment of protein-protein interactions members in expression data.
Lindberg et al. 2008	Natalizumab	11 patients	Undefined[1] 2 years	PBMC samples before the infusion of drug or placebo and at 6 additional time points, (Affymetrix GeneChip array more than 22,000 sequences)	DEGs related to T cells, regulating B cells, neutrophil and erythrocyte functions.

[1]undefined – drug effects were measured response phenotype not determined

[2]controls – expression profiles of MS patients (disregarding responsiveness) were compared to those of healthy controls

CXCL10 (gene for C-X-C motif chemokine 10) and more, were detected by all platforms, whole genome arrays, as well as custom-made arrays (Goertsches, Hecker, and Zettl 2008). The expression of mRNA of *MxA* (*MX1*), the most robust type I interferon response gene, has been suggested as a surrogate marker for the development of neutralizing antibodies (NAbs) to IFN therapy, as well as a marker of treatment responsiveness. NAbs to IFN therapy develop in up to 30 percent of IFN-treated patients within six to twelve months following initiation of therapy, and are among the factors that limit IFN therapy efficacy. Bioassays have been developed that rely on in vitro measurement of neutralization of IFN activity, such as virus neutralization. However, these assays are time consuming, expensive, and difficult to standardize among different laboratories. The predictive value of the MxA assay and the clinical usefulness of MxA mRNA level are still controversial. While some researchers found MxA mRNA increases in significant correlation with IFN-β bioreactivity and treatment response(Malucchi et al. 2008; Van der Voort, Visser et al. 2009; Van der Voort, Kok et al. 2009), others found the marker to have a variable baseline level and a short window of time for measurement, making it a difficult marker with which to work. The baseline level of MxA mRNA increases rapidly during the first four hours after treatment initiation, but decreases rapidly thereafter. Hence, changes in the time interval between injection and sampling will affect the assay results (Hemmer and Berthele 2009). The utility and value of MxA transcript measurement remain to be determined in a large-scale, well-designed clinical trial.

Additional Markers for IFN-β Therapy

Several other transcripts have been evaluated as predictors for IFN-β treatment, such as IFN-α/β receptor (IFNAR), which seems to influence both the magnitude and the nature of the biologic response to IFN-β (Gilli et al. 2008), and β2-microglobulin mRNA, a widely used inflammatory marker (Weinstock-Guttman et al. 2008). Recently, IFN-α inducible protein 27 (IFI27) was suggested as a predictor for response to IFN-β treatment, since it was found to be reduced in patients with low NAb levels and lost in patients with intermediate/high NAb levels (Sellebjerg et al. 2009). Vandenbroeck and his group identified a specific set of fifteen IFN response genes, the expression levels of which before treatment initiation were negatively correlated with the IFN-β pharmacological response. In addition, the negative correlation with response was maintained at three and six months after treatment commenced. Moreover, this result was replicated in an independent group of patients, suggesting that the expression levels of these fifteen genes could serve as biomarkers for the clinical response to IFN-β (Van Baarsen et al. 2008; Vandenbroeck et al. 2009).

The advantage of this specific set of markers is that the differential expression levels seem to be maintained over a long period, making the time window for measurement more reasonable for a clinical situation. In addition, despite large baseline variation among patients, the expression levels seem to correlate well with long-term drug response. Despite these encouraging results, the predictive value of these transcribed markers remains to be validated and determined in a double-blind clinical trial.

Future Directions and Work

Currently, there are too few standardized approaches to analyzing data. This, in turn, has resulted in the identification of a multitude of biomarker candidates that predict poor drug response. Most studies have used a predetermined set of mRNA genes, therefore creating bias for the obvious candidate genes, such as cytokine, adhesion molecule, and IFN-β pathway genes. For most results, replication and validation have been lacking.

It is now clear that to improve MS treatment and develop personalized medicine, a multidisciplinary approach will be required. The scope and depth of such efforts exceeds the capacity of individual research groups; hence, numerous research groups, as well as multidisciplinary teams, will be required to collaborate on these projects. Such a research initiative was organized in 2009 in Europe and funded by the Seventh Framework Program of the European Commission (Vandenbroeck et al. 2009). In addition, the declining cost of high-throughput, genome-wide methodology justifies research that complements hypothesis-drawn research with assumption-free approaches that are data drive. This approach will most likely highlight many novel candidates, never before considered in MS treatment response. Obviously, these studies will have to be followed by functional studies to reveal the molecular mode of action, as well as by large-scale clinical trials to determine the utility of these markers. We hope that this approach will result in better predictive algorithms for personalized medicine in MS, as well as lower treatment costs. However, several years of intensive research and collaboration will be required before such integrative results will be available.

DNA Markers for Pharmacogenetics

Experimental Approaches to DNA Markers

Drug response is multifactorial and is influenced by genetic, epigenetic, and environmental components, as well as gene–environment interactions. As previously described, RNA and protein markers are quantitative, and their levels can be affected by age, gender, disease state, and environmental factors. In contrast, DNA markers remain constant during one's lifetime. DNA material is easy to obtain and inexpensive to isolate. The interpretation of DNA marker results is relatively straightforward, making adaptation of DNA markers tests to clinical practice simple. However, genetic polymorphisms that have a large effect on drug response are rare (Takeuchi et al. 2009; Daly et al. 2009). For most drugs, including those used in MS treatments, multiple genes with modest effects appear to be the rule, making the pharmacogenetics (PGx) discovery and translation to medicine a long and laborious road.

Earlier PGx studies for MS therapies were performed on small groups of selected candidate genes, based on drug mechanism of action and disease mechanisms. More recently, with the decline of whole genome scan costs, research has shifted toward hypothesis-free approaches, where hundreds of thousands markers spanning the entire human genome are interrogated for association with drug response. The drawback of this approach is that it requires large cohorts, both to accumulate power to detect modest effects and to obtain statistical significance and replication after multiple test corrections.

A list of the PGx studies published for MS drug treatments is presented in Tables 11-5 and 11-6 with a concise summary of the study design and main results.

Interferon-β Pharmacogenetic Studies

The majority of MS-related PGx studies have focused on IFN-β therapies, the major prescribed immunomodulatory treatment, for which more information is available on its mechanism of action. Initial studies focused on *HLA DRB1* gene, as one of the genes with a major effect on disease susceptibility and progression (Oksenberg et al. 2008; Svejgaard 2008). However, no association between *HLA* genes with IFN-β response could be established (Fusco et al. 2001; Villoslada et al. 2002; Fernandez et al. 2005; Comabella,

Table 11-5 Summary of Pharmacogenetics Studies for IFN-β Therapy Published Through August 2009

Study	*Study design*	*Sample size* **(Good responders%) (Poor responders %)**	*Response definition: Main measures (follow-up time)*	*Reported associated genes*	*Comments and functional significance of positive results*
Candidate gene studies					
HLA genes					
Fusco et al. 2001	HLA class II *DRB1* and *DQB1* genes	39 patients **(22 56.4%)** **(17 43.6%)**	Relapse no. EDSS (2 years)	NO	
Villoslada et al. 2002	HLA class I A and B and HLA class II *DRB1* and *DQB1* genes	154 patients[1] **(77 38.9%)** **(57 28.8%)**	Relapse no. EDSS (2 years)	NO	
Fernandez et al. 2005	HLA class II genes *DRB1, DQA1, DQB1*	96 patients (66 69%) (30 31%)	Relapse no. EDSS (1 year)	NO	
Comabella et al. 2009b	*HLA-A, -B, -C, -DRB1, -DQA1*, and *-DQB1*	149 patients[1] 74 responders 75 non-responders	Relapse no. EDSS (2 years)	NO	
Other candidate genes					
Sriram et al. 2003	8 SNPs in *IFNRA1* & *IFNRA2* genes	147 patients[1] **(57 38.8%)** **(48 32.7%)**	Relapse no. EDSS (2 years)	NO	
Leyva et al. 2005	2 SNPs in *IFNAR1* and one in *IFNAR2*	147 patients **(104 70.7%)** **(43 29.3%)**	Relapse no. EDSS (1 year)	NO	

(*continued*)

Table 11-5 (continued)

Cunningham et al. 2005	Focus on genes with ISRE IFN stimulated response elements in their promoters. 2-stage design with pooled samples in first stage	230 patients[1] **(94 40.9%)** **(68 29.6%)**	Relapse no. EDSS (6-9 months)	*IFNRA1, LMP7, CTSS, MX1*	IFN type 1 receptor subunit, proteasome, protease and viral response pathway genes.
Wergeland et al. 2005	Comparison of clinical parameters for treatment groups stratified by a 3 SNP-based haplotype in IL10 promoter	25 patients stratified into 2 haplotype groups: GCC or non-GCC	Responder groups not defined. Relapse no. MRI, EDSS, evaluated in 2 haplotype groups (6 months)	Larger no. of new enhancing lesions in GCC haplotype group.	Study includes also 38 patients treated with IFN-α2a in addition to the IFN-β group. IFN-β group is very small.
Martinez et al. 2006	CA repeat in *IFNG*	110 patients **(71 64.5%)** **(39 35.5%)**	Relapse no. (2 years)	*IFNG*	Lack of replication of previous results.
Weinstock-Guttman et al. 2007	2 SNPs in promoter of *MX1*	37 patients **(19 51.4%)** **(18 48.6%)**	Relapse no. MRI (2 years)	NO	Viral response gene
Byun et al. 2008	Reanalysis of SNPs in 100 candidate genes from previous studies (Cunningham et al. 2005) using pooled genotyping data from original cohort (see below), in GWA studies	206 patients[1] **(99 48.1%)** **(107 51.9%)**	Relapse no. EDSS (2 years)	Associations with SNPs in or near candidate genes including: *ADAR, PTGS2, CF1, SYN2, IFNG, TRAF6*	Lack of replication of any of the previous candidate gene associations, with the exception of *IFNG*.

(*continued*)

Table 11-5 (continued)

Study	*Study design*	*Sample size* **(Good responders%) (Poor responders %)**	*Response definition: Main measures (follow-up time)*	*Reported associated genes*	*Comments and functional significance of positive results*
GWA studies					
Byun et al. 2008	Multistage study: 1. Original cohort–samples pooled into groups defined by response pools genotyped on Affymetrix 100K array. 2. Validation by individual sample genotyping of original cohort and join cohort on 35 candidate SNPs, ranked by p value and cluster analyses	1. Original: 206 patients[1] **(99 48.1%)** **(107 51.9%)** 2. Joint validation: 285 patients **(143 49.7%)** **(142 50.3%)**	Relapse no. EDSS (2 years)	Association with 18 SNPs in GWA, genes including *GPC5, HAPLN1, FLJ32978, COL25A1, CAST, TAFA1, NPAS3.*	None of the previous candidate gene associations were replicated.
Comabella et al. 2009a	1. Initial cohort - Pooled samples; all responders versus all non-responders genotyping with Affymetrix 500K array. 2. Validation cohort: genotyping for 383 SNPs selected by rank of silhouette statistics and cluster analysis	1. Initial cohort: 106[1] **(53 50%)** **(53 50%)** 2. Validation cohort: 94[1] **(49 52.1%)** **(45 47.9%)**	Relapse no. EDSS (2 years)	*GRIA3, CIT, ADAR, IFNAR2,ZFAT, ZFHX4, STARD13* And several SNPs in non-genic regions, Uncorrected P values	SNPs within genes related to functional activity of IFN-β and glutamatergic pathways. *ADAR* association from Byun et al. candidate gene study is replicated here.
Replication studies					
Cénit et al. 2009	3 SNPs in *GPC5* and 3 SNPs in *HAPLN1*	199 patients[1] **(55 28%)** **(79 40%)**	Relapse no. EDSS (2 years)	One SNP in *GPC5* associated with response. NO for *HAPLN1*	

EDSS – Expanded Disability Status Scale (Kurtzke 1983); NO - None observed; GWA - Genome wide association

[1]Intermediate response phenotype excluded from genotype analyses

Craig et al. 2009), highlighting the dichotomy between genes involved in disease susceptibility and those affecting the drug response phenotype. Other PGx studies focused on DNA polymorphisms in major players of the IFN-β pathway, including the IFN type 1 receptor subunits, and viral response genes such as *MX1*, however here, too, no clear association was found (Sriram et al. 2003; Fernandez et al. 2005; Weinstock-Guttman et al. 2007; Wergeland et al. 2005).

The 2005 PGx study by Cunningham and colleagues adopted a wider search approach by focusing on genes whose promoter sequences contain IFN response elements, and by increasing to over sixty the number of patients in each response group (Cunningham et al. 2005). They reported several significant associations between drug response and *IFNRA1* and *MX1*, genes known from the IFN-β signaling pathway, but also associations with genes involved in protein breakdown such as *CTSS*, encoding cathepsin S, a cystein proteinase, and *LMP7*, encoding the β subunit of the proteosome subunit. However, other studies failed to replicate these results (Byun et al. 2008).

Following the growing disillusion from the candidate gene study design, a shift in PGx research approaches toward an unbiased search for predictive markers has emerged, as exemplified by two recent studies (see Table 11-5; (Byun et al. 2008; Comabella, Craig et al. 2009). Both studies employ a staged design, employing whole-genome SNP array for the exploratory stage, followed by a validation stage comprising of selected SNPs. Both studies used pooled DNA samples for the initial stage, followed by individual genotyping for the validation stage. Surprisingly, although both studies identified overrepresentation in similar pathways, such as pathways related to immune activity and IFN-β response, or related to neurotransmitter receptors and ion channels, no overlap was found between the gene-lists of the two studies. Nevertheless, the importance of these studies lies in their ability to highlight genes never before associated with the IFN-β pathway that may provide insight into drug mechanism of action.

At present, only a handful of DNA polymorphisms have emerged in at least two independent studies as associated with IFN-β response. These include *IFNG* encoding interferon γ, whose signaling pathway overlaps in part the IFN-β pathway, and is considered as one of the major cytokines involved in MS disease activity (Martinez et al. 2006; Byun et al. 2008); *ADAR*, an interferon inducible gene whose product, adenosine deaminase, performs RNA editing as part of the anti viral cellular response, as well as modulation of functional activity of cellular transcripts (Byun et al. 2008; Comabella, Craig et al. 2009); and *GCP5*, encoding glypican 5, a cell surface proteoglycan expressed in neurons, that may play a role in sequestering of cytokines, and has recently also been implicated as a risk factor for MS (Byun et al. 2008; Cénit et al. 2009; Baranzini, Wang et al. 2009). These results emphasize that response to IFN-β is not likely to be governed by a few genes with major effects, but more likely by numerous genes having modest effect. Hence, further validation of the above genes as well as identification of additional PGx markers will be required in the future.

Pharmacogenetics for Other MS Therapies

More than a decade after the initial approval of glatiramer acetate (GA) for marketing, only 2 PGx studies with this drug have been published (see Table 11-6; (Fusco et al. 2001; Grossman et al. 2007). The earlier study, by Fusco and colleagues, reported an association between drug response phenotype and the *DRB1*1501*, the major MS susceptibility allele (Fusco et al. 2001). In a larger candidate genes study that used a subset of the original clinical trial cohorts, the association with 27 candidate genes was assessed including *DRB1*1501* (Comi, Filippi, and Wolinsky 2001; Johnson et al. 1998). No association with *DRB1*1501*

Table 11-6 Summary of Pharmacogenetics Studies for Other MS-Related Therapies Published Through August 2009

Study	*Study design*	*Sample size* **(Good responders%) (Poor responders %)**	*Response definition: Main measures (follow-up time)*	*Reported associated genes*	*Comments and functional significance of positive results*
Glatiramer acetate					
Fusco et al. 2001	HLA class II *DRB1* and *DQB1* genes	44 patients **(22 50%)** **(22 50%)**	Relapse no.; EDSS (2 years)	Association with *DRB1*1501*	
Grossman et al. 2007	Sixty-three SNPs in 27 candidate genes, and *DRB1*1501* genotype analyses. Employed two clinical trial cohorts, including a placebo group analysis (included to distinguish drug- and disease-related effects).	Cohort #1: 49 GA treated **(17 34.7%)** **(32 65.3%)** 52 placebo **(10 19.2%)** **(42 80.8%)** Cohort #2: 36 GA treated **(14 38.9%)** **(22 61.1%)** 37 placebo **(11 29.8%)** **(26 70.2%)**	Cohort #1: Relapse no.; MRI (9 months) Cohort #2: Relapse no.; EDSS (2 years)	Associations detected with *TCRB@, CTSS, MBP, CD86, IL1R1, IL12RB2*	Did not replicate *DRB1*1501* association.

(*continued*)

Table 11-6 (continued)

Study	*Study design*	*Sample size* **(Good responders%) (Poor responders %)**	*Response definition: Main measures (follow-up time)*	*Reported associated genes*	*Comments and functional significance of positive results*
Study	*Study design*	*Sample size* **(Good responders%) (Poor responders %)**	*Response definition: Main measures (follow-up time)*	*Reported associated genes*	*Comments and functional significance of positive results*
Mitoxantrone					
Cotte et al. 2009	Genotypic analysis of the multiple drug transporters genes *ABCB1* and *ABCG2*, based on genotypes defined by functional *in vitro* analyses as high, intermediate or low efflux mediators. Analysis of a 2 gene genotype utilizing 2 SNPs in each gene was done for association with either monotherapy of mitoxantrone, or combination therapy of mitoxantrone and glucocorticoids.	1. Monotherapy: 155 patients **(121 78.1%)** **(34 21.9%)** 2. Combination therapy with glucocorticoids: 154 patients **(91 59.1%)** **(63 40.9%)**	One or more of: Relapse no.; EDSS; MSFC; MRI 9, 12, and 21-24 months, (depending on data availability)	*ABCB1/ABCG2* genotypic associations with response in monotherapy group. NO for combination therapy.	

EDSS – Expanded Disability Status Scale (Kurtzke 1983); NO - None observed; MSFC: Multiple Sclerosis Functional Composite

[1]Clinical trial cohorts

was found, however several other genes, including *CTSS* and *TCRBa,* were found to be associated with GA response phenotype (Grossman et al. 2007).

A PGx study for two candidate genes using mitoxantrone, a second-line drug therapy for MS, has recently been reported (Cotte et al. 2009). Both candidate genes encode members of the ATP binding cassette (ABC) transporters, *ABCB1* and *ABCG2*, known to mediate efflux of mitoxantrone from cells. The combined genotype of both transporters was found to significantly contribute to mitoxantrone response with an odds ratio (OR) of 3.5. Moreover, researchers were able to establish that this combined genotype correlates with mitoxantrone-related cell death in immune cells. Because of the severe adverse events occurring in mitoxantrone therapy, the identification of additional predictive markers to guide prescription of this potentially toxic drug is a must, requiring larger cohorts and a greater number of genes to be studied.

Limitations of PGx MS Studies

The previous sections delineated the large variability of results obtained across the various PGx studies on MS drugs. In retrospect, several shortcomings can be recognized, which are likely to have compromised detection of genetic factors associated with MS drug response and to contribute to this variability. These shortcomings include small group size, differences in response definitions, and associations that might reflect a population-specific effect (Xie et al. 2001). Finally, a major limitation inherent in PGx studies is the distinction between drug-related effects and disease-course effects. Thus, a benign disease course may be scored as good response, whereas a poor response may be the outcome of a malignant MS form. To allow for the distinction between drug-specific effects and disease-related effects, a non-treated group of patients is required for comparison, such as the placebo group in the setting of a clinical trial. Indeed, in the only MS PGx study that was based on a clinical trial population (Grossman et al. 2007), the placebo group analysis allowed identification of genes related to the disease course effect, such as *SPP1* that contributed to "good response" in the placebo group, as well as in the drug-treated group. However, placebo-controlled trials are becoming the exception rather than the rule, as new treatments are being compared to established therapies as a control for efficacy. It is therefore important for the scientific community, together with the pharmaceutical industry, to consider sharing genetic data from placebo groups from clinical trials, contingent on the appropriate ethical approvals and anonymization of sample sources, to allow correct identification of drug-related effects.

Future Prospects in MS Pharmacogenetics

The lack of positive associations from the candidate gene and genome-wide association studies (GWAS) indicates that drug response in MS, at least for IFN-β, is probably a polygenic effect. It is now apparent that the candidate-gene-study design is not likely to deliver a major breakthrough, but rather that a combined approach, using GWAS and validation studies on selected candidate genes, is required for identification of a set of genes whose polymorphisms will be of predictive value. However, the success of these analyses depends on the ability to amass a large collaborative effort of research groups and clinics, to allow collection of a large number of MS-affected individuals, whose response phenotype can be clearly defined.

Unification of response criteria, with a clear definition of good and poor response, is required to allow comparison and reproduction of methodologies in different research settings. The recent establishment of a European network of research groups to promote

MS PGx research (the United Europeans for the Development of Pharmacogenomics in Multiple Sclerosis Network, UEPHA*MS) (Vandenbroeck et al. 2009), funded by the European Commission's (EC) Seventh Framework Program, is a large step forward toward achieving these aims. Expansion of the network to include non-European study groups is likely to allow the application of this effort to other populations worldwide, with the goal of identification of predictive DNA markers for drug prescription for most of the individuals with MS.

Environmental and Epigenetic Roles for MS Predisposition and Drug Response

While genetics shape the population's overall MS susceptibility, observed epidemiological patterns strongly suggest a role for the environment in disease initiation and modulation. Findings from studies on seasonality in MS patients' birth, disease onset, and exacerbations, as well as apparent temporal trends in incidence and gender ratio, support the importance of metabolic and lifestyle factors, vitamin D status, and smoking on MS risk, which may explain such epidemiological patterns.

Vitamin D and MS Predisposition

Evidence from epidemiological studies of geographic distribution, sun exposure, and vitamin D intake, as well as experimental animal models of MS, indicate a possible influence of vitamin D on disease susceptibility and progression. Vitamin D deficiency is highly prevalent in MS patients and dietary intake of vitamin D is below the recommended level in 80 percent of patients (Nieves et al. 1994). In humans, vitamin D is obtained through two distinct pathways: vitamin D synthesis in the skin, and dietary intake. Vitamin D3 is biologically inert and requires hydroxylation in the liver to 25-hydroxyvitamin D3 (25(OH) D). Once formed, this major circulating form of vitamin D3 is further hydroxylated in the kidney to its biologically active form, 1,25-dihydroxyvitamin D3 (1,25$(OH)_2$D). Most biological effects of 1,25-$(OH)_2$D or calcitriol, are mediated by the vitamin D receptor (VDR). This receptor is a member of the steroid receptor super-family and influences the rate of transcription of vitamin D responsive genes by acting as a ligand-activated transcription factor that binds to vitamin D response elements (VDREs) in gene promoters. The vitamin D receptor is expressed on antigen-presenting cells (APCs), activated lymphocytes,and enhanced anti-inflammatory response (Smolders et al. 2009). The genetic association of *VDR* gene polymorphisms with MS was studied in several populations with mixed results, with some finding a weak association and others failing to replicate this result (Smolders et al. 2009) . However, the recent finding of a highly conserved *VDRE* in the promoter region of the *HLA-DRB1*1501* allele has raised the possibility of a functional interaction between vitamin D and *HLA-DRB1*1501* (Oksenberg et al. 2008; Svejgaard 2008; Ramagopalan et al. 2009). In experimental autoimmune encephalomyelitis (EAE) mice, a well-established animal model for MS, oral or intraperitoneal administration of 1,25$(OH)_2$D before EAE-induction with myelin proteins completely prevented the appearance of any symptoms. Furthermore, after immunization, but before the appearance of clinical symptoms, 1,25$(OH)_2$D was still able to prevent disease (Muthian et al. 2006). In the animals in which EAE was not prevented, the disease was milder and the animals survived longer (Muthian et al. 2006). Potential effects of oral vitamin D intake on MS susceptibility have been shown in a Norwegian case-control study, where the protective effect of regular consumption of fish and cod-liver oil—both foods rich in vitamin D—was found,

(Kampman and Brustad 2008). In addition, there are also some indications for disease-modifying effects of vitamin D in MS, where the vitamin D-treated group experienced fewer relapses compared to the untreated group (Myhr 2009). Based on the anti-inflammatory effects of vitamin D, relapsing remitting MS patients may be a good target population for vitamin D intervention. The benefit of vitamin D supplementation and the optimal dosage and frequency of dosage remain to be determined.

Epigenetic Effect on MS Predisposition

The gender prevalence of MS, the low-level concordance in monozygotic twins, and the low effect size of the few genetic loci found to be associated with MS predisposition all suggest an epigenetic component in this autoimmune disease (Casaccia-Bonnefil, Pandozy, and Mastronardi 2008; Kurtuncu and Tuzun 2008) . Epigenetic mechanisms have been shown to modulate gene expression by modifying DNA without mutations (Ptashne 2007), and include adding or removing methyl, acetyl, phospho, ribosyl and ubiquitin groups to or from histones, or by adding methyl groups to cytosines located in CpG islands of the promoters, thereby activating or silencing the genes. These modifications have been shown to be tightly regulated by specific enzymes during development and cell differentiation (Casaccia-Bonnefil, Pandozy, and Mastronardi 2008).

Acetylation of selected lysine residues in the tails of nucleosomal histones represents an efficient way to regulate gene expression; the process is regulated by two families of enzymes: histone acetyltransferases (HATs) and histone deacetylases (HDACs). The removal of the negative acetyl groups from the histone tails has been associated with transcriptional repression. HDAC inhibitors have been proposed as an alternative immune modulator treatment for MS. This concept was supported by the finding that the action of the most potent anti-inflammatory agents, the glucocorticoids, is partly due to HDAC activity. It has been shown that glucocorticoid treatment recruited HDAC to the promoter of pro-inflammatory genes, and that specific knockdown of *HDAC2* by RNA interference inhibited the anti-inflammatory response (Ito et al. 2006). In addition, HDAC inhibitors have been shown to ameliorate disability in the relapsing phase of EAE mice, the animal model for MS (Camelo et al. 2005). Other studies reported that treatment with HDAC inhibitors had a negative impact on oligodendrocyte precursor cell (OPC) differentiation (Lyssiotis et al. 2007). The proper re-expression of histone deacetylation in OPC progenitors is required for OPC differentiation, myelin production and axon repair. However, it was found that the MS OPC failed to differentiate into mature myelin-producing OPCs and maintained OPC progenitor markers. This result would indicate a defective establishment of the epigenetic identity in the MS brain, similar to patterns observed in aging (Shen et al. 2008).

Methylation, which occurs on CpG sequences in the promoter regions of genes, is an important regulatory mechanism for gene expression. Methylation of cytosine in CpG sequences turns off promoter activity and thereby decreases gene expression, since methylation of DNA affects both DNA structure and the binding of transcription factors. This event is regulated by the balance between DNA methyltransferase and DNA demethylases. Decreased methylation of cytosines in CpG islands in MS brains is thought to be caused by increased DNA demethylase activity. DNA isolated from white matter of MS brains contained only about one-third of the amount of methyl cytosine found in DNA from normal subjects. This abnormal methylation pattern was found to be specific to MS (Mastronardi et al. 2007). The hypomethylation in the MS brain is thought to be caused by overexpression of protein-arginine deiminase type-2 (PAD2) enzyme in mature oligodendrocytes, and results in hypercitrullination of the MBP, an MS autoantigen (Mastronardi et al. 2007).

Citrullination could increase auto-cleavage of the protein, or disrupt the physical and chemical properties of MBP and, thereby, affect its localization within the myelin membrane (Musse, Boggs, and Harauz 2006).

The impact of epigenetic modification on individuals predisposed to MS, as well as the possible effects of drugs that might modulate methylation or acetylation patterns on brain cells (Franklin and Ffrench-Constant 2008), remain to be determined. Next-generation sequencing technology is emerging as the method of choice for genomic and epigenomic profiling, allowing more comprehensive understanding of epigenetic contributions to brain diseases (Gargiulo and Minucci 2009). In addition, a systems-biology approach employing microarray analyses of gene expression and methylation patterns can lead to a better estimation of drug safety and efficacy (Csoka and Szyf 2009). Epigenetic assays ought to be incorporated into assessment of new pharmaceuticals; this new approach has been termed "phamacoepigenomics."

Integrative Medicine Tailored for Patients with MS

Integrative medicine (IM) is an emerging field of health care that seeks to combine the best approaches of conventional and alternative medicine, to provide patients with evidence-based, proven, safe, and cost-effective treatments, while activating their innate capacity for healing. As the healing response varies within and between patients at various stages of their illness processes— especially in chronic diseases such as MS—a major challenge of IM is not only to address the question, "Which treatment is the best?" but more importantly, to ask, "Best or better for whom, when and why?" (Caspi and Bell 2004). Thus, the fundamental tenets and premises of IM converge well with those of another emerging field in modern health care: personalized medicine.

Complementary, Alternative, and Integrative Medicine

While the term "complementary" implies the use of unconventional therapies alongside conventional medicine, the term "alternative" is reserved for cases in which such unconventional medicine is used in lieu of conventional medicine. As most alternative modalities are based on theoretical frameworks that are incongruent with allopathic medicine (i.e., conventional medicine), their use, including in treatment of MS, comes almost always as an add-on to conventional medicine. This segment of health care is often referred to as "complementary and alternative medicine" or "CAM." The NIH now hosts two CAM sections: the National Center for Complementary and Alternative Medicine (NCCAM; http://nccam.nih.gov/) and the Office of Cancer Complementary and Alternative Medicine (OCCAM; http://www.cancer.gov/cam/).

Integrative medicine (IM), on the other hand, is an approach to health care that goes beyond CAM. (Caspi 2001). Committed to the practice of good medicine, whether its origins are conventional or alternative, IM recognizes that good medicine must always be based on good science that is inquiry-driven and open to new paradigms. It neither rejects conventional medicine nor uncritically accepts alternative practices. Thus, IM is a healing-oriented medicine that takes account of the whole person (body, mind, and spirit), including all aspects of lifestyle. It emphasizes the therapeutic relationship and makes use of all appropriate therapies, both conventional and alternative (Gaudet 1998).

Table 11-7 Worldwide Surveys of CAM Use among MS patients

Reference	*Country*	*Sample*	*Comments*
Page et al. 2003	Canada	70% of 440	75% women
Nayak et al. 2003	USA	57.1% of 3140	herbs (26.6%), chiropractic (25.5%), massage (23.3%) and acupuncture (19.9%). Women > men, Caucasians > non-Caucasians
Stuifbergen & Harrison 2004	USA	50% of 621	
Pucci et al. 2004	Italy	35.7% of 109	positive attitude towards CAM - 39.4% perceived benefit - 61.5% of cases
Sastre-Garriga et al. 2003	Spain	41% of 140	low levels of satisfaction with allopathy and high disability
Somerset et al. 2001	UK	47% of 318	many needs are unsatisfied

CAM Use in MS

Worldwide surveys show that CAM use among MS patients is common: between 36 percent to more than 70 percent of all MS patients use it at some time (Table 11-7) (Page et al. 2003; Nayak et al. 2003; Apel, Greim, and Zettl 2005; Yadav and Bourdette 2006; Carlson and Krahn 2006; Zur et al. 2002; Sparber and Wootton 2002; Apel et al. 2006; Campbell et al. 2006). Most patients use it for pragmatic reasons—to improve functioning and to enhance quality of life—rather than as result of a New Age-type ideology (Olsen 2009). Shinto et al. (2006) used a cross-sectional survey and the SF-12 (An Even Shorter Health Survey) to study demographic and health-related factors in a cohort of 1,700 US adult MS patients(Shinto et al. 2006). They found that female gender, high education level, longer MS duration, lower physical well-being, and lack of use of disease-modifying drugs were independent factors associated with CAM use. Marrie and colleagues examined 20,778 MS patients enrolled in the North American Research Consortium on Multiple Sclerosis (NARCOMS) Patient Registry and found that chiropractors (51 percent), massage therapists (34 percent), and nutritionists (24 percent) were the most commonly used CAM providers. Their analyses indicate that demographic factors play only a minimal role in predicting the use of CAM in the MS population, while disease factors play a slightly stronger role. They postulated that there must be other factors involved in CAM use in MS that may include accessibility, social acceptability, and cultural factors (Marrie, Hadjimichael, and Vollmer 2003). Indeed, as with other chronic diseases, research findings confirm that CAM use in MS can be understood as a critical component of self-care management in general, and not as a rejection of conventional medicine or an unrealistic search for cure (Thorne et al. 2002).

A recent study by Schwarz et al., looking at MS patients in Germany, found that 70 percent reported lifetime use of at least one CAM method, including diet modification (41 percent); omega-3 fatty acids (37 percent); removal of amalgam fillings (28 percent); taking vitamins E (28 percent), B (36 percent), and C (28 percent); homeopathy (26 percent), and taking selenium (24 percent). Sixty-nine percent of respondents were satisfied with the effects of CAM (Schwarz et al. 2008). Another study, from the United Kingdom, found that the six most-used CAM therapies in MS were reflexology, massage, yoga, relaxation and meditation, acupuncture, and aromatherapy, with each rated by a quarter to two-fifths of patients as "extremely helpful" (Esmonde and Long 2008). Does the use of these CAM methods represent smart health care decisions? What are patients with MS really happy about with respect to CAM? What does the evidence suggest (Yadav and Bourdette 2006)? Replication and validation in double-blind placebo controlled studies are required to resolve their benefits.

Evidence-Based CAM for MS Patients

A recent systematic review of randomized controlled trials (RCTs) found that there are very few rigorous studies assessing the efficacy of CAM therapies for MS (Huntley 2006). Although there is no evidence that any single CAM therapy can be a substitute for DMDs for MS, the findings from these studies indicate that some CAM therapies may provide beneficial effects for some symptoms of MS. Examples of evidence-based CAM treatments that are possibly effective and relatively safe include the following.

- **Omega-3 fatty acids.** Based on the known anti-inflammatory effects of omega-3 fatty acids, a few studies examined the effect of linoleic acid as a treatment for MS, both in remitting–relapsing and chronic progressive patients. Although these studies are imperfect, they suggest that linoleic acid may be of benefit to MS patients (Dworkin et al. 1984).
- **Naturopathic medicine combined with usual care**. Naturopathic medicine combined with usual care for MS showed a trend in improvement in the General Health subscale of the SF-36, timed walk, and neurological impairment (Shinto et al. 2008). The naturopathic intervention included daily supplementation of multivitamin/mineral without iron, vitamin C, vitamin E, fish oil, alpha-lipoic acid, intramuscular vitamin B12 once a week, and four levels of anti-inflammatory plant-based diets tailored for each patient.
- **CAM modalities**. Several CAM modalities, including massage (Hernandez-Reif et al. 1998), reflexology (Mackereth et al. 2009), and guided imagery (Maguire 1996) have been shown to reduce anxiety level and depression in MS patients. In some patients, these changes have translated into significantly improved self-esteem, better body image, better image of disease progression, and enhanced social functional status.
- **Craniosacral therapy**. A recent prospective small cohort study in MS patients showed that craniosacral therapy is effective and safe for reducing lower urinary tract symptoms (LUTS)—such as urinary frequency and urgency, using objective measures of post-voiding residual volume, subjective ratings of LUTS, and quality of life (Raviv et al. 2009).
- **Magnet therapy.** Several sham-controlled RCTs examined the effect of magnet therapy on symptoms and quality of life in patients with MS. In one of these trials, patients were randomly assigned to wear either an active magnetic pulsing device or a sham device on one of three acupuncture points, for ten to twenty-four hours a day for two months. The results showed a significant improvement in the combined rating for bladder control, cognitive function, fatigue level, mobility, spasticity, and vision (Richards et al. 1997).

It seems that, although MS patients report significant benefit from conventional therapies and providers, they may seek CAM providers for emotional support (Shinto et al. 2005). There have been significant concerns in both the conventional and alternative medical worlds that the various benefits of CAM and IM represent merely enhanced placebo effects. While a recent position paper questioned the ethics of placebo research in MS

(Polman et al. 2008), it remains crucial to ascertain the extent to which CAM has specific effects in MS.

From CAM to IM: Tailoring Treatments for Patients with MS

Despite the popularity of CAM among MS patients, the findings from an Italian survey (Pucci et al. 2004) showed that the referral source to CAM was a physician in only 12.8 percent of cases, that the neurologist was not consulted in 82 percent of cases, and the generalist was not consulted in 67 percent of cases. Of sixty-one CAM interventions, twenty-one were expected to be disease modifying (despite lack of convincing evidence). Furthermore, the use of CAM is costly (and typically paid for out of pocket). The aforementioned are all alarming. This is because when two systems of medicine (allopathy and CAM) exist side-by-side rather than integrated into one health care system, the possibility for treatment oversights and lapses, unwanted drug–herb–nutrient interactions, and unwise utilization of scarce healthcare resources all increase dramatically. Furthermore, CAM is not always safe (Levinson and Chinn 2001). Therefore, the following five main recommendations can be considered:

- Since many MS patients do not tell their physicians that they are using CAM, especially if they feel the physician will be judgmental (Murray 2006), neurologists should actively ask their patients about CAM use in an open and nonjudgmental way. MS patients who ask for advice on CAM use should be referred to IM experts, preferably neurologists, for further consultation.
- Since "natural" does not mean safe, patients might be better advised to refrain from making any health care decisions based solely on information on the Internet, because of potential conflict of interest related to hidden commercial motivations.
- There is no evidence that CAM can cure MS. Therefore, patients who express an interest in exploring CAM to improve quality of life, or patients who might benefit from CAM for symptom management, should be instructed to use it alongside, and not in lieu of, conventional medicine, while continuing close medical follow-up.
- Many health care decisions require trade-offs. Even evidence-based medicine (EBM) admits that "evidence alone is never sufficient to make a clinical decision. Decision makers must always trade the benefits and risks, inconvenience, and costs associated with alternative management strategies, and in doing so consider the patient's values" (Haynes et al. 1996). Therefore, as the evidence is integrated, best practice evidence can be resolved Eisenberg (Eisenberg 2002). It is crucial to understand the patient's needs and worldview, and to involve him or her in a shared decision-making process, just as EBM suggests.
- The success of tailoring treatments, especially for the chronically ill, rests on accurate characterization of two complementary parameters: disease factors and humanistic factors. Personalized medicine, with its emphasis on the host biology, is therefore not enough. Since IM goes beyond CAM and allopathy in putting the patient, rather than the disease, in the foreground, it offers an opportunity to re-humanize medicine (Hartzband and Groopman 2009). For that to happen, priority areas of future research should include: testing packages of care with multiple outcomes, rather than individual therapies, from a whole systems approach (Caspi and Bell 2004); finding the balance between approaching the whole person and treating the disease (Federoff and Gostin 2009); focusing on individual differences, rather than on homogeneity of subjects (Heng 2008); and applying comparative

effectiveness and cost-benefit/cost-utility analyses for health care policy decision-making (Garber and Tunis 2009).

Tailored Information for Patients with MS

Computerized visual interfaces offer an additional dimension to personalized medicine, allowing the development of information sources that are tailored to the individual needs of the patient. Moreover, since MS is characterized by phases of exacerbation and remission, associated with changes in function and presenting symptoms, information needs are likely to vary over time—depending on the time duration since diagnosis, stage of the illness, and disability, among other factors—emphasizing the need for tailoring information to the individual .(Hepworth M 2002; Hepworth Mark 2004; MacLean and Russell 2005).

In addition, any physical issues being suffered by the patient, such as fatigue, loss of concentration, and limited mobility, needs to be taken into account, since it affects his or her information-acquiring and processing behavior, including via the Internet.

Information provision, therefore, needs to be sensitive to the impact of these symptoms on the person with MS.

Information for Newly Diagnosed MS Patients

People newly diagnosed with MS and their families have an increased demand for receiving information. When people with a chronic disease recognize a specific gap in their knowledge or understanding, they seek information to help them make sense of their situation, solve a problem, or make informed decisions. Their information-seeking behaviours represent an attempt to maintain some control over their lives, which is vital for people with long-term disability or illness. In a retrospective study by Wollin, newly diagnosed MS patients stated their wish to be provided with a range of information reflective of their individual needs. The topics of interest were: therapies, counseling, support services, and research. The preferable format for information delivery was in person, in both group or individual sessions (Wollin et al. 2000).Not only do people have various informational needs, but each person's response is likely to be either information-seeking or information-blocking. What remains important is for people with MS and their families to know where and how to seek information, so that they can obtain it when they wish. Wollin's results support the view that providing accurate, relevant, and timely information soon after diagnosis of a serious disease, to both patients and their family members, is an essential management strategy (Wollin et al. 2000).

The Internet as a Tool for Providing Health Information

In recent years, the Internet has become one of the major sources of health information. A study of seventeen MS patients on their use of the Internet found that MS patients felt they use the Internet more than their healthy peers. Patients reported that they usually look for information before and after medical visits and find the Internet useful for helping them understand medical terms. On the other hand, patients reported that the information could be overwhelming, and they often have difficulty finding the information they want (Atreja et al. 2005).

In another study, fifty out of the sixty-one (82 percent) MS participants looked for information on the Internet previous to their first visit to an MS clinic (Hay et al. 2008).

Internet activity was correlated with income but not with education, marital status, physician assessment of health, or gender. Even though large number of patients reported looking for online data on MS, this information was not intended to replace the information from the physician. One of the surprising results was that two-thirds of the patients seemed to be reluctant to discuss information obtained by online search with their physician. It is possible that patients were concerned that this online search might be perceived as a lack of confidence in the physician's skills and jeopardize the doctor-patient relationship. Supporting this hypothesis was the fact that patients with longer disease duration, and hence more dependent on their physician, tended to discuss less information found on the Internet (Hay et al. 2008).

A recent study at the Carmel Medical Center in Haifa, Israel on the use of the Internet by MS patients found a correlation between age and disease severity and Internet usage. On average, younger patients (N=61) used the Internet more than older patients (N=17) (40.9 and 47.7, respectively, p= 0.016). In addition, patients with more disability, as indicated by high EDSS score, were more interested in support groups and contact with other patients through the Internet than people with lower EDSS. It is possible that as patients' quality of life deteriorates and mobility decreases, they become more dependent on the Internet for communication and socialization (Lejbkowicz et al. 2010).

As part of an effort to plan a Web-based portal and communication system for MS patients, a survey that included fifteen patients was performed (Atreja et al. 2005). Most patients believe that such a portal will be valuable. Their requests from such a portal included options to self-monitor symptoms, request prescriptions, retrieve lab results, take part in online patient education, and receive the latest research updates. However, the most valuable service would be timely communication with the medical team. Similar results were found in a recent Canadian study, where 263 MS patients were surveyed regarding their use of online health services. (Wardell L 2009). The findings of this study indicate that a vast majority of MS patients are highly interested in online health care services. Skilled Internet users, Internet surfers, and health-information seekers report the greatest likelihood of using online services.

Web Chat as a Communication Device Between MS Patients and the Health Care Team

One possible method of patient-doctor communication is Internet chat. Patients can interact with physicians without the need to go to the clinic, thereby facilitating access to medical care, especially for patients with physical disabilities. The Clalit health maintenance organization *(HMO)* in Israel has just launched such a service in pediatric medicine. Data gathered from this chat service will help guide the development and implementation of an online communication system for MS patients. Teaching sessions and technical support must be provided to patients with lower levels of Internet proficiency and literacy, who nevertheless express an interest in using online health care services. One of the problems faced by MS patients using the Internet is vision impairment (Atreja et al. 2005). One of the possible solutions could be to develop customizable Web sites using scripting language, which can allow users to switch between different style sheets and dramatically alter the appearance of a Web page's layout. It remains to be seen if this approach will become a standard feature for online health services.

Some Web sites, such as that of the MS Centers of Excellence created by the U.S. Veterans Health Administration (http://www4.va.gov/ms/), go beyond merely offering information, and actually support self-monitoring of the disease (Hatzakis et al. 2006). Personalized registration of symptoms and signs by MS patients can also contribute to tailored care. MS patients experience a vast array of symptoms that may worsen or decrease over time, making it difficult for those patients, who also may suffer from memory impairment, to keep track of their symptoms. Detailed registration of the symptoms may improve significantly management of the disease, by allowing more accurate reports to the clinician. Another example of such a system was developed by Lowe-Strong and McCullagh, who built a computerized visual interface for self-recording of pain associated with MS (Lowe-Strong and McCullagh 2005).

The quality of Web sites available for the MS patient has been found to be variable, but some Web sites meet nearly all of the information needs of people with MS, and it was concluded that it is important to assess the quality of the Web sites with MS-specific tools (Harland and Bath 2007).

In conclusion, Web sites and online tools are well accepted by MS patients and can offer tailored information and tailored self-management support to them. The studies of information needs of MS patients reported in recent years should be the basis for the development of additional portals and systems. Patients' visual impairment and other disabilities should also be taken into consideration. Dedicated health portals and management support tools will assist patients and their families in making informed decisions, and contribute to the overall improvement of personalized care of MS patients.

References

Achiron, A., M. Gurevich, N. Friedman, N. Kaminski, and M. Mandel. 2004. Blood transcriptional signatures of multiple sclerosis: unique gene expression of disease activity. *Ann Neurol* 55 (3):410–7.

Alvarez-Cermeno, J. C., T. Gasalla, and L. M. Villar. 2008. Value of oligoclonal band study in clinically isolated syndromes and multiple sclerosis. *Expert Rev Neurother* 8 (9):1279–80.

Andersson, M., J. Alvarez-Cermeno, G. Bernardi, I. Cogato, P. Fredman, J. Frederiksen, S. Fredrikson, P. Gallo, L. M. Grimaldi, M. Gronning, and et al. 1994. Cerebrospinal fluid in the diagnosis of multiple sclerosis: a consensus report. J Neurol Neurosurg Psychiatry 57 (8):897–902.

Anthony, D. C., B. Ferguson, M. K. Matyzak, K. M. Miller, M. M. Esiri, and V. H. Perry. 1997. Differential matrix metalloproteinase expression in cases of multiple sclerosis and stroke. *Neuropathol Appl Neurobiol* 23 (5):406–15.

Apel, A., B. Greim, N. Konig, and U.K. Zettl. 2006. Frequency of current utilisation of complementary and alternative medicine by patients with multiple sclerosis. *Journal of Neurology* 253:1331–6.

Apel, A., B. Greim, and U.K. Zettl. 2005. How frequently do patients with multiple sclerosis use complementary and alternative medicine? *Complementary Therapies in Medicine* 13:258–63.

Atreja, A., N. Mehta, D. Miller, S. Moore, K. Nichols, H. Miller, and C. M. Harris. 2005. One size does not fit all: using qualitative methods to inform the development of an Internet portal for multiple sclerosis patients. *AMIA Annu Symp Proc*:16–20.

Bagert, B. A. 2009. Epstein-Barr virus in multiple sclerosis. *Curr Neurol Neurosci Rep* 9 (5): 405–10.

Baranzini, S. E., N. W. Galwey, J. Wang, P. Khankhanian, R. Lindberg, D. Pelletier, W. Wu, B. M. Uitdehaag, L. Kappos, C. H. Polman, P. M. Matthews, S. L. Hauser, R. A. Gibson, J. R. Oksenberg, and M. R. Barnes. 2009. Pathway and network-based analysis of genome-wide association studies in multiple sclerosis. *Hum Mol Genet* 18 (11):2078–90.

Baranzini, S. E., P. Mousavi, J. Rio, S. J. Caillier, A. Stillman, P. Villoslada, M. M. Wyatt, M. Comabella, L. D. Greller, R. Somogyi, X. Montalban, and J. R. Oksenberg. 2005. Transcription-based prediction of response to IFNbeta using supervised computational methods. *PLoS Biol* 3 (1):e2.

Baranzini, S. E., J. Wang, R. A. Gibson, N. Galwey, Y. Naegelin, F. Barkhof, E. W. Radue, R. L. Lindberg, B. M. Uitdehaag, M. R. Johnson, A. Angelakopoulou, L. Hall, J. C. Richardson,

R. K. Prinjha, A. Gass, J. J. Geurts, J. Kragt, M. Sombekke, H. Vrenken, P. Qualley, R. R. Lincoln, R. Gomez, S. J. Caillier, M. F. George, H. Mousavi, R. Guerrero, D. T. Okuda, B. A. Cree, A. J. Green, E. Waubant, D. S. Goodin, D. Pelletier, P. M. Matthews, S. L. Hauser, L. Kappos, C. H. Polman, and J. R. Oksenberg. 2009. Genome-wide association analysis of susceptibility and clinical phenotype in multiple sclerosis. *Hum Mol Genet* 18 (4):767–78.

Bielekova, B., and R. Martin. 2004. Development of biomarkers in multiple sclerosis. *Brain* 127 (Pt 7):1463–78.

Bjartmar, C., and B. D. Trapp. 2001. Axonal and neuronal degeneration in multiple sclerosis: mechanisms and functional consequences. *Curr Opin Neurol* 14 (3):271–8.

Blanco, Y., E. A. Moral, M. Costa, M. Gomez-Choco, J. F. Torres-Peraza, L. Alonso-Magdalena, J. Alberch, D. Jaraquemada, T. Arbizu, F. Graus, and A. Saiz. 2006. Effect of glatiramer acetate (Copaxone) on the immunophenotypic and cytokine profile and BDNF production in multiple sclerosis: a longitudinal study. *Neurosci Lett* 406 (3):270–5.

Blanco, Y., A. Saiz, M. Costa, J. F. Torres-Peraza, E. Carreras, J. Alberch, D. Jaraquemada, and F. Graus. 2005. Evolution of brain-derived neurotrophic factor levels after autologous hematopietic stem cell transplantation in multiple sclerosis. *Neurosci Lett* 380 (1–2):122–6.

Braitch, M., R. Nunan, G. Niepel, L. J. Edwards, and C. S. Constantinescu. 2008. Increased osteopontin levels in the cerebrospinal fluid of patients with multiple sclerosis. *Arch Neurol* 65 (5):633–5.

Brettschneider, J., M. Maier, S. Arda, S. Claus, S. D. Sussmuth, J. Kassubek, 2005. Tumani, H. 2005. Tau protein level in cerebrospinal fluid is increased in patients with early multiple sclerosis. *Mult Scler* 11 (3):261–5.

Brettschneider, J., A. Petzold, A. Junker, and H. Tumani. 2006. Axonal damage markers in the cerebrospinal fluid of patients with clinically isolated syndrome improve predicting conversion to definite multiple sclerosis. *Mult Scler* 12 (2):143–8.

Buttgereit, F., G. Doering, A. Schaeffler, S. Witte, S. Sierakowski, E. Gromnica-Ihle, S. Jeka, K. Krueger, J. Szechinski, and R. Alten. 2008. Efficacy of modified-release versus standard prednisone to reduce duration of morning stiffness of the joints in rheumatoid arthritis (CAPRA-1): a double-blind, randomised controlled trial. *Lancet* 371 (9608):205–14.

Buttmann, M., C. Merzyn, H. H. Hofstetter, and P. Rieckmann. 2007. TRAIL, CXCL10 and CCL2 plasma levels during long-term Interferon-beta treatment of patients with multiple sclerosis correlate with flu-like adverse effects but do not predict therapeutic response. *J Neuroimmunol* 190 (1–2):170–6.

Byun, E., S. J. Caillier, X. Montalban, P. Villoslada, O. Fernandez, D. Brassat, M. Comabella, J. Wang, L. F. Barcellos, S. E. Baranzini, and J. R. Oksenberg. 2008. Genome-wide pharmacogenomic analysis of the response to interferon beta therapy in multiple sclerosis. *Arch Neurol* 65 (3):337–44.

Calabresi, P. A., G. Giovannoni, C. Confavreux, S. L. Galetta, E. Havrdova, M. Hutchinson, L. Kappos, D. H. Miller, P. W. O'Connor, J. T. Phillips, C. H. Polman, E. W. Radue, R. A. Rudick, W. H. Stuart, F. D. Lublin, A. Wajgt, B. Weinstock-Guttman, D. R. Wynn, F. Lynn, and M. A. Panzara. 2007. The incidence and significance of anti-natalizumab antibodies: results from AFFIRM and SENTINEL. *Neurology* 69 (14):1391–403.

Camelo, S., A. H. Iglesias, D. Hwang, B. Due, H. Ryu, K. Smith, S. G. Gray, J. Imitola, G. Duran, B. Assaf, B. Langley, S. J. Khoury, G. Stephanopoulos, U. De Girolami, R. R. Ratan, R. J. Ferrante, and F. Dangond. 2005. Transcriptional therapy with the histone deacetylase inhibitor trichostatin A ameliorates experimental autoimmune encephalomyelitis. *J Neuroimmunol* 164 (1–2):10–21.

Campbell, D.G., A.P. Turner, R.M. Williams, M. Hatzakis, Jr., J.D. Bowen, A. Rodriguez, and J.K. Haselkorn. 2006. Complementary and alternative medicine use in veterans with multiple sclerosis: Prevalence and demographic associations. *Journal of Rehabilitation Research & Development* 43:99–110.

Carlson, M.J., and G. Krahn. 2006. Use of complementary and alternative medicine practitioners by people with physical disabilities: estimates from a National US Survey. *Disability & Rehabilitation* 28:505–13.

Carrieri, P. B., P. Ladogana, G. Di Spigna, M. F. de Leva, M. Petracca, S. Montella, L. Buonavolonta, C. Florio, and L. Postiglione. 2008. Interleukin-10 and interleukin-12 modulation in patients with relapsing-remitting multiple sclerosis on therapy with interferon-beta 1a: differences in responders and non responders. *Immunopharmacol Immunotoxicol* 30 (4):1–9.

Casaccia-Bonnefil, P., G. Pandozy, and F. Mastronardi. 2008. Evaluating epigenetic landmarks in the brain of multiple sclerosis patients: a contribution to the current debate on disease pathogenesis. *Prog Neurobiol* 86 (4):368–78.

Caspi, O. 2001. Integrated medicine: orthodox meets alternative. Bringing complementary and alternative medicine (CAM) into mainstream is not integration. *Bmj* 322 (7279):168.

Caspi, O., and IR. Bell. 2004. One size does not fit all: Aptitude x Treatment Interaction (ATI) as a conceptual framework for CAM outcome research. Part II - Research designs and their applications. Journal of Alternative & Complementary *Medicine* 10:698–705.

Cénit, M. D., F. Blanco-Kelly, V. de las Heras, M. Bartolome, E. G. de la Concha, E. Urcelay, R. Arroyo, and A. Martinez. 2009. Glypican 5 is an interferon-beta response gene: a replication study. *Mult Scler* 15 (8):913–7.

Chawla-Sarkar, M., D. W. Leaman, and E. C. Borden. 2001. Preferential induction of apoptosis by interferon (IFN)-beta compared with IFN-alpha2: correlation with TRAIL/Apo2L induction in melanoma cell lines. *Clin Cancer Res* 7 (6):1821–31.

Cheang, M. C., M. van de Rijn, and T. O. Nielsen. 2008. Gene expression profiling of breast cancer. *Annu Rev Pathol* 3:67–97.

Chiasserini, D., M. Di Filippo, A. Candeliere, F. Susta, P. L. Orvietani, P. Calabresi, L. Binaglia, and P. Sarchielli. 2008. CSF proteome analysis in multiple sclerosis patients by two-dimensional electrophoresis. *Eur J Neurol* 15 (9):998–1001.

Chowdhury, S. A., J. Lin, and S. A. Sadiq. 2008. Specificity and correlation with disease activity of cerebrospinal fluid osteopontin levels in patients with multiple sclerosis. *Arch Neurol* 65 (2):232–5.

Comabella, M., D. W. Craig, C. Morcillo-Suarez, J. Rio, A. Navarro, M. Fernandez, R. Martin, and X. Montalban. 2009. Genome-wide scan of 500,000 single-nucleotide polymorphisms among responders and nonresponders to interferon beta therapy in multiple sclerosis. *Arch Neurol* 66 (8):972–8.

Comabella, M., J. Rio, C. Espejo, M. Ruiz de Villa, H. Al-Zayat, C. Nos, F. Deisenhammer, S. E. Baranzini, L. Nonell, C. Lopez, E. Julia, J. R. Oksenberg, and X. Montalban. 2009. Changes in matrix metalloproteinases and their inhibitors during interferon-beta treatment in multiple sclerosis. *Clin Immunol* 130 (2):145–50.

Comi, G., M. Filippi, and J. S. Wolinsky. 2001. European/Canadian multicenter, double-blind, randomized, placebo-controlled study of the effects of glatiramer acetate on magnetic resonance imaging--measured disease activity and burden in patients with relapsing multiple sclerosis. European/Canadian Glatiramer Acetate Study Group. *Ann Neurol* 49 (3):290–7.

Compston, A., G. C. Ebers, H. Lassmann, J. McDonald, PM. Matthews, and H. Wekerle. 1998. *McAlpine's multiple sclerosis*. 3rd ed,. London: Churchill Livingstone.

Cotte, S., N. von Ahsen, N. Kruse, B. Huber, A. Winkelmann, U. K. Zettl, M. Starck, N. Konig, N. Tellez, J. Dorr, F. Paul, F. Zipp, F. Luhder, H. Koepsell, H. Pannek, X. Montalban, R. Gold, and A. Chan. 2009. ABC-transporter gene-polymorphisms are potential pharmacogenetic markers for mitoxantrone response in multiple sclerosis. *Brain* 132 (9):2517–30.

Cross, S. A. 2007. Rethinking neuromyelitis optica (Devic disease). *J Neuroophthalmol* 27 (1):57–60.

Csoka, A. B., and M. Szyf. 2009. Epigenetic side-effects of common pharmaceuticals: A potential new field in medicine and pharmacology. *Med Hypotheses* 73 (5):770–80.

Cunningham, S., C. Graham, M. Hutchinson, A. Droogan, K. O'Rourke, C. Patterson, G. McDonnell, S. Hawkins, and K. Vandenbroeck. 2005. Pharmacogenomics of responsiveness to interferon IFN-beta treatment in multiple sclerosis: a genetic screen of 100 type I interferon-inducible genes. *Clin Pharmacol Ther* 78 (6):635–46.

Cuzner, M. L., D. Gveric, C. Strand, A. J. Loughlin, L. Paemen, G. Opdenakker, and J. Newcombe. 1996. The expression of tissue-type plasminogen activator, matrix metalloproteases and endogenous inhibitors in the central nervous system in multiple sclerosis: comparison of stages in lesion evolution. *J Neuropathol Exp Neurol* 55 (12):1194–204.

Daly, A. K., P. T. Donaldson, P. Bhatnagar, Y. Shen, I. Pe'er, A. Floratos, M. J. Daly, D. B. Goldstein, S. John, M. R. Nelson, J. Graham, B. K. Park, J. F. Dillon, W. Bernal, H. J. Cordell, M. Pirmohamed, G. P. Aithal, and C. P. Day. 2009. HLA-B*5701 genotype is a major determinant of drug-induced liver injury due to flucloxacillin. *Nat Genet* 41 (7):816–9.

De Jager, P. L., X. Jia, J. Wang, P. I. de Bakker, L. Ottoboni, N. T. Aggarwal, L. Piccio, S. Raychaudhuri, D. Tran, C. Aubin, R. Briskin, S. Romano, S. E. Baranzini, J. L. McCauley, M. A. Pericak-Vance, J. L. Haines, R. A. Gibson, Y. Naeglin, B. Uitdehaag, P. M. Matthews, L. Kappos, C. Polman, W. L. McArdle, D. P. Strachan, D. Evans, A. H. Cross, M. J. Daly, A. Compston, S. J. Sawcer, H. L. Weiner, S. L. Hauser, D. A. Hafler, and J. R. Oksenberg. 2009. Meta-analysis of genome scans and replication identify CD6, IRF8 and TNFRSF1A as new multiple sclerosis susceptibility loci. *Nat Genet* 41 (7):776–82.

Dworkin, R.H., D. Bates, J.H. Millar, and D.W. Paty. 1984. Linoleic acid and multiple sclerosis: a reanalysis of three double-blind trials. *Neurology* 34:1441–5.

Ehling, R., A. Lutterotti, J. Wanschitz, M. Khalil, C. Gneiss, F. Deisenhammer, M. Reindl, and T. Berger. 2004. Increased frequencies of serum antibodies to neurofilament light in patients with primary chronic progressive multiple sclerosis. *Mult Scler* 10 (6):601–6.

Eisenberg, J.M. 2002. Globalize the evidence, localize the decision: evidence-based medicine and international diversity. *Health Affairs* 21:166–8.

Esmonde, L., and A.F. Long. 2008. Complementary therapy use by persons with multiple sclerosis: benefits and research priorities. *Complementary Therapies in Clinical Practice* 14:176–84.

Farina, C., S. Wagenpfeil, and R. Hohlfeld. 2002. Immunological assay for assessing the efficacy of glatiramer acetate (Copaxone) in multiple sclerosis. A pilot study. *J Neurol* 249 (11): 1587–92.

Farina, C., M. S. Weber, E. Meinl, H. Wekerle, and R. Hohlfeld. 2005. Glatiramer acetate in multiple sclerosis: update on potential mechanisms of action. *Lancet Neurol* 4 (9):567–75.

Federoff, H.J., and L.O. Gostin. 2009. Evolving from reductionism to holism: is there a future for systems medicine? *JAMA* 302:994–6.

Felgenhauer, K., H. J. Schadlich, M. Nekic, and R. Ackermann. 1985. Cerebrospinal fluid virus antibodies. A diagnostic indicator for multiple sclerosis? *J Neurol Sci* 71 (2–3):291–9.

Fernald, G. H., S. Knott, A. Pachner, S. J. Caillier, K. Narayan, J. R. Oksenberg, P. Mousavi, and S. E. Baranzini. 2007. Genome-wide network analysis reveals the global properties of IFN-beta immediate transcriptional effects in humans. *J Immunol* 178 (8):5076–85.

Fernandez, O., V. Fernandez, C. Mayorga, M. Guerrero, A. Leon, J. A. Tamayo, A. Alonso, F. Romero, L. Leyva, G. Luque, and E. de Ramon. 2005. HLA class II and response to interferon-beta in multiple sclerosis. *Acta Neurol Scand* 112 (6):391–4.

Fox, E. J., T. K. Vartanian, and S. S. Zamvil. 2007. The immunogenicity of disease-modifying therapies for multiple sclerosis: clinical implications for neurologists. *Neurologist* 13 (6): 355–62.

Franklin, R. J., and C. Ffrench-Constant. 2008. Remyelination in the CNS: from biology to therapy. *Nat Rev Neurosci* 9 (11):839–55.

Freedman, M. S., J. Laks, N. Dotan, R. T. Altstock, A. Dukler, and C. J. Sindic. 2009. Anti-alpha-glucose-based glycan IgM antibodies predict relapse activity in multiple sclerosis after the first neurological event. *Mult Scler* 15 (4):422–30.

Fusco, C., V. Andreone, G. Coppola, V. Luongo, F. Guerini, E. Pace, C. Florio, G. Pirozzi, R. Lanzillo, P. Ferrante, P. Vivo, M. Mini, M. Macri, G. Orefice, and M. L. Lombardi. 2001. HLA-DRB1*1501 and response to copolymer-1 therapy in relapsing-remitting multiple sclerosis. *Neurology* 57 (11):1976–9.

Galboiz, Y., S. Shapiro, N. Lahat, H. Rawashdeh, and A. Miller. 2001. Matrix metalloproteinases and their tissue inhibitors as markers of disease subtype and response to interferon-beta therapy in relapsing and secondary-progressive multiple sclerosis patients. *Ann Neurol* 50 (4):443–51.

Garber, A.M., and S.R. Tunis. 2009. Does comparative-effectiveness research threaten personalized medicine? *New England Journal of Medicine* 360:1925–7.

Gargiulo, G., and S. Minucci. 2009. Epigenomic profiling of cancer cells. *Int J Biochem Cell Biol* 41 (1):127–35.

Gaudet, T.W. 1998. Integrative Medicine: The evoluation of a new approach to medicine and to medical education. *Integrative Medicine* 1:67–73.

Gilli, F., A. Bertolotto, A. Sala, F. Hoffmann, M. Capobianco, S. Malucchi, T. Glass, L. Kappos, R. Lindberg, and D. Leppert. 2004. Neutralizing antibodies against IFN-beta in multiple sclerosis: antagonization of IFN-beta mediated suppression of MMPs. *Brain* 127 (Pt2):259–68.

Gilli, F., P. Valentino, M. Caldano, L. Granieri, M. Capobianco, S. Malucchi, S. Glass, F. Marnetto, and A. Bertolotto. 2008. Expression and regulation of IFNalpha/beta receptor in IFNbeta-treated patients with multiple sclerosis. *Neurology* 71 (24):1940–7.

Glass-Marmor, L., T. Paperna, Y. Galboiz, and A. Miller. 2009. Immunomodulation by chronobiologically-based glucocorticoids treatment for multiple sclerosis relapses. *J Neuroimmunol* 210 (1–2):124–7.

Goertsches, R. H., M. Hecker, and U. K. Zettl. 2008. Monitoring of multiple sclerosis immunotherapy: from single candidates to biomarker networks. *J Neurol* 255 Suppl 6:48–57.

Grossman, I., N. Avidan, C. Singer, D. Goldstaub, L. Hayardeny, E. Eyal, E. Ben-Asher, T. Paperna, I. Pe'er, D. Lancet, J. S. Beckmann, and A. Miller. 2007. Pharmacogenetics of glatiramer acetate therapy for multiple sclerosis reveals drug-response markers. *Pharmacogenet Genomics* 17 (8):657–66.

Grossman, I., and A. Miller. 2010. Multiple Sclerosis Pharmacogenetics: Personalized approach towards Tailored Therapeutics. *The EPMA Journal* In Press.

Hammack, B. N., K. Y. Fung, S. W. Hunsucker, M. W. Duncan, M. P. Burgoon, G. P. Owens, and D. H. Gilden. 2004. Proteomic analysis of multiple sclerosis cerebrospinal fluid. *Mult Scler* 10 (3):245–60.

Harland, J., and P. Bath. 2007. Assessing the quality of websites providing information on multiple sclerosis: evaluating tools and comparing sites. *Health Informatics J* 13 (3):207–21.

Hartzband, P., and J. Groopman. 2009. Keeping the patient in the equation--humanism and health care reform. *New England Journal of Medicine* 361:554–5.

Hatzakis, M. J., Jr., C. Allen, M. Haselkorn, S. M. Anderson, P. Nichol, C. Lai, and J. K. Haselkorn. 2006. Use of medical informatics for management of multiple sclerosis using a chronic-care model. *J Rehabil Res Dev* 43 (1):1–16.

Hay, M. C., C. Strathmann, E. Lieber, K. Wick, and B. Giesser. 2008. Why patients go online: multiple sclerosis, the internet, and physician-patient communication. *Neurologist* 14 (6):374–81.

Haynes, R.B., R.B. Sackett, J.M.A. Gray, D.C. Cook, and G.H. Guyatt. 1996. Transferring evidence from research into practice, 1: the role of clinical care research evidence in clinical decisions. *ACP Journal Club* 125:A-14–15.

Hemmer, B., and A. Berthele. 2009. Should we measure the bioavailability of interferon beta in vivo in patients with multiple sclerosis? *Nat Clin Pract Neurol* 5 (3):126–7.

Heng, H.H. 2008. The conflict between complex systems and reductionism. *JAMA* 300:1580–1.

Hepworth M, Harrison J, James N. 2002. *The Information needs of people with Multiple Sclerosis (MS)*. Loughborough University and MS Trust [cited. Available from http://www.mstrust.org.uk/downloads/inareport.pdf.

Hepworth Mark, Harrison Hanet. 2004. A Survey of the Information Needs of People with Multiple Sclerosis. *Health Informatics J* 10 (1):49–69.

Hernandez-Reif, M., T. Field, T. Field, and H. Theakston. 1998. Multiple sclerosis patients benefit from massage therapy. *Journal of Bodywork & Movement Therapies* 2:168–174.

Hesse, D. and P. S. Sorensen. 2007. Using measurements of neutralizing antibodies: the challenge of IFN-beta therapy. *Eur J Neurol* 14 (8):850–9.

Hesse, D., F. Sellebjerg, and P. S. Sorensen. 2009. Absence of MxA induction by interferon beta in patients with MS reflects complete loss of bioactivity. *Neurology* 73 (5):372–7.

Hohlfeld, R. 2004. Immunologic factors in primary progressive multiple sclerosis. *Mult Scler* 10 Suppl 1:S16–21; discussion S21–2.

Huntley, A. 2006. A review of the evidence for efficacy of complementary and alternative medicines in MS. *International MS Journal* 13:5–12.

Iglesias, A. H., S. Camelo, D. Hwang, R. Villanueva, G. Stephanopoulos, and F. Dangond. 2004. Microarray detection of E2F pathway activation and other targets in multiple sclerosis peripheral blood mononuclear cells. *J Neuroimmunol* 150 (1–2):163–77.

Ito, K., S. Yamamura, S. Essilfie-Quaye, B. Cosio, M. Ito, P. J. Barnes, and I. M. Adcock. 2006. Histone deacetylase 2-mediated deacetylation of the glucocorticoid receptor enables NF-kappaB suppression. *J Exp Med* 203 (1):7–13.

Jarius, S., P. Eichhorn, C. Jacobi, B. Wildemann, M. Wick, and R. Voltz. 2009. The intrathecal, polyspecific antiviral immune response: Specific for MS or a general marker of CNS autoimmunity? *J Neurol Sci* 280 (1–2):98–100.

Jarius, S., D. Franciotta, R. Bergamaschi, S. Rauer, K. P. Wandinger, H. F. Petereit, M. Maurer, H. Tumani, A. Vincent, P. Eichhorn, B. Wildemann, M. Wick, and R. Voltz. 2008. Polyspecific, antiviral immune response distinguishes multiple sclerosis and neuromyelitis optica. *J Neurol Neurosurg Psychiatry* 79 (10):1134–6.

Jarius, S., D. Franciotta, R. Bergamaschi, H. Wright, E. Littleton, J. Palace, R. Hohlfeld, and A. Vincent. 2007. NMO-IgG in the diagnosis of neuromyelitis optica. *Neurology* 68 (13):1076–7.

Jasperse, B., C. Jakobs, M. J. Eikelenboom, C. D. Dijkstra, B. M. Uitdehaag, F. Barkhof, C. H. Polman, and C. E. Teunissen. 2007. N-acetylaspartic acid in cerebrospinal fluid of multiple sclerosis patients determined by gas-chromatography-mass spectrometry. *J Neurol* 254 (5):631–7.

Johnson, K. P., B. R. Brooks, J. A. Cohen, C. C. Ford, J. Goldstein, R. P. Lisak, L. W. Myers, H. S. Panitch, J. W. Rose, R. B. Schiffer, T. Vollmer, L. P. Weiner, and J. S. Wolinsky. 1998. Extended use of glatiramer acetate (Copaxone) is well tolerated and maintains its clinical effect on multiple sclerosis relapse rate and degree of disability. Copolymer 1 Multiple Sclerosis Study Group. *Neurology* 50 (3):701–8.

Joseph, F. G., C. L. Hirst, T. P. Pickersgill, Y. Ben-Shlomo, N. P. Robertson, and N. J. Scolding. 2009. CSF oligoclonal band status informs prognosis in multiple sclerosis: a case control study of 100 patients. *J Neurol Neurosurg Psychiatry* 80 (3):292–6.

Kampman, M. T., and M. Brustad. 2008. Vitamin D: a candidate for the environmental effect in multiple sclerosis - observations from Norway. *Neuroepidemiology* 30 (3):140–6.

Kanabrocki, E. L., M. D. Ryan, D. Lathers, N. Achille, M. R. Young, J. V. Cauteren, S. Foley, M. C. Johnson, N. C. Friedman, G. Siegel, and B. A. Nemchausky. 2007. Circadian distribution of serum cytokines in multiple sclerosis. *Clin Ter* 158 (2):157–62.

Khademi, M., L. Bornsen, F. Rafatnia, M. Andersson, L. Brundin, F. Piehl, F. Sellebjerg, and T. Olsson. 2009. The effects of natalizumab on inflammatory mediators in multiple sclerosis: prospects for treatment-sensitive biomarkers. *Eur J Neurol* 16 (4):528–36.

Kirstein-Grossman, I., J. S. Beckmann, D. Lancet, and A. Miller. 2002. Pharmacogenetic Development of Personalized Medicine: Multiple Sclerosis Treatment as a Model. *Drug News Perspect* 15 (9):558–567.

Kivisakk, P., B. C. Healy, V. Viglietta, F. J. Quintana, M. A. Hootstein, H. L. Weiner, and S. J. Khoury. 2009. Natalizumab treatment is associated with peripheral sequestration of proinflammatory T cells. *Neurology* 72 (22):1922–30.

Koike, F., J. Satoh, S. Miyake, T. Yamamoto, M. Kawai, S. Kikuchi, K. Nomura, K. Yokoyama, K. Ota, T. Kanda, T. Fukazawa, and T. Yamamura. 2003. Microarray analysis identifies interferon beta-regulated genes in multiple sclerosis. *J Neuroimmunol* 139 (1–2):109–18.

Kracke, A., P. von Wussow, A. N. Al-Masri, G. Dalley, A. Windhagen, and F. Heidenreich. 2000. Mx proteins in blood leukocytes for monitoring interferon beta-1b therapy in patients with MS. *Neurology* 54 (1):193–9.

Kuhle, J., C. Pohl, M. Mehling, G. Edan, M. S. Freedman, H. P. Hartung, C. H. Polman, D. H. Miller, X. Montalban, F. Barkhof, L. Bauer, S. Dahms, R. Lindberg, L. Kappos, and R. Sandbrink. 2007. Lack of association between antimyelin antibodies and progression to multiple sclerosis. *N Engl J Med* 356 (4):371–8.

Kurtuncu, M., and E. Tuzun. 2008. Multiple sclerosis: could it be an epigenetic disease? *Med Hypotheses* 71 (6):945–7.

Lamers, K. J., H. P. de Reus, and P. J. Jongen. 1998. Myelin basic protein in CSF as indicator of disease activity in multiple sclerosis. *Mult Scler* 4 (3):124–6.

Lamers, K. J., B. G. van Engelen, F. J. Gabreels, O. R. Hommes, G. F. Borm, and R. A. Wevers. 1995. Cerebrospinal neuron-specific enolase, S-100 and myelin basic protein in neurological disorders. *Acta Neurol Scand* 92 (3):247–51.

Lehmensiek, V., S. D. Sussmuth, G. Tauscher, J. Brettschneider, S. Felk, F. Gillardon, and H. Tumani. 2007. Cerebrospinal fluid proteome profile in multiple sclerosis. *Mult Scler* 13 (7):840–9.

Lejbkowicz, I., T. Paperna, N. Stein, S. Dishon, and A. Miller. 2010. Internet usage by patients with multiple sclerosis: Implications to participatory medicine and personalized healthcare. *Multiple Sclerosis International* article ID: 640749, doi:10.1155/2010/6407.

Lennon, V. A., D. M. Wingerchuk, T. J. Kryzer, S. J. Pittock, C. F. Lucchinetti, K. Fujihara, I. Nakashima, and B. G. Weinshenker. 2004. A serum autoantibody marker of neuromyelitis optica: distinction from multiple sclerosis. *Lancet* 364 (9451):2106–12.

Levinson, A., and J. Chinn. 2001. Lead poisoning from complementary and alternative medicine in multiple sclerosis. *Journal of Neurology, Neurosurgery & Psychiatry* 71: 281–2.

Lichtinghagen, R., T. Seifert, A. Kracke, S. Marckmann, U. Wurster, and F. Heidenreich. 1999. Expression of matrix metalloproteinase-9 and its inhibitors in mononuclear blood cells of patients with multiple sclerosis. *J Neuroimmunol* 99 (1):19–26.

Lindberg, R. L., L. Achtnichts, F. Hoffmann, J. Kuhle, and L. Kappos. 2008. Natalizumab alters transcriptional expression profiles of blood cell subpopulations of multiple sclerosis patients. *J Neuroimmunol* 194 (1–2):153–64.

Lowe-Strong, A., and P. J. McCullagh. 2005. Monitoring of symptoms and interventions associated with multiple sclerosis. *Stud Health Technol Inform* 117:223–8.

Lycke, J. N., J. E. Karlsson, O. Andersen, and L. E. Rosengren. 1998. Neurofilament protein in cerebrospinal fluid: a potential marker of activity in multiple sclerosis. *J Neurol Neurosurg Psychiatry* 64 (3):402–4.

Lyssiotis, C. A., J. Walker, C. Wu, T. Kondo, P. G. Schultz, and X. Wu. 2007. Inhibition of histone deacetylase activity induces developmental plasticity in oligodendrocyte precursor cells. *Proc Natl Acad Sci U S A* 104 (38):14982–7.

Mackereth, P.A., K. Booth, V.F Hillier, and A-L. Caress. 2009. Reflexology and progressive muscle relaxation training for people with multiple sclerosis: A crossover trial. *Complementary Therapies in Clinical Practice* 15:14–21.

MacLean, R., and A. Russell. 2005. Innovative ways of responding to the information needs of people with MS. *Br J Nurs* 14 (14):754–7.

Maguire, B.L. 1996. The effects of imagery on attitudes and moods in multiple sclerosis patients. *Alternative Therapies in Health & Medicine* 2:75–79.

Malmestrom, C., S. Haghighi, L. Rosengren, O. Andersen, and J. Lycke. 2003. Neurofilament light protein and glial fibrillary acidic protein as biological markers in MS. *Neurology* 61 (12):1720–5.

Malucchi, S., F. Gilli, M. Caldano, F. Marnetto, P. Valentino, L. Granieri, A. Sala, M. Capobianco, and A. Bertolotto. 2008. Predictive markers for response to interferon therapy in patients with multiple sclerosis. *Neurology* 70 (13 Pt 2):1119–27.

Malucchi, S., A. Sala, F. Gilli, R. Bottero, A. Di Sapio, M. Capobianco, and A. Bertolotto. 2004. Neutralizing antibodies reduce the efficacy of betaIFN during treatment of multiple sclerosis. *Neurology* 62 (11):2031–7.

Marchionni, L., R. F. Wilson, A. C. Wolff, S. Marinopoulos, G. Parmigiani, E. B. Bass, and S. N. Goodman. 2008. Systematic review: gene expression profiling assays in early-stage breast cancer. *Ann Intern Med* 148 (5):358–69.

Marrie, R.A., O. Hadjimichael, and T. Vollmer. 2003. Predictors of alternative medicine use by multiple sclerosis patients. *Multiple Sclerosis* 9:461–6.

Martinez, A., V. de las Heras, A. Mas Fontao, M. Bartolome, E. G. de la Concha, E. Urcelay, and R. Arroyo. 2006. An IFNG polymorphism is associated with interferon-beta response in Spanish MS patients. *J Neuroimmunol* 173 (1–2):196–9.

Masjuan, J., J. C. Alvarez-Cermeno, N. Garcia-Barragan, M. Diaz-Sanchez, M. Espino, M. C. Sadaba, P. Gonzalez-Porque, J. Martinez San Millan, and L. M. Villar. 2006. Clinically isolated syndromes: a new oligoclonal band test accurately predicts conversion to MS. *Neurology* 66 (4):576–8.

Massaro, A. R. 2002. The role of NCAM in remyelination. *Neurol Sci* 22 (6):429–35.

Massaro, A. R., F. Michetti, A. Laudisio, and P. Bergonzi. 1985. Myelin basic protein and S-100 antigen in cerebrospinal fluid of patients with multiple sclerosis in the acute phase. *Ital J Neurol Sci* 6 (1):53–6.

Massaro, A. R., C. Soranzo, and A. Carnevale. 1997. Cerebrospinal-fluid ciliary neurotrophic factor in neurological patients. *Eur Neurol* 37 (4):243–6.

Mastronardi, F. G., A. Noor, D. D. Wood, T. Paton, and M. A. Moscarello. 2007. Peptidyl argininedeiminase 2 CpG island in multiple sclerosis white matter is hypomethylated. *J Neurosci Res* 85 (9):2006–16.

McDonald, W. I., A. Compston, G. Edan, D. Goodkin, H. P. Hartung, F. D. Lublin, H. F. McFarland, D. W. Paty, C. H. Polman, S. C. Reingold, M. Sandberg-Wollheim, W. Sibley, A. Thompson, S. van den Noort, B. Y. Weinshenker, and J. S. Wolinsky. 2001. Recommended diagnostic criteria for multiple sclerosis: guidelines from the International Panel on the diagnosis of multiple sclerosis. *Ann Neurol* 50 (1):121–7.

Miller, A., N. Avidan, N. Tzunz-Henig, L. Glass-Marmor, I. Lejbkowicz, R. Y. Pinter, and T. Paperna. 2008. Translation towards personalized medicine in Multiple Sclerosis. *J Neurol Sci* 274 (1–2):68–75.

Murray, T.J. 2006. Complementary and alternative medicine for MS. *International MS Journal* 13:3.

Musse, A. A., J. M. Boggs, and G. Harauz. 2006. Deimination of membrane-bound myelin basic protein in multiple sclerosis exposes an immunodominant epitope. *Proc Natl Acad Sci U S A* 103 (12):4422–7.

Muthian, G., H. P. Raikwar, J. Rajasingh, and J. J. Bright. 2006. 1,25 Dihydroxyvitamin-D3 modulates JAK-STAT pathway in IL-12/IFNgamma axis leading to Th1 response in experimental allergic encephalomyelitis. *J Neurosci Res* 83 (7):1299–309.

Myhr, K. M. 2009. Vitamin D treatment in multiple sclerosis. *J Neurol Sci* 286 (1–2):104–8.

Nayak, S., R.J. Matheis, N.E. Schoenberger, and S.C. Shiflett. 2003. Use of unconventional therapies by individuals with multiple sclerosis. *Clinical Rehabilitation* 17:181–91.

Nieves, J., F. Cosman, J. Herbert, V. Shen, and R. Lindsay. 1994. High prevalence of vitamin D deficiency and reduced bone mass in multiple sclerosis. *Neurology* 44 (9):1687–92.

Oksenberg, J. R., S. E. Baranzini, S. Sawcer, and S. L. Hauser. 2008. The genetics of multiple sclerosis: SNPs to pathways to pathogenesis. *Nat Rev Genet* 9 (7):516–26.

Olsen, S.A. 2009. A review of complementary and alternative medicine (CAM) by people with multiple sclerosis. *Occupational Therapy International* 16:57–70.

Osako, T., Y. Ito, S. Takahashi, N. Tokudome, T. Iwase, and K. Hatake. 2008. Efficacy and safety of trastuzumab plus capecitabine in heavily pretreated patients with HER2-positive metastatic breast cancer. *Cancer Chemother Pharmacol* 62 (1):159–64.

Otten, U., and R. A. Gadient. 1995. Neurotrophins and cytokines--intermediaries between the immune and nervous systems. *Int J Dev Neurosci* 13 (3–4):147–51.

Ozenci, V., L. Rinaldi, N. Teleshova, D. Matusevicius, P. Kivisakk, M. Kouwenhoven, and H. Link. 1999. Metalloproteinases and their tissue inhibitors in multiple sclerosis. *J Autoimmun* 12 (4):297–303.

Page, S.A., M.J. Verhoef, R.A. Stebbins, L.M. Metz, and J.C. Levy. 2003. The use of complementary and alternative therapies by people with multiple sclerosis. *Chronic Diseases in Canada* 24:75–9.

Paik, S. 2006. Molecular profiling of breast cancer. *Curr Opin Obstet Gynecol* 18 (1):59–63.

Petereit, H. F., S. Nolden, S. Schoppe, S. Bamborschke, R. Pukrop, and W. D. Heiss. 2002. Low interferon gamma producers are better treatment responders: a two-year follow-up of interferon beta-treated multiple sclerosis patients. *Mult Scler* 8 (6):492–4.

Petzold, A., M. J. Eikelenboom, D. Gveric, G. Keir, M. Chapman, R. H. Lazeron, M. L. Cuzner, C. H. Polman, B. M. Uitdehaag, E. J. Thompson, and G. Giovannoni. 2002. Markers for different glial cell responses in multiple sclerosis: clinical and pathological correlations. *Brain* 125 (Pt 7):1462–73.

Petzold, A., M. J. Eikelenboom, G. Keir, D. Grant, R. H. Lazeron, C. H. Polman, B. M. Uitdehaag, E. J. Thompson, and G. Giovannoni. 2005. Axonal damage accumulates in the progressive phase of multiple sclerosis: three year follow up study. *J Neurol Neurosurg Psychiatry* 76 (2):206–11.

Polman, C. H., S. C. Reingold, G. Edan, M. Filippi, H. P. Hartung, L. Kappos, F. D. Lublin, L. M. Metz, H. F. McFarland, P. W. O'Connor, M. Sandberg-Wollheim, A. J. Thompson, B. G. Weinshenker, and J. S. Wolinsky. 2005. Diagnostic criteria for multiple sclerosis: 2005 revisions to the "McDonald Criteria". *Ann Neurol* 58 (6):840–6.

Polman, C.H., S.C. Reingold, F. Barkhof, P.A. Calabresi, M. Clanet, and J.A. Cohen. 2008. Ethics of placebo-controlled clinical trials in multiple sclerosis: A reassessment. *Neurology* 70:1134–40.

Pozzilli, C., and L. Prosperini. 2008. Clinical markers of therapeutic response to disease modifying drugs. *Neurol Sci* 29 Suppl 2:S211–3.

Ptashne, M. 2007. On the use of the word 'epigenetic'. *Curr Biol* 17 (7):R233–6.

Pucci, E., E. Cartechini, C. Taus, and G. Giuliani. 2004. Why physicians need to look more closely at the use of complementary and alternative medicine by multiple sclerosis patients. *European Journal of Neurology* 11:263–7.

Qin, Z., Y. Qin, and S. Liu. 2009. Alteration of DBP levels in CSF of patients with MS by proteomics analysis. *Cell Mol Neurobiol* 29 (2):203–10.

Quintana, F. J., M. F. Farez, V. Viglietta, A. H. Iglesias, Y. Merbl, G. Izquierdo, M. Lucas, A. S. Basso, S. J. Khoury, C. F. Lucchinetti, I. R. Cohen, and H. L. Weiner. 2008. Antigen microarrays identify unique serum autoantibody signatures in clinical and pathologic subtypes of multiple sclerosis. *Proc Natl Acad Sci U S A* 105 (48):18889–94.

Ramagopalan, S. V., N. J. Maugeri, L. Handunnetthi, M. R. Lincoln, S. M. Orton, D. A. Dyment, G. C. Deluca, B. M. Herrera, M. J. Chao, A. D. Sadovnick, G. C. Ebers, and J. C. Knight. 2009. Expression of the multiple sclerosis-associated MHC class II Allele HLA-DRB1*1501 is regulated by vitamin D. *PLoS Genet* 5 (2):e1000369.

Raviv, G., S. Shefi, D. Nizani, and A. Achiron. 2009. Effect of craniosacral therapy on lower urinary tract signs and symptoms in multiple sclerosis. *Complementary Therapies in Clinical Practice* 15:72–75.

Reiber, H. 1998. Cerebrospinal fluid--physiology, analysis and interpretation of protein patterns for diagnosis of neurological diseases. *Mult Scler* 4 (3):99–107.

Richards, T.L., M.S. Lappin, J. Acosta-Urquidi, G.H. Kraft, A.C. Heide, and F.W. Lawrie. 1997. Double-blind study of pulsing magnetic field effects on multiple sclerosis. *Journal of Alternative & Complementary Medicine* 3:21–29.

Rieckmann, P., B. Altenhofen, A. Riegel, B. Kallmann, and K. Felgenhauer. 1998. Correlation of soluble adhesion molecules in blood and cerebrospinal fluid with magnetic resonance imaging activity in patients with multiple sclerosis. *Mult Scler* 4 (3):178–82.

Robinson-Agramonte, M., H. Reiber, J. A. Cabrera-Gomez, and R. Galvizu. 2007. Intrathecal polyspecific immune response to neurotropic viruses in multiple sclerosis: a comparative report from Cuban patients. *Acta Neurol Scand* 115 (5):312–8.

Rosengren, L. E., J. Lycke, and O. Andersen. 1995. Glial fibrillary acidic protein in CSF of multiple sclerosis patients: relation to neurological deficit. *J Neurol Sci* 133 (1–2):61–5.

Rot, U., A. Sominanda, A. Fogdell-Hahn, and J. Hillert. 2008. Impression of clinical worsening fails to predict interferon-beta neutralizing antibody status. *J Int Med Res* 36 (6):1418–25.

Rudick, R. A., and C. H. Polman. 2009. Current approaches to the identification and management of breakthrough disease in patients with multiple sclerosis. *Lancet Neurol* 8 (6):545–59.

Rudick, R. A., R. M. Ransohoff, J. C. Lee, R. Peppler, M. Yu, P. M. Mathisen, and V. K. Tuohy. 1998. In vivo effects of interferon beta-1a on immunosuppressive cytokines in multiple sclerosis. *Neurology* 50 (5):1294–300.

Salama, H. H., J. Hong, Y. C. Zang, A. El-Mongui, and J. Zhang. 2003. Blocking effects of serum reactive antibodies induced by glatiramer acetate treatment in multiple sclerosis. *Brain* 126 (Pt 12):2638–47.

Santos, R., B. Weinstock-Guttman, M. Tamano-Blanco, D. Badgett, R. Zivadinov, T. Justinger, F. Munschauer, 3rd, and M. Ramanathan. 2006. Dynamics of interferon-beta modulated mRNA biomarkers in multiple sclerosis patients with anti-interferon-beta neutralizing antibodies. *J Neuroimmunol* 176 (1–2):125–33.

Sarchielli, P., L. Greco, A. Stipa, A. Floridi, and V. Gallai. 2002. Brain-derived neurotrophic factor in patients with multiple sclerosis. *J Neuroimmunol* 132 (1–2):180–8.

Sarchielli, P., M. Zaffaroni, A. Floridi, L. Greco, A. Candeliere, A. Mattioni, S. Tenaglia, M. Di Filippo, and P. Calabresi. 2007. Production of brain-derived neurotrophic factor by mononuclear cells of patients with multiple sclerosis treated with glatiramer acetate, interferon-beta 1a, and high doses of immunoglobulins. *Mult Scler* 13 (3):313–31.

Satoh, J., M. Nakanishi, F. Koike, S. Miyake, T. Yamamoto, M. Kawai, S. Kikuchi, K. Nomura, K. Yokoyama, K. Ota, T. Kanda, T. Fukazawa, and T. Yamamura. 2005. Microarray analysis identifies an aberrant expression of apoptosis and DNA damage-regulatory genes in multiple sclerosis. *Neurobiol Dis* 18 (3):537–50.

Scagnolari, C., P. Duda, F. Bagnato, G. De Vito, A. Alberelli, V. Lavolpe, E. Girardi, V. Durastanti, M. Trojano, L. Kappos, and G. Antonelli. 2007. Pharmacodynamics of interferon beta in multiple sclerosis patients with or without serum neutralizing antibodies. *J Neurol* 254 (5):597–604.

Schwarz, S., C. Knorr, H. Geiger, and P. Flachenecker. 2008. Complementary and alternative medicine for multiple sclerosis. *Multiple Sclerosis* 14:1113–9.

Sega, S., A. Horvat, U. Rot, B. Wraber, and A. Ihan. 2008. Could cytokines levels or adhesion molecules expression be predictor of IFN-beta treatment response in multiple sclerosis patients? *Clin Neurol Neurosurg* 110 (9):947–50.

Sellebjerg, F., P. Datta, J. Larsen, K. Rieneck, I. Alsing, A. Oturai, A. Svejgaard, P. Soelberg Sorensen, and L. P. Ryder. 2008. Gene expression analysis of interferon-{beta} treatment in multiple sclerosis. *Multiple Sclerosis* 14 (5):615–21.

Sellebjerg, F., M. Krakauer, D. Hesse, L. P. Ryder, I. Alsing, P. E. Jensen, N. Koch-Henriksen, A. Svejgaard, and P. Soelberg Sorensen. 2009. Identification of new sensitive biomarkers for the in vivo response to interferon-beta treatment in multiple sclerosis using DNA-array evaluation. *Eur J Neurol* 16 (12):1291–8.

Shen, S., J. Sandoval, V. A. Swiss, J. Li, J. Dupree, R. J. Franklin, and P. Casaccia-Bonnefil. 2008. Age-dependent epigenetic control of differentiation inhibitors is critical for remyelination efficiency. *Nat Neurosci* 11 (9):1024–34.

Shinto, L., C. Calabrese, C. Morris, V. Yadav, D. Griffith, R. Frank, B.S. Oken, S. Baldauf-Wagner, and D. Bourdette. 2008. A randomized pilot study of naturopathic medicine in multiple sclerosis. *Journal of Alternative & Complementary Medicine* 14:489–96.

Shinto, L., V. Yadav, C. Morris, J.A. Lapidus, A. Senders, and D. Bourdette. 2005. The perceived benefit and satisfaction from conventional and complementary and alternative medicine (CAM) in people with multiple sclerosis. *Complementary Therapies in Medicine* 13:264–72.

———. 2006. Demographic and health-related factors associated with complementary and alternative medicine (CAM) use in multiple sclerosis. *Multiple Sclerosis* 12:94–100.

Singh, M. K., T. F. Scott, W. A. LaFramboise, F. Z. Hu, J. C. Post, and G. D. Ehrlich. 2007. Gene expression changes in peripheral blood mononuclear cells from multiple sclerosis patients undergoing beta-interferon therapy. *J Neurol Sci* 258 (1–2):52–9.

Smith, C. H., E. Waubant, and A. Langer-Gould. 2009. Absence of neuromyelitis optica IgG antibody in an active relapsing-remitting multiple sclerosis population. *J Neuroophthalmol* 29 (2):104–6.

Smolders, J., E. Peelen, M. Thewissen, P. Menheere, J. W. Cohen Tervaert, R. Hupperts, and J. Damoiseaux. 2009. The relevance of vitamin D receptor gene polymorphisms for vitamin D research in multiple sclerosis. *Autoimmun Rev* 8 (7):621–6.

Soilu-Hanninen, M., M. Laaksonen, A. Hanninen, J. P. Eralinna, and M. Panelius. 2005. Downregulation of VLA-4 on T cells as a marker of long term treatment response to interferon beta-1a in MS. *J Neuroimmunol* 167 (1–2):175–82.

Sorensen, P., S. Ross, C. Clemmesen, K. M. Bendtzen, K. Frederiksen, J. L. Jensen, K. Kristensen, O. Petersen, T. Rasmussen, S. Ravnborg, M. Stenager, E. Koch-Henriksen, N. 2003. Clinical importance of neutralising antibodies against interferon beta in patients with relapsing-remitting multiple sclerosis. *Lancet* 362 (9391):1184–91 .

Sorensen, P. S., N. Koch-Henriksen, C. Ross, K. M. Clemmesen, and K. Bendtzen. 2005. Appearance and disappearance of neutralizing antibodies during interferon-beta therapy. *Neurology* 65 (1):33–9.

Sparber, A., and J.C. Wootton. 2002. Surveys of complementary and alternative medicine: Part V. Use of alternative and complementary therapies for psychiatric and neurologic diseases. *Journal of Alternative & Complementary Medicine* 8:93–6.

Sriram, U., L. F. Barcellos, P. Villoslada, J. Rio, S. E. Baranzini, S. Caillier, A. Stillman, S. L. Hauser, X. Montalban, and J. R. Oksenberg. 2003. Pharmacogenomic analysis of interferon receptor polymorphisms in multiple sclerosis. *Genes Immun* 4 (2):147–52.

Stoop, M. P., L. J. Dekker, M. K. Titulaer, P. C. Burgers, P. A. Sillevis Smitt, T. M. Luider, and R. Q. Hintzen. 2008. Multiple sclerosis-related proteins identified in cerebrospinal fluid by advanced mass spectrometry. *Proteomics* 8 (8):1576–85.

Sturzebecher, S., K. P. Wandinger, A. Rosenwald, M. Sathyamoorthy, A. Tzou, P. Mattar, J. A. Frank, L. Staudt, R. Martin, and H. F. McFarland. 2003. Expression profiling identifies responder and non-responder phenotypes to interferon-beta in multiple sclerosis. *Brain* 126 (Pt 6):1419–29.

Svejgaard, A. 2008. The immunogenetics of multiple sclerosis. *Immunogenetics* 60 (6):275–86.

Takeuchi, F., R. McGinnis, S. Bourgeois, C. Barnes, N. Eriksson, N. Soranzo, P. Whittaker, V. Ranganath, V. Kumanduri, W. McLaren, L. Holm, J. Lindh, A. Rane, M. Wadelius, and P. Deloukas. 2009. A genome-wide association study confirms VKORC1, CYP2C9, and CYP4F2 as principal genetic determinants of warfarin dose. *PLoS Genet* 5 (3):e1000433.

Teunissen, C. E., C. Dijkstra, and C. Polman. 2005. Biological markers in CSF and blood for axonal degeneration in multiple sclerosis. *Lancet Neurol* 4 (1):32–41.

Teunissen, C. E., E. Iacobaeus, M. Khademi, L. Brundin, N. Norgren, M. J. Koel-Simmelink, M. Schepens, F. Bouwman, H. A. Twaalfhoven, H. J. Blom, C. Jakobs, and C. D. Dijkstra. 2009. Combination of CSF N-acetylaspartate and neurofilaments in multiple sclerosis. *Neurology* 72 (15):1322–9.

Thorne, S., B. Paterson, C. Russell, and A. Schultz. 2002. Complementary/alternative medicine in chronic illness as informed self-care decision making. *International Journal of Nursing Studies* 39:671–83.

Trapp, B. D., and K. A. Nave. 2008. Multiple sclerosis: an immune or neurodegenerative disorder? *Annu Rev Neurosci* 31:247–69.

Trojano, M., C. Avolio, G. M. Liuzzi, M. Ruggieri, G. Defazio, M. Liguori, M. P. Santacroce, D. Paolicelli, F. Giuliani, P. Riccio, and P. Livrea. 1999. Changes of serum sICAM-1 and MMP-9 induced by rIFNbeta-1b treatment in relapsing-remitting MS. *Neurology* 53 (7):1402–8.

Tumani, H., H. P. Hartung, B. Hemmer, C. Teunissen, F. Deisenhammer, G. Giovannoni, and U. K. Zettl. 2009. Cerebrospinal fluid biomarkers in multiple sclerosis. *Neurobiol Dis* 35 (2):117–127.

Tumani, H., V. Lehmensiek, D. Rau, I. Guttmann, G. Tauscher, H. Mogel, C. Palm, V. Hirt, S. D. Suessmuth, I. Sapunova-Meier, A. C. Ludolph, and J. Brettschneider. 2009. CSF proteome analysis in clinically isolated syndrome (CIS): candidate markers for conversion to definite multiple sclerosis. *Neurosci Lett* 452 (2):214–7.

Valis, M., R. Talab, P. Stourac, C. Andrys, and J. Masopust. 2008. Tau protein, phosphorylated tau protein and beta-amyloid42 in the cerebrospinal fluid of multiple sclerosis patients. *Neuro Endocrinol Lett* 29 (6):971–6.

Vallittu, A. M., M. Halminen, J. Peltoniemi, J. Ilonen, I. Julkunen, A. Salmi, and J. P. Eralinna. 2002. Neutralizing antibodies reduce MxA protein induction in interferon-beta-1a-treated MS patients. *Neurology* 58 (12):1786–90.

Van Baarsen, L. G., S. Vosslamber, M. Tijssen, J. M. Baggen, L. F. Van der Voort, J. Killestein, T. C. van der Pouw Kraan, C. H. Polman, and C. L. Verweij. 2008. Pharmacogenomics of interferon-beta therapy in multiple sclerosis: baseline IFN signature determines pharmacological differences between patients. *PLoS One* 3 (4):e1927.

Van Boxel-Dezaire, A. H., S. C. van Trigt-Hoff, J. Killestein, H. M. Schrijver, J. C. van Houwelingen, C. H. Polman, and L. Nagelkerken. 2000. Contrasting responses to interferon beta-1b treatment in relapsing-remitting multiple sclerosis: does baseline interleukin-12p35 messenger RNA predict the efficacy of treatment? *Ann Neurol* 48 (3):313–22.

Van der Voort, L. F., A. Kok, A. Visser, C. B. Oudejans, M. Caldano, F. Gilli, A. Bertolotto, C. H. Polman, and J. Killestein. 2009. Interferon-beta bioactivity measurement in multiple sclerosis: feasibility for routine clinical practice. *Mult Scler* 15 (2):212–8.

Van der Voort, L. F., A. Visser, D. L. Knol, C. B. Oudejans, C. H. Polman, and J. Killestein. 2009. Lack of Interferon-beta bioactivity is associated with the occurrence of relapses in multiple sclerosis. *Eur J Neurol* 16 (9):1049–52.

Vandenbroeck, K., M. Comabella, E. Tolosa, R. Goertsches, D. Brassat, R. Hintzen, C. Infante-Duarte, A. Favorov, S. Escorza, R. Palacios, J. R. Oksenberg, and P. Villoslada. 2009. United Europeans for development of pharmacogenomics in multiple sclerosis network. *Pharmacogenomics* 10 (5):885–94.

Villar, L. M., N. Garcia-Barragan, M. C. Sadaba, M. Espino, J. Gomez-Rial, J. Martinez-San Millan, P. Gonzalez-Porque, and J. C. Alvarez-Cermeno. 2008. Accuracy of CSF and MRI criteria for dissemination in space in the diagnosis of multiple sclerosis. *J Neurol Sci* 266 (1–2):34–7.

Villar, L. M., J. Masjuan, P. Gonzalez-Porque, J. Plaza, M. C. Sadaba, E. Roldan, A. Bootello, and J. C. Alvarez-Cermeno. 2002. Intrathecal IgM synthesis predicts the onset of new relapses and a worse disease course in MS. *Neurology* 59 (4):555–9.

Villar, L. M., J. Masjuan, P. Gonzalez-Porque, J. Plaza, M. C. Sadaba, E. Roldan, A. Bootello, and J. C. Alvarez-Cermeno.. 2003. Intrathecal IgM synthesis is a prognostic factor in multiple sclerosis. *Ann Neurol* 53 (2):222–6.

Villar, L. M., T. Masterman, B. Casanova, J. Gomez-Rial, M. Espino, M. C. Sadaba, P. Gonzalez-Porque, F. Coret, and J. C. Alvarez-Cermeno. 2009. CSF oligoclonal band patterns reveal disease heterogeneity in multiple sclerosis. *J Neuroimmunol* 211 (1–2):101–4.

Villoslada, P., L. F. Barcellos, J. Rio, A. B. Begovich, M. Tintore, J. Sastre-Garriga, S. E. Baranzini, P. Casquero, S. L. Hauser, X. Montalban, and J. R. Oksenberg. 2002. The HLA locus and multiple sclerosis in Spain. Role in disease susceptibility, clinical course and response to interferon-beta. *J Neuroimmunol* 130 (1–2):194–201.

Vogt, M. H., S. Floris, J. Killestein, D. L. Knol, M. Smits, F. Barkhof, C. H. Polman, and L. Nagelkerken. 2004. Osteopontin levels and increased disease activity in relapsing-remitting multiple sclerosis patients. *J Neuroimmunol* 155 (1–2):155–60.

Wandinger, K. P., J. D. Lunemann, O. Wengert, J. Bellmann-Strobl, O. Aktas, A. Weber, E. Grundstrom, S. Ehrlich, K. D. Wernecke, H. D. Volk, and F. Zipp. 2003. TNF-related apoptosis inducing ligand (TRAIL) as a potential response marker for interferon-beta treatment in multiple sclerosis. *Lancet* 361 (9374):2036–43.

Wardell L, Stanley H, Laizner AM and Lapierre Y. 2009. Multiple Sclerosis Patients Interest in and Likelihood of Using Online Health-Care Services. *Int J MS Care [Serial Online]* 11:79–89.

Waubant, E., D. Goodkin, A. Bostrom, P. Bacchetti, J. Hietpas, R. Lindberg, and D. Leppert. 2003. IFNbeta lowers MMP-9/TIMP-1 ratio, which predicts new enhancing lesions in patients with SPMS. *Neurology* 60 (1):52–7.

Waubant, E., D. E. Goodkin, L. Gee, P. Bacchetti, R. Sloan, T. Stewart, P. B. Andersson, G. Stabler, and K. Miller. 1999. Serum MMP-9 and TIMP-1 levels are related to MRI activity in relapsing multiple sclerosis. *Neurology* 53 (7):1397–401.

Weder, C., G. M. Baltariu, K. A. Wyler, H. J. Gober, C. Lienert, M. Schluep, E. W. Radu, G. De Libero, L. Kappos, and P. W. Duda. 2005. Clinical and immune responses correlate in glatiramer acetate therapy of multiple sclerosis. *Eur J Neurol* 12 (11):869–78.

Weiner, H. L. 2009. The challenge of multiple sclerosis: how do we cure a chronic heterogeneous disease? *Ann Neurol* 65 (3):239–48.

Weinstock-Guttman, B., D. Badgett, K. Patrick, L. Hartrich, R. Santos, D. Hall, M. Baier, J. Feichter, and M. Ramanathan. 2003. Genomic effects of IFN-beta in multiple sclerosis patients. *J Immunol* 171 (5):2694–702.

Weinstock-Guttman, B., K. Bhasi, D. Badgett, M. Tamano-Blanco, M. Minhas, J. Feichter, K. Patrick, F. Munschauer, R. Bakshi, and M. Ramanathan. 2008. Genomic effects of once-weekly, intramuscular interferon-beta1a treatment after the first dose and on chronic dosing: Relationships to 5-year clinical outcomes in multiple sclerosis patients. *J Neuroimmunol* 205 (1–2):113–25.

Weinstock-Guttman, B., M. Tamano-Blanco, K. Bhasi, R. Zivadinov, and M. Ramanathan. 2007. Pharmacogenetics of MXA SNPs in interferon-beta treated multiple sclerosis patients. *J Neuroimmunol* 182 (1–2):236–9.

Wergeland, S., A. Beiske, H. Nyland, H. Hovdal, D. Jensen, J. P. Larsen, T. H. Maroy, A. I. Smievoll, C. A. Vedeler, and K. M. Myhr. 2005. IL-10 promoter haplotype influence on interferon treatment response in multiple sclerosis. *Eur J Neurol* 12 (3):171–5.

Wiesemann, E., M. Deb, C. Trebst, B. Hemmer, M. Stangel, and A. Windhagen. 2008. Effects of interferon-beta on co-signaling molecules: upregulation of CD40, CD86 and PD-L2 on monocytes in relation to clinical response to interferon-beta treatment in patients with multiple sclerosis. *Mult Scler* 14 (2):166–76.

Wiesemann, E., J. Klatt, C. Wenzel, F. Heidenreich, and A. Windhagen. 2003. Correlation of serum IL-13 and IL-5 levels with clinical response to Glatiramer acetate in patients with multiple sclerosis. *Clin Exp Immunol* 133 (3):454–60.

Wingerchuk, D. M., V. A. Lennon, S. J. Pittock, C. F. Lucchinetti, and B. G. Weinshenker. 2006. Revised diagnostic criteria for neuromyelitis optica. *Neurology* 66 (10):1485–9.

Wollin, J, H. Dale, N. Spenser, and A. Walsh. 2000. What people with newly diagnosed MS (and their families and friends) need to know. *International Journal of MS Care* 2 (3):22–29.

Xie, H. G., R. B. Kim, A. J. Wood, and C. M. Stein. 2001. Molecular basis of ethnic differences in drug disposition and response. *Annu Rev Pharmacol Toxicol* 41:815–50.

Yadav, V., and D. Bourdette. 2006. Complementary and alternative medicine: is there a role in multiple sclerosis? *Current Neurology & Neuroscience Reports* 6:259–67.

Yamaguchi, K. D., D. L. Ruderman, E. Croze, T. C. Wagner, S. Velichko, A. T. Reder, and H. Salamon. 2008. IFN-beta-regulated genes show abnormal expression in therapy-naive relapsing-remitting MS mononuclear cells: Gene expression analysis employing all reported protein-protein interactions. *J Neuroimmunol* 195 (1–2):116–20.

Yong, V. W. 2009. Prospects of repair in multiple sclerosis. *J Neurol Sci* 277 Suppl 1:S16–8.

Zur, K., K. Ginzburg, Y. Yagil, J. Chapman, M. Drori, and A. Korczyn. 2002. The use of alternative medicine by multiple sclerosis patients. *Harefuah* 141:611–6.

Section 4

Personalized Medicine and Health Care

12 Brain-Related Health Care

New Models for Personalized Medicine in Psychiatry

David Whitehouse, MD

Introduction

At the center of the wisdom imparted in the preceding chapters is a belief that the insights presented can be turned into practical, relevant, and effective approaches to relieving human suffering and maximizing human potential. We expect that new studies and approaches will yield better biological classifications of pathology, which will provide a deeper understanding of the etiologies of disorders. With this understanding will come an improved awareness of the unique mechanisms that help create a particular phenotypic presentation, and from these, improved, focused, and targeted interventions will be developed. As we discover and better understand the various components that create sickness or promulgate health—whether they are genetic in origin, or brain instabilities based in individual behavior, societal stressors, or environmental factors—we more fully appreciate how sensitive individual adaptation is.

In the last few years, we have witnessed an exponential growth in studies that provide insights into and understanding of how the human brain works. Biology, biochemistry, psychology, neuroimaging, neuroscience, cognitive science, psychiatry, nosology, human development, neurology, pharmacology, sociology, philosophy, and anthropology have all added innovative and often differing perspectives. Each field at various times has wanted to claim the preeminent contribution to advancing understanding and treatment. And while everyone hoped that the human gene code project would accelerate our wisdom in ways we could not expect, that road may be longer in fulfilling those ambitions than many hoped. Even as our knowledge of genetics, brain function, and disease etiology increases, confounding factors—including the influence of genes at more than one locus, and the effect of environmental factors on genetic expression—leave us to understand that genes alone will not explain the cause of all illness.

Thus, we still need other tools in our armamentarium to allow better diagnosis and treatment of brain diseases now. Furthermore, as managed health care figures more prominently in patient service, requiring a dual focus on keeping health care costs down while helping patients make the right decisions, we need new, pragmatic, and realistic paradigms to better assist the patient in selecting the best therapeutic modality or modalities for them, individualized clinical decision support that goes beyond the algorithms we have today. Web-based assessments tools seem to offer particular promise in this regard; my own work with the WebNeuro tool will be presented later in the chapter as an example.

While advances in neuroscience and disease management approaches seemed to offer such promise, that same promise has often disappointed because of one important fact. We forget that we are dealing with human beings, with all their foibles, imperfections, backgrounds, comorbidities, and cultural attributes. As doctors and other health care providers, however right, wise and compassionate we are, we forget that we are human, too, and we cannot make people do things just because we tell them these things will make their lives better, without showing them how, exactly, that will be achieved. Increased neuroscientific understanding of health and disease states must be matched by equal advances in health psychology, social marketing and the insights that behavioural economics sheds on how reward systems drive our irrational decision processes.

Our hope is that the tools to personalize medicine—some of which may be more convenient for patients to utilize—will lead to patients becoming more willing to participate in and take control of their own care (perhaps under the guidance of a health care professional), and to engage proactively in healthier lifestyles, habits, and behaviors.

Personalized Medicine and Brain Health

This new era of personalized medicine presents particular opportunities and challenges to the field of brain health and psychiatric disease. If no single science, or approach truly holds the unique key to unravelling the mysteries of brain health and pathology our best chance at developing a truly personalized approach may come from looking at brain and behavior through an "integrated" neuroscience perspective. I use the term here to refer to an approach to assessment that examines how variations in genetics, results of functional magnetic resonance imaging (fMRI), electrophysiological recordings, and tests of general and social cognition simultaneously recorded while the brain is at work around particular tasks can create a more comprehensive fuctional whole brain assessment that offers a unique perspective. This perspective can then be greatly enhanced when those same integrated assessments are compared to a huge database of healthy and disease state norms whose data has similarly been collected in a standardized manner.

The urgency for this type of personalized medicine approach comes from the frustrations that easily arise from failings in the system as it exists today. Our current classification system for diagnosis, which is the first step toward treatment recommendations, is still based exclusively on a patient's self-report of symptoms, along with the clinician's subjective interpretation of that self-reporting. Treatment recommendations are based on this diagnosis. In every other area of medicine, in addition to patient self-report, we can make some attempt to assess the functional capacity of the organ or organ system through laboratory or more extensive tests, such as echocardiography and stress tests, or renal function tests, These assessments provide us with comparable objective measures, and they sometimes contradict the self-report. Part of the value of a stress test is that while the patient may huff and puff and say they feel no pain, the EKG can show that despite being asymptomatic the patient has significant S-T changes. The patient "feels"okay but the heart is functionally impaired. Are there equivalent functional assessments for the seat of emotional and behavioral states and actions—that is, the brain? Yes. There are electrophysiological recordings, such as EEGs and ERPs, functional MRIs, neuropsychological tests, and other psychological batteries. But making these assessments a regular part of decision-making support for primary care physicians (PCPs), who deliver 70 percent of all the behavioral treatment in the U. S., would be highly impractical, as well as cost prohibitive. Integrated neuroscience based Web-based assessments sitting on top of a huge comparative database might just hold the key.

Even when we improve our assessment capabilities we must remain cautious about casting solutions into definitive approaches to be applied universally to all without reflection. If Evidence Based Practice can show us the statistically most valid path to a particular outcome at times Practice Based Evidence can equally inform us about what happens when research truths and real world chaos collide. This was amply demonstrated when the former Agency for Health Care Policy and Research (AHCPR, now the U. S. Department of Health and Human Services [HHS], published their first recommendations for antidepressant use (U.S. Dept. Health and Human Services, 1993). The Agency stated that all antidepressants were equally efficacious: "No one antidepressant medication is clearly more effective than another, and no single medication results in remission for all patients." It was then assumed by some insurance companies that if this was the case, then a justifiable approach would be to insist that all new cases of depression should be started on the lowest-cost agent, unless there were serious clinical reasons to override that decision. At that time, amitryptiline was the cheapest agent. As a result, protocols were put in place to make amitryptiline the first-line agent. One of the problems with this decision was that, while the information on the success of amitryptiline in the AHCPR guidelines was accurate in the controlled circumstances of a study, where patients had their medications presented to them and they were encouraged to take them, in the real world, the side-effect profile of the drug led many to discontinue taking it. Thus, while the initial AHCPR guidance was "true," the application of scientific truth in establishing not just efficacy, but practical effectiveness, required a reevaluation in light of real-world experience. This should serve as a warning that doctors and therapists need to make use of many modalities to make an appropriate drug recommendation to patients. Adherence is one of the most significant challenges faced by any intervention. An effective intervention should not cause a problem with patient adherence, but if the treatment is ineffective and/or has significant side effects, adherence to the treatment then becomes a problem. In this regard much rides on the impact of our new investment in "comparative effectiveness" research.

Web-Based Brain Assessment

So what about looking to see if there might be particular value in Web-based assessments? This would certainly be easier in terms of cost and availability than extensive laboratory, electrophysiological, and imaging testing. But would it be as valuable as more extensive, traditional testing? The key to the usefulness of such is the number of patients assessed, and the integration of the results from the Web-assessment tool with other data. In developing WebNeuro (described in detail below), a whole series of cases, including normal individuals with no pathology and various patient cohorts with specific disease categories, underwent the Web-based assessment, including standard neuropsychological assessments together with assessments of social cognition, covering the range of brain structure and functions; these findings were integrated with EEG datasets, ERP data, fMRIs, MRI, and genetic data. These data were collected in precisely the same standardized manner. Because of the amount of data collected and their correlation, each assessment then can be compared with an age-matched, gender-matched, and education-matched set of normals, as well as with cohorts with specific diagnoses.

For any individual data variable (and potential marker), there are millions of possible correlations. It is from these individual data elements that constellations of the strongest correlations emerge (after controlling for multiple comparisons), and are applied in the most pragmatic manner, such as via the Internet. These are susceptibility markers that not only point the way toward selection of specific treatments, but also offer insights into the underlying mechanisms of certain endophenotypes (Gatt et al. 2009). WebNeuro and other Web-based assessment tools provide the possibility of rapidly, conveniently, and accurately bringing a greater level of decision support based on published scientific consensus. Without an external level of support, the challenge of selecting a particular antidepressant agent is currently based on signs and symptoms and related guideline support, and considerations of drug-drug interactions, side effects, and comorbidities—in other words,on the personal knowledge base of the clinician.

Psychiatrists prescribe medications for off-label uses with the greatest frequency of any specialty. There is some general guidance and research on these uses available, however, the discovery and promulgation of new insights takes place so rapidly, and across such a large range of journals, papers, and conferences, that it is almost impossible to make that knowledge available in any practical way to influence clinical decision-making in the near term. One estimate is that it takes up to seventeen years to integrate only 30 percent of clinical recommendations into practice (Committee on Quality of Health Care in America 2001). Furthermore, 50 percent of the published guidelines are obsolete within six years.

Given that 75 percent of mental health care is delivered by primary care physicians, having a clinical decision support system in place is even more important. When we examined the pharmarcy utilisation patterns of many Fortune 100 companies, we found that not only do antidepressants frequently rank first or second in any employer's list of drug categories by volume or cost, but between fifteen and twenty-five percent of these prescriptions are filled only once. There may be multiple reasons for this, ranging from inappropriate initial prescribing, problems with side effects, or failure to explain how the drugs work or when to expect a response. In addition, while alternative drugs may be offered should the first agent fail, this may be considered likely to increase the patient's hope, while lowering the patient's confidence that the physician knows how to select the most effective agent. This challenge is only compounded when considering as well what we know from the National Institute of Mental Health's (NIMH) Sequenced Treatment Alternatives to Relieve Depression (STAR*D) Study: namely, that when a particular SSRI is chosen and used as a single agent, because the various guidelines suggest that no particular agent is considered superior, then the chance of a successful first response may be around only 37 percent (Rush et al. 2006). If it became clear that there are various endophenotypes within the depressive spectrum and that identification of "susceptibility markers," determined using a simple Web-based assessment, may aid in selection of the correct drugs, there could be huge cost savings, improved outcomes, improved patient adherence and satisfaction, decreased medical and comorbidity costs; and improved productivity. Such guidance can only improve patient treatment, response, and compliance.

There are four features that are essential for clinical decision support systems to improve clinical practice (1) automatic provision of decision support as part of clinician workflow; (2) making recommendations in addition to assessments; (3) provision of decision support at the time and location of decision-making, and; (4) offering computer-based assistance (Kawamoto et al. 2005). This is now all available through WebNeuro.

WebNeuro is currently being tested, validated, and used at Optum Health Behavioral Services. It is an online battery of assessments of social and general cognition intended to achieve a personalized perspective to help guide the use and interpretation of diagnostics, as well as choices in therapies and prevention (Optum Health 2008).

The first portion of the assessment is a brief online survey instrument that asks a series of questions and takes only ten minutes to complete. This instrument is called the Brain Resource Inventory of Social Cognition (BRISC). The survey is easy to administer. It is

Web-based, but also touchscreen-capable and easily translated to application software (or "app" as it is more commonly known) or a personal digital assistant (or PDA). Its particular value derives from the simplicity of the insights it provides to care managers. Because it is backed by sufficient validation, it allows care managers to feel confident in using it to help determine the adaptive strengths and limitations of each individual patient. The BRISC measures three components: negativity bias, social skills, and emotional resilience.

Negativity bias measures the triad of: a) experienced symptoms (depression, anxiety, and somatic features); b) heritable aspects of cognition and temperament (negative cognitive set and vulnerability to stress versus resilience); and c) learned coping that determines disorders, such as depression. Thus, negativity bias complements a Patient Health Questionnaire-9-(PHQ-9) measure of depression/anxiety symptoms alone, by providing the extent of their hypersensitivity and likelihood to over-react to threat. A borderline or deficit negativity bias confirms the need to undertake the full WebNeuro Cognition, the second more extensive portion of the full Web-Neuro which looks at a comprehensive battery of computer based tests of general and social cognition.

In this context, "social skills" measures the ability to foster positive social relationships, develop empathy, and express emotional intelligence. These are traits that are vital to workplace functioning and also are impaired in most mental illness presentations. Social skills measures combine the behavioral (social skills and fluidity, relationship building), empathetic, and temperamental (extraversion) aspects of social functioning. In the absence of a substantial disorder, it indicates deficits that would benefit from the gamut of social skills training for all interpersonal relationships, through to superior skills suited to specific employment benchmarks.

Emotional resilience is a marker of adaptive coping. It reflects the capacity for regulation of emotion, and for the skill of self-monitoring, which is essential to the development and maintenance of self-esteem and emotional stability, which are also impaired in many disorders. These capacities may also be deficient in some normal individuals, who may be expected to rapidly lose control in very stressful interpersonal situations. Emotional resilience thus assesses behavioral (self-esteem, self-efficacy), temperamental (emotional lability, conscientiousness), and cognitive (ability to focus, think clearly) aspects of emotion regulation. It captures the spectrum from emotional instabilities, which characterize psychopathology, to normal variation in resilience to the stressors of daily life.

In practical terms, negativity bias serves as a "thermometer" for brain functioning, letting the physician or therapist know if the emotional and cognitive systems are well or malfunctioning. If they are malfunctioning compared to age-matched, gender-matched, and education-matched norms, there is a need for a more comprehensive evaluation. This would be achieved by taking a more complex functional assessment of general and social cognition based on tried and tested neuropsychological assessment tools, as well as the markers of face emotion recognition in the social cognition dimension of WebNeuro. This assessment takes an additional forty minutes beyond the first 10 minutes devoted to completing the survey. Because BRISC screens for general disturbance, whether caused by stress, depression, anxiety, or organic deficits, it provides a more global test than disease-specific diagnostic instruments, such as the PHQ-9, the Hamilton Rating Scale for Depression (HAM-D), the Hamilton Anxiety Scale (HAM-A), or the Yale-Brown Obsessive Compulsive Scale (Y-BOCS).

But what is even more important is the fact that the additional perspectives added by the social skills and emotional resilience elements can play critical roles in "engaging" the patient. The social skills number can give us a quick read on how well the individual is going to be able to make use of other human beings in their immediate environment (on the phone, or face to face) to help him or her create solutions. For example, if the mere presence of others creates anxiety, it is unlikely that a lengthy dialogue, however empathic, is going to create a real benefit. Feeling that even limited social interaction is an invasion of their private boundary, some patients cannot "hear" what is being said. It is far better if they score low to keep face time or telephone time to a minimum, and rely instead on brief interactions, during which material is offered that they can take away and think about privately in their own time and space. The health care professional can then check in with them later for further clarification or help. Interactions might need to be brief, rather than extended, to minimize the threat or anxiety felt.

Likewise, the emotional resilience measurement can give the clinician a perspective on the individual's capacity to overcome challenges or obstacles that he or she may encounter daily. While some individuals begin to fall apart the minute they encounter the slightest obstacle, others find they can weather all sorts of difficulties if only they are given practical advice, or make a plan in advance in consultation with their doctor or therapist. Some people just don't do life well, or a certain aspect of it, whether it is relationships, school, or work, while others have an innate capacity to overcome all sorts of hardships. Some people can be given a plan, guidance, or education and then they are ready face life's challenges. Others will need continual hand holding, encouragement, and support at all stages to ensure success. Disease management that is based purely on severity of illness criteria, without taking personality and character into account, misses these individual nuances at its peril.

Following a screening that can help identify broadly the at-risk population and how to approach them, the next challenge is to determine, via traditional diagnosis, the particular "disease" the person might have. But it also is important to determine whether there are particular biomarkers that suggest that this individual belongs to a sub-cohort within the large group with the same diagnosis, for whom some interventions may be better than others. This process creates the possibility of personalized clinical decision support. While some of the focus in this area is heavily on the particular genetic variants (genotype), it is now clear that creating personalized approaches from genetic information alone will be insufficient. The multifactorial nature of most disease phenotypes, the heterogeneity of individuals who share those phenotypes, and the impact of environmental factors on gene expression and phenotype itself cannot be ignored. It is very likely that a broader range of susceptibility markers that suggest particular endophenotypes will yield distinctions that will support decisions for particular sub-groups.

Thus, an integrated neuroscience approach supported by a standardized methodology of data collection and comparison will most likely give us some of the most reliable insights. It is with this in mind that we have made WebNeuro available, not to replace current clinical wisdom, but rather to empower clinicians and supplement that same knowledge with new insights from a different perspective that otherwise would not be available, in the hope that it will yield a richer, more informed level of decision-making.

Brain Exercises and Online Cognitive Therapy Programs

If we can improve identification of disease, can we also do more to improve prevention, or augment current therapies by supporting neurogenesis and synaptogenesis with "brain training"? Although there are computer-based games purporting to increase brain health, in fact, most are direct-to-consumer marketing ploys with no solid scientific data behind them, that merely match a personalized assessment to a range of training exercises (an example that does match personalized assessment to a holistic set of brain training exercises is www.MyBrainSolutions.com). Because validated data take a long time to collect, and the effort-to-benefit ratio of clinical trials takes a long time to establish, it may be a while before scientifically proven brain training programs become widely accepted in the marketplace. Nevertheless, they are becoming increasingly popular due to their noninvasive nature, early studies of their benefits, and the fact that they offer an alternative to medication. These programs are also already taking advantage of the latest delivery technologies, such as cell phones, iPods, and other portable personal technologies.

In addition to computer-based brain training programs, there is now an increasing array of online programs based on cognitive-behavioral therapy. These programs take the best of what we know from cognitive therapy and make it conveniently available to individuals, with 24-hour online access. In the United Kingdom, for example, the National Institute for Health and Clinical Excellence, part of the National Health Service, offers "Beating the Blues" and many other programs geared toward depression, as well as "BT Steps," for obsessive-compulsive disorder. The programs are self-paced, sometimes with the assistance of an online coach, and offer a much more readily accessible alternative to face-to-face to therapy.

Another area that lends itself to individual, self-paced instruction and practice is heart rate variability (HRV) training, a more scientifically presented and personalized approach to breathing techniques that optimally reduce stress. The technology allows individuals to see the impact of particular breathing rates on the ability to control HRV. This technique not only decreases stress, but seems to affect a whole variety of medical conditions that appear to be mediated, in part, by overactivation of the sympathetic nervous system (SNS), which actives the "fight or flight" response. The practice of mindfulness meditation also offers individuals the opportunity to become more active in taking greater control of their well-being.

Also emerging are low-cost, simple feedback instruments that will allow individual training using real-time feedback of electrophysiological body activity (B\biofeedback) and tracking of brain-wave activity (neurofeedback) to increase particular patterns to aid focus or improve relaxation.

With these advances in neuroscience, early intervention and personalized treatment are possible in a much more systematic way.

Better Clinical Outcomes, Better Financial and Workplace Outcomes

Frequently, the discussion about benefit design or effectiveness has focused on expense management. What are the conditions or diseases that affect the population, and what should we do to manage them, to achieve better clinical and financial outcomes? This approach sees health care as an expense, and it is a fiduciary responsibility to manage that expense as effectively as possible.

However, that is only part of the story. The other part comes into play when you expand your view to ask another question: What are the health issues that, if we were to resolve them, would make us more profitable and more successful as a company or more productive as a society? This question shifts the focus to the best clinical solutions for improving productivity, incidental absence, turnover, and the whole realm of metrics that represent the effectiveness of clinical practice for maximizing human capital investment in the workplace and the quality of life in society. (Human capital investment covers costs stemming from investment in hiring, brainpower, manpower, decision-making, leadership, morale, alignment, and productivity.) Frequently, companies invest in adding productivity questionnaires to their regular health risk assessment (such as the WHO Health and Work Performance Questionnaire [HPQ] by Ron Kessler, or the Work Limitations Questionnaire

[WLQ] by Debra Lerner). The results of these questionnaires shift the focus from diseases (such as back pain, cardiac conditions, and cancer), which are critical to driving medical expenses, to stress, depression, anxiety, and insomnia, which drive productivity loss. The results clearly vary by industry. But in the world of "service" industries and "knowledge-based" industries, where the value of the human capital is gauged by intellectual output, creativity, engagement, social connectedness, or decision-making acumen, the importance of cognitive and emotional health becomes even more critical.

Companies that embrace this approach, such as CISCO or Nationwide Insurance, are leaders in realizing how investing in brain health can further enhance the productivity and output of their human capital. In initial studies with Nationwide, we have shown how brain training, especially in the areas of "feeling" and "self-regulation," enhanced productivity, not only for those who were at average performance, but also for those who were already peak performers. By far, the greatest impact of detecting and treating illnesses, such as depression, lies in their impact on productivity, absenteeism, and the effect that those illnesses have on coworkers and team production.

Maximizing the cognitive and emotional brain health of individuals could have a huge social and economic impact. Foresight, part of the U. K. Government Office for Science, produced a report, "Mental Capital and Wellbeing," (2008) that challenges us to look at the full societal impact of mental health for a country over the entire life span of its citizenry. It begins with challenges faced in the peripartum period, where the future responsiveness of an individual's neurons to brain-derived neurotrophic factor (BDNF) to drive neurogenesis can be forever limited due to exposure to chronic stress. It continues through the educational period of the developing brain, and then goes into the workplace, where life- and work-related stressors wreak their own havoc on the brain. Our increased life spans suggest that we have to develop new understandings about human capital potential, which can be lost to society through retirement, especially early retirement. A report like this challenges us to think differently. We need to consider social programs that support the peripartum period, emphasize inclusion of social and emotional competence training in schools, and maintain an ongoing focus on enhancing resilience and hardiness in the workplace. This will maximize and preserve brain health, not just through interventions for sickness, but through a wealth of programs and approaches, such as exercise, nutrition, brain training, meditation, and HRV enhancement. These are programs that individuals can engage in alone, or with minimal support, to increase their brain health.

Alternative, and more controversial, shortcuts to this process have been hinted at in some research, such as the article published in *Nature*, in which a group of neuroscientists advocates for greater acceptance of "mind-enhancing" drugs, such as Ritalin (methylphenidate) and Provigil® (modafinil), and suggests that they be made more widely available (Maher 2008). The argument is that maximizing our biological system is ethically acceptable. After all, we do it for the immune system with vaccinations, so why not do it for the mental functioning systems? At the same time, the military is studying new approaches to accelerated learning by using transcranial direct cortical stimulation (tDCS) to improve learning and executive function (Smith & Clithero, 2009). All of these approaches must be evaluated for their ethical and practical implications.

Conclusions

In the United Kingdom, the movement behind the IAPT (Increasing Access to Psychological Therapies) initiative sees the need to reach a much wider population of the stressed and distressed, beyond the severely psychiatrically impaired. Those involved in the initiative understand that to try and help this broader population with the current tools of medication and individual one-to-one therapy would be impractical and financially impossible. A potential means of making optimal brain health available to everyone is through the dissemination of brain health activities, individual empowerment, online cognitive therapies, and other tools—backed with coaching as necessary—as well as with the face-to-face use of "cognitive therapy extenders." These are individuals who are thoroughly trained in delivering cognitive therapy, although they are not PhD-holding psychologists. These tools could expand the possibility of meeting the need for brain health for a generation of individuals who have no concept of brain health, and for whom resilience development, if it were to happen, would be, at best, accidental. Once we embrace creative and forward-looking approaches like this we will really be able to begin to meet the ever-growing needs of our society.

What is at stake? Every seventeen minutes, someone commits suicide in the United States. Every year, employers suffer millions of dollars in losses in productivity to employee brain health problems. There is widespread, unattended, and untreated comorbidity of illnesses, such as depression, in chronic diseases. Many people who lack training in resilience and hardiness cannot manage effectively the anxiety and stress of daily life, leaving them vulnerable to a rapidly changing world, and inundated with multiple stressors 24/7. We have the possibility of significantly improving our chances to maximize human capital and well-being at all points along life's axis.

Our health care system can continue to use the psychopharmacological and psychotherapeutic interventions we currently employ in 90 percent of our mental health care today. Or, it can embrace new partnerships: where evidence-based practice and practice-based evidence complement each other; where the understanding of individuals as people with lives, values, dreams, and concerns is integrated into the personalization of a treatment plan; where the findings and discoveries of neuroscience become a stimulus, not just to improve brain ill-health, but to support brain health, resilience, and hardiness across the life span; and where integrated neuroscience and standardized databases are valued as powerful tools in accelerating our understanding of gene-brain markers and their incorporation into pragmatic clinical decision support systems that become part of a "continuum of care," rather than fragmented solutions.

Reimbursement of new personalized medicine markers that predict treatment outcomes will require validation and demonstration of cost-effectiveness. Their adoption will also be accelerated if they readily interface with existing clinical decision support systems, such as WebNeuro.

References

Committee on Quality of Health Care in America, Institute of Medicine. 2001. *Crossing the quality chasm: A new health system for the 21st century.* Washington, DC: National Academy Press.

Gatt, J. M., C. B. Nemeroff, C. Dobson-Stone, R. H. Paul, R. A. Bryant, P. R. Schofield, E. G. A. Kemp, and L. M. Williams. 2009. Interactions between BDNF Val66Met polymorphism and early life stress predict brain and arousal pathways to syndromal depression and anxiety. *Molecular Psychiatry* 14: 681–95.

Hogan, M., and B. Altevogt. 2008. *From molecules to minds: Challenges for the 21st century: Workshop summary.* Washington, D.C.: The National Academies Press.

Kawamoto, K., C. A. Houlihan, E. A. Balas, and D. F. Lobach. 2005. Information in practice. Improving clinical practice using clinical decision support systems: A systematic review of trials to identify features critical to success. *British Medical Journal* 11: 1–8.

Kinnally, E. L., L. A. Lyons, K. Abel, S. Mendoza, and J. P. Capitanio. 2007. Effects of early experience and genotype on serotonin transporter regulation in infant rhesus macaques. *Genes, Brain and Behavior* 7: 481–86.

Maher, B. 2008. Poll results: Look who's doping. *Nature* 452: 674–75.

Optum Health. 2008. *WebNeuro user manual*. pp. 12–13.

Rush, A. J., M. H. Trivedi, S. R. Wisniewski, A. A. Nierenberg, J. W. Stewart, D. Warden, G. Niederehe, M. E. Thase, P. W. Lavori, B. D. Lebowitz, P. J. McGrath, J. F. Rosenbaum, H. A. Sackeim, D. J. Kupfer, J. Luther, and M. Fava. 2006. Acute and longer-term outcomes in depressed outpatients requiring one or several treatment steps: A STAR*D report. *American Journal of Psychiatry* 163: 1905–17.

Smith, D. V., and J. A. Clithero. 2009. Manipulating executive function with transcranial direct current stimulation. *Frontiers in Integrative Neuroscience* 3: 26.

U. K. Government Office for Science. Foresight. 2008. *Mental Capital and Wellbeing*. London: UK Government Office for Science. http://www.foresight.gov.uk/OurWork/ActiveProjects/Mental%20Capital/Welcome.asp (accessed June 16, 2010).

U. S. Department of Health and Human Services, Agency for Health Care Policy and Research. 1993. *Treatment of Major Depression, Volume 2: Clinical Practice Guideline Number 5*. AHCPR Publication No. 93–0551: National Library of Medicine DOCLINE Information: CAT/9411422.

13

Clinical Decision Support in Employee Assistance Programs

Personalizing the Therapeutic Approach

Eugene Baker, PhD

Introduction

Employers help finance the majority of health insurance in the United States, and they select programs that are the most cost-effective for their employees. A basic function of an Employee Assistance Program (EAP) is to offer assessment and treatment referral for employees or dependents with various forms of behavioral health issues. While this is by no means the only function of an EAP, being able to effectively and efficiently refer people for appropriate treatment services is a key value of the EAP. Depression is the most common complaint of individuals seeking mental health treatment services (Kessler et al. 2005). In most cases, depression can effectively be treated using psychotherapy, medication, or even self-directed learning (www.helpguide.org). The question, then, becomes, "What is the best personalized approach to use for a particular client?" The focus is not on the markers, but on the most effective treatment programs. Personalized medicine will in the future be incorporated into such programs.

Treatment is often based on the professional background of the evaluator, rather than the specific needs of the individual. Counselors are often likely to recommend psychotherapy, while physicians frequently prescribe medication. General literature supports the use of both medication and psychotherapy in dealing with depression, and there are even debates about which modality to use first (Antonuccio et al. 1995; NICE 2009). The decision to use psychotherapy, which is less expensive and has a lower relapse rate than medication alone, is of great importance to both the individual as well as the functioning healthcare delivery system (Antonuccio et al. 1995).

The Clinical Decision Support System

"Clinical decision support (CDS) provides clinicians, staff, patients or other individuals with knowledge and person-specific information, intelligently filtered or presented at appropriate times, to enhance health and health care" (American Medical Informatics Association 2006). Could a clinical decision support system (CDSS) help EAP professionals manage the range of problems presented in EAP settings? A CDSS has been effectively used in many other areas of the health care delivery system. Through their meta-analysis, Kawamoto and colleagues (2005) identified characteristics of a CDSS found to improve clinical outcome:

- Automatic provision of decision support as part of clinician workflow ($P < 0.00001$);
- Provision of recommendations, rather than just assessments ($P = 0.0187$);
- Provision of decision support at the time and location of decision-making ($P = 0.0263$); and
- Computer-based decision support ($P = 0.0294$).

WebNeuro Assessment Tool

The WebNeuro Assessment, introduced in chapter 12, is an exemplar CDSS. It is a 40-minute, Web-based tool that assesses general and social cognition and could provide information in a CDSS setting. Figure 13-1 shows a sample report of self-report and cognition marker scores measured by WebNeuro.

These markers are linked to insights and correlations in a standardized and integrative international database (see www.BrainResource.com; www.BRAINnet.com). These markers and their current applications point the way forward as to how they could be extended, as new marker discoveries that emerge from personalized medicine.

Scores on each marker vary from one toten, representing stens scores. Individual scores are compared to similar age, education, and gender scores in the Brain Resource International Database (http://www.brainnet.net/about/brain-resource-international-database/). The markers are named based on the principal components analysis of the individual measures, and have a logical and empirical connection with brain function. Each report contains a summary with "information for consideration" based on clinical consensus guidelines. The clinician can use the report in conjunction with other information gained from the patient's assessment to determine which approach would be most beneficial to each individual patient.

Personalization of treatment requires information based on standardized datasets, and sufficiently large databases, to provide robust insights into the selection of appropriate treatments. As a CDSS tool, WebNeuro is an exemplar of a positive step in the move toward personalized medicine—matching individuals to effective treatments based on objective neurocognitive markers (in this case). In the absence of widely agreed-upon standards for determining the most appropriate treatment for each patient, WebNeuro provides an objective standardized assessment tool based on a large data set (>5000 subjects) across the life span.

The WebNeuro CDSS screens for a mental health problem (Yes/No). It then produces objective marker information for consideration of treatment options or referral to a specialist. Finally, it encourages reassessment to objectively determine the extent of improvement.

The first component of WebNeuro is the Brain Resource Inventory of Social Cognition(BRISC), which screens for the presence of problems, while the Cognition assessment component provides additional data for determining individualized, appropriate treatment. In essence, this tool provides the standard (DSM-based) PHQ9, plus more objective brain-based cognition markers for depression (and there are versions for ADHD and schizophrenia). Figure 13-2 provides a pictorial summary of how this process works.

Most mental health clinicians are not familiar with the concept of the CDSS. For many, it is a novel approach, and we (at OptumHealth) have noted that clinicians do not know how to effectively use the information as they would, for example, in cardiology CDSS systems.

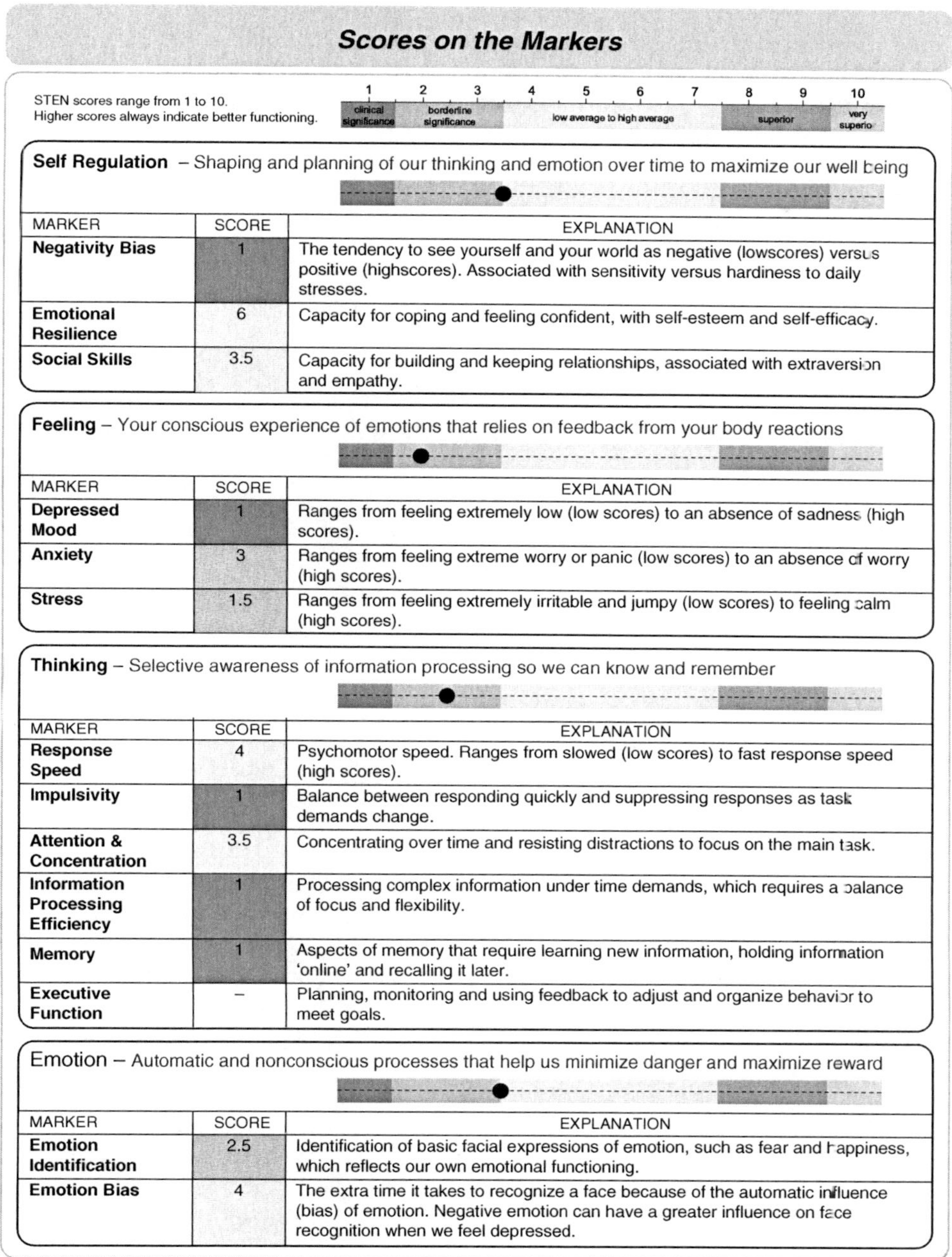

Scores on the Markers

STEN scores range from 1 to 10.
Higher scores always indicate better functioning.

Self Regulation – Shaping and planning of our thinking and emotion over time to maximize our well being

MARKER	SCORE	EXPLANATION
Negativity Bias	1	The tendency to see yourself and your world as negative (lowscores) versus positive (highscores). Associated with sensitivity versus hardiness to daily stresses.
Emotional Resilience	6	Capacity for coping and feeling confident, with self-esteem and self-efficacy.
Social Skills	3.5	Capacity for building and keeping relationships, associated with extraversion and empathy.

Feeling – Your conscious experience of emotions that relies on feedback from your body reactions

MARKER	SCORE	EXPLANATION
Depressed Mood	1	Ranges from feeling extremely low (low scores) to an absence of sadness (high scores).
Anxiety	3	Ranges from feeling extreme worry or panic (low scores) to an absence of worry (high scores).
Stress	1.5	Ranges from feeling extremely irritable and jumpy (low scores) to feeling calm (high scores).

Thinking – Selective awareness of information processing so we can know and remember

MARKER	SCORE	EXPLANATION
Response Speed	4	Psychomotor speed. Ranges from slowed (low scores) to fast response speed (high scores).
Impulsivity	1	Balance between responding quickly and suppressing responses as task demands change.
Attention & Concentration	3.5	Concentrating over time and resisting distractions to focus on the main task.
Information Processing Efficiency	1	Processing complex information under time demands, which requires a balance of focus and flexibility.
Memory	1	Aspects of memory that require learning new information, holding information 'online' and recalling it later.
Executive Function	–	Planning, monitoring and using feedback to adjust and organize behavior to meet goals.

Emotion – Automatic and nonconscious processes that help us minimize danger and maximize reward

MARKER	SCORE	EXPLANATION
Emotion Identification	2.5	Identification of basic facial expressions of emotion, such as fear and happiness, which reflects our own emotional functioning.
Emotion Bias	4	The extra time it takes to recognize a face because of the automatic influence (bias) of emotion. Negative emotion can have a greater influence on face recognition when we feel depressed.

Figure 13-1 Summary of Markers from WebNeuro Report.

Further, while treatment options follow consensus guidelines, many clinicians are not familiar with how common clinical decision support systems are in mainstream medicine, or understand how the guidelines were established. The guidelines are simply a summary of the scientific literature, with levels of evidence made explicit and delivered in an integrated and simple manner via the Internet.

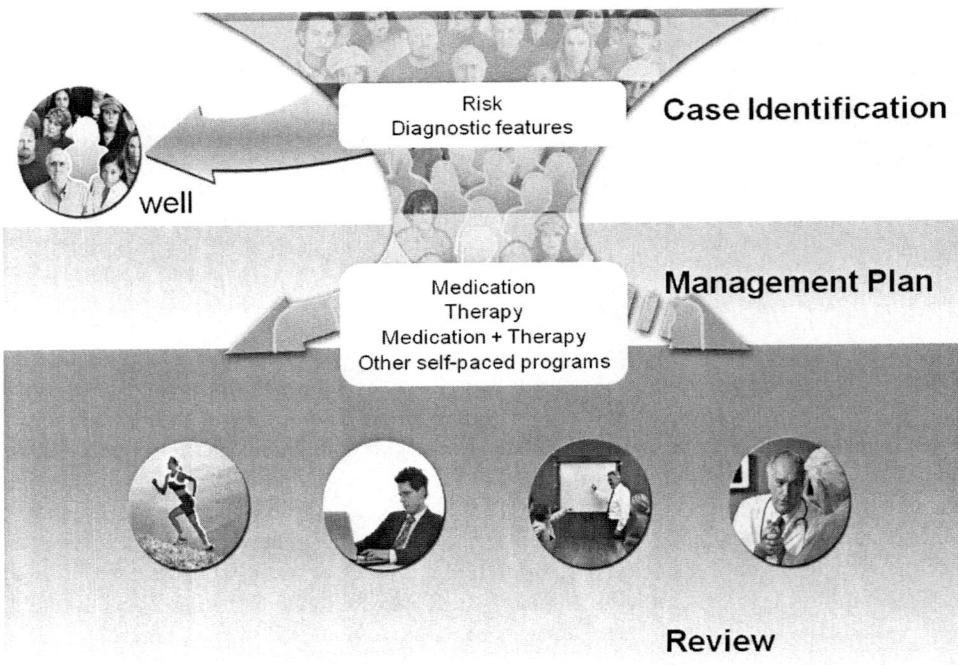

Figure 13-2 WebNeuro as a clinical decision support tool.

We have found it useful to train clinicians in how to use the assessment tool. We developed an orientation training program, and an accompanying monthly case seminar, to facilitate learning for a small group of our own full-time, on-site EAP counselors. This enabled our counselors to understand the logical structure of the tool. There was ample opportunity to discuss how clinical observation and CDSS worked together and, therefore, treatment options made more sense. Training materials are available on our practitioner Web site (www.ubhonline.com).

This CDSS uses the negativity bias marker as an indicator of risk (Williams et al. 2009). A clinician can use the PHQ9 data as they normally do, but also add more objective cognitive markers, such as impulsivity or a slow response, for determining the potential class of medication. The emotion recognition and emotion bias markers can be used to determine the potential usefulness of medication. Additionally, the client's self-reported depression, stress, and anxiety provide information, as do markers of thinking (i.e., attention, concentration, memory, and executive functioning). Potentially, the EAP professional can suggest which treatment strategy is best suited for each client—further medication evaluation with a physician, psychotherapy, perhaps both in a sequential manner, or alternatively, self-directed online methods (for example, see http://mybrainsolutions.com).

The feedback we received from our EAP counselors has been instructive. They note that WebNeuro assists in clinical decision-making; it helps to solidify clinical understanding of problem areas so that clinicians can treat or refer clients according to the standard of practice for addressing these issues. Also, WebNeuro is used in conjunction with other sources of information, including clinicians' own clinical assessment, as well as communication with patients to guide clinical decision-making. Reported benefits for clinicians include the ability to:

(i) Identify strengths as well as weaknesses of each patient;
(ii) Provide credible, reliable, valid, scientific, and objective feedback;
(iii) Inform clinical perceptions/impressions, and develop clinical decision–making tools;
(iv) Identify specific areas needing immediate, urgent clinical focus, or uncover previously undetected areas requiring further assessment (e.g., cognitive or emotion functioning);
(v) Provide "talking points" to discuss with patients;
(vi) Expand the treatment options clinicians can offer patients;
(vii) Determine when the complexity of the case warrants referral to a specialist; and
(viii) Objectively reassess the patient's recovery, or lack thereof.

In an actual practice setting, a major benefit of WebNeuro is its ability to promote conversations between the patient and therapist to discuss the advantages and disadvantages of different treatments, thereby facilitating the therapeutic alliance. This opportunity to discuss alternatives is key to the personalization of treatment. Beyond simply identifying what treatment modalities might be helpful, WebNeuro offers both clinician and patient the opportunity to jointly consider what areas to address, and what other resources might be relevant to improving those functions. While personalized medicine is still often thought to be based on genomics and brain markers to predict best medication interventions, clinical decision support systems are a prelude to personalized medicine in clinical practice and include psychosocial treatments to address cognitive functioning, which has an important relationship to the underlying biology. The research addressing cognitive remediation to treat schizophrenia in conjunction with medication provides another example of the personalization of treatment beyond genomic-brain markers (McGurk et al. 2007).

Conclusion

Because of the Mental Health Parity and Addiction Equity Act of 2008, cost-effective treatment for depression is critical. At present, the decision to initiate psychotherapy or use medication to treat mental health issues is not typically based on CDSS or neurocognitive markers. More clinicians need to be trained in CDSS tools such as WebNeuro, so that the impact the CDSS has on the outcome and efficiency of mental health treatment can be rigorously studied, by comparing therapeutic courses for patients whose clinicians have used a CDSS versus those who have not. Furthermore, researchers can carry out comparative effectiveness studies on treatment with medication versus psychotherapy using data for persons, and their profiles, on WebNeuro.

To date, WebNeuro has been shown to be a promising and effective tool for applying a standardized objective method to facilitate the most appropriate treatment approach. It serves as an exemplar CDSS that can be readily extended to include new personalized medicine markers as they emerge, are validated, and show high clinical utility and cost-effectiveness.

References

American Medical Informatics Association. 2006. A Roadmap for National Action on Clinical Decision Support. https://www.amia.org/files/cdsroadmap.pdf

Antonuccio, D. O., W. G. Danton, and G. Y. DeNelsky. 1995. Psychotherapy versus medication for depression: Challenging the conventional wisdom with data. *Professional Psychology: Research and Practice* 26: 574–85.

http://www.brainresource.com/brain_health_solutions/

http://helpguide.org/mental/treatment_strategies_depression.htm

Kawamoto, K., C. A. Houlihan, E. A. Balas, and D. F. Lobach. 2005. Improving clinical practice using clinical decision support systems: A systematic review of trials to identify features critical to success. *BMJ* 330(7494): 765.

Kessler, R. C., W. T. Chiu, O. Demler, and E. E. Walters. 2005. Prevalence, severity, and comorbidity of twelve-month DSM-IV disorders in the National Comorbidity Survey Replication (NCS-R). *Archives of General Psychiatry* 62: 617–27.

McGurk, S. R., E. W. Twamley, D. I. Sitzer, G. J. McHugo, and K. T. Mueser. 2007. A meta-analysis of cognitive remediation in schizophrenia. *American Journal of Psychiatry* 164: 1791–1802.

NICE National Institute for Health and Clinical Excellence. National Clinical Practice Guidelilne 90.Depression in Adults. Depression: the treatment and management of depression in adults. October 2009. http://wwww.nice.org.uk/nicemedia/live/12329/45896/45896.pdt

Williams, L. M., J. M. Gatt, P. R. Schofield, G. Oivieri, G. Perduto, and E. Gordon. 2009. Negativity bias in risk for depression and anxiety: Brain-body fear circuitry correlates, 5-HTT-LPR and early life stress. *Neuroimage* 47: 804–14.

14

Economic Impacts of the Personalized Medicine Tsunami

Dan Segal, MSc and David E. Williams, MBA

Information is really the lifeblood of medicine. Health information technology is its circulatory system.

—Dr. David Blumenthal, National Health IT Coordinator
Office of the National Coordinator for Health Information Technology
Office of the Secretary for the U.S. Department of Health and Human Services

We propose to describe clinical decision support as health information technology functionality that builds upon the foundation of an EHR [electronic health record] to provide persons involved in care processes with general and person specific information, intelligently filtered and organized, at appropriate times, to enhance health and health care.

—Meaningful Use Rule, Department of Health and Human Services (HHS)
Centers for Medicare & Medicaid

Introduction

We are currently experiencing one of the most exciting periods in the evolution of health care. Never before have so many variables that directly impact the core of patient care been simultaneously in transition. This confluence of changes—including technology, policy, legislation, and consumer attitudes—involves and affects everyone. This confluence is based upon the following seminal transitions (which are also presented in Table 14-1).

First, any of the above factors, acting in isolation, would be sufficient to drive industry transformation. In combination, we are looking at a tsunami of change. This chapter sets out to evaluate these parameters and their associated economic outcomes.

Second, the evolution of these drivers could not be timelier. Most health care systems are at the breaking point, even without taking into account the increasing challenges facing our aging population. The United States, in particular, has multiple tensions on its health care system. Passage of the Affordable Health Care for America Act in 2010, ensures that some form of health coverage will be made available to more than 47 million uninsured people currently living outside the public safety net. Furthermore, Mental Health Parity will also have a widespread impact on existing service providers. Thus, not only does the U. S. have to control costs, it has to do this while increasing both the number of individuals

Table 14-1 Contrasting the Key Features of the Current Approach to Health Care with that of Personalized Medicine

Current Healthcare	*Personalized Medicine*
1. Highly reliant on subjective assessment to complement the science	Data centric, driven by new technologies and utilizing the data to drive better health care outcomes
2. Paper based records	Electronic Health Records
3. One size fits all "blockbuster" model	Personalized models
4. Payment for activity	Payment for value, including factoring comparative effectiveness
5. US Regulation - uninsured with no public safety net and discrimination against coverage for Mental Health	Universal coverage and Mental Health Parity
6. Disempowered consumers	Consumer centric
7. Mental Health – with the stigma of psychiatric labels and focus on disability	Brain Health – Destigmatized focus on brain function (e.g., a hypersensitivity to stress or trouble with memory) and on the entire spectrum of health needs, from disability to well-being/prevention
8. Nothing measured in the course of diagnosis or treatment for psychiatric illnesses	Evidence-based measures, including genomics, biology and brain function

covered, and the extent of that coverage. Accordingly, the benefits of the digital wave are an essential component to ensure these challenges are met.

As a note on terminology, in our view, personalized medicine (PM) is simply the utilization of technology to improve health care. It places the consumer at the center of the health care process, recognizing that individual needs differ, and addresses the full spectrum of personal needs, from disability to well-being and prevention. We believe this definition better captures the above forces and, just as importantly, the interrelationship between them.

Specific Health Care Challenges that Personalized Medicine Can Help Address

The following are specific areas in health care that are in need of productivity gains.

Health care spending in the U. S. is estimated to reach above $4 trillion by 2016, and accounts for just under 20 percent of the gross domestic product (US Department of Health & Human Services, National Health Expenditure Data) (including the growing proportion of the aging population)—a 4 percent increase from the current 16 percent (National Health Expenditures 2008 Highlights). This growth is clearly unaffordable and unsustainable. In a March 5, 2009 statement at the White House Forum on Health Reform, President Obama said, "the greatest threat to America's fiscal health is not Social Security, though that is a significant challenge; and it is not the investments we've made to rescue our economy; it is the skyrocketing cost of health care." The breakdown of total spending (shown in Figure 14-1) is noteworthy. There are also other breakdowns of this spending that show administration costs at 35 percent (Burrill 2009).

Declining pharmaceutical company research and development (R & D) productivity has led to depleted patent-protected portfolios. Data from a PriceWaterhouseCoopers report

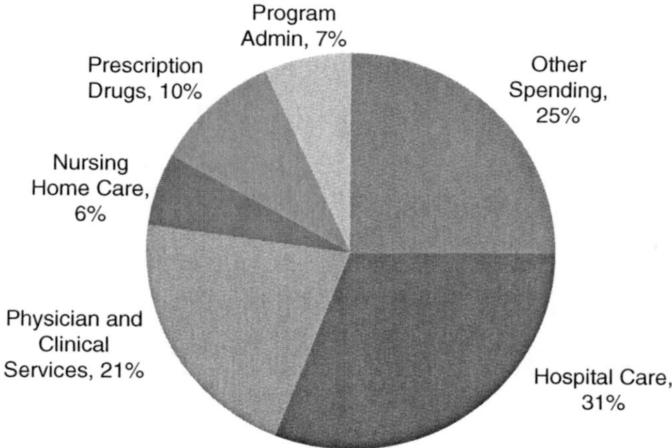

Figure 14-1 Health care spending breakdown.

show that R & D spending continues to grow disproportionately to new drug approval, while new drugs are not being developed at a pace necessary to replace those whose patents have expired. For example, the report shows that Pfizer revenues will be significantly impacted if it remains on its current path. The Approvals trends highlighted in that report have continued, with approvals in 2007 and 2008 of only eighteen and twenty new molecular entities, respectively.

The United Kingdom's National Institute for Health and Clinical Excellence (NICE) agreed to purchase a cancer treatment made by Johnson & Johnson on the condition that NICE would pay only when the drug was effective, since it was expected to have little effect on one-third of patients. "For those who may get a full response or a partial response . . . it's a cost-effective intervention for the National Health Service. Narrowing down to patients getting the best response makes it worth it," said Andrew Dillon, NICE chief executive (Loftus, *Wall Street Journal,* October 10, 2007). Currently, pharmaceutical companies receive payment irrespective of whether or not the treatment works.

On April 21, 2009, P. Lee, executive director of the Pacific Business Group on Health, told a U. S. Senate Committee on Finance Roundtable Discussions on Health Reform, "Our health care system is broken: Quality of care varies dramatically between doctors and hospitals, but those differences are invisible to patients. Payments reward quantity over quality and fixing problems over prevention. Lack of standardized performance measures makes it impossible to know which providers are doing a good job, and those who are not. Consumers lack information to make the choices that are right for them."

Adverse events continue to plague the U. S. health care system. A recent study has shown that from 1998 through 2005, reported serious adverse drug events almost tripled, to almost 90,000 patients (Moore, Cohen, and Furberg 2007).

At the time of this writing, the Mental Health Parity and Addiction Equity Act of 2008 (MHP) was still the subject of much deliberation, with the associated regulations newly released. The key issues included:

- What is the collective increase in mental health coverage required to align current coverage with parity, and to what extent will this increase in coverage translate into increase service utilization?
- What does parity mean in relation to coverage and reimbursement?

According to a Harvard Medical School study, over 35 million people—about 14 percent of the U. S. adult population—suffer from a moderate or severe mental health disorder in any given year (Kessler 2005). More days of impaired and lost work result from mental illness than from any other chronic condition, including diabetes, asthma, and arthritis (Stewart et al. 2003). Employees with depression cost employers an estimated $44 billion per year in lost productivity (Marlowe 2002). Individuals who have depression (the most frequently diagnosed mental illness), but are not receiving care, use two to four times the health care resources of other enrolees (Watson and Wyatt 1998). Minutes are allocated to treating depression in a primary care setting (Tai-Seale et al. 2007). In primary care settings, the prevalence of major depressive disorder (MDD) ranges from 5 to 13 percent in adults and from 6 to 9 percent in older adults. Many of the treated depression cases are managed in primary care; roughly one-third to one-half of non-elderly adults, and almost two-thirds of older adults who are treated for depression, are treated in primary care (O'Connor 2009).

Personalized Medicine: The Next Phase of the "Digital Wave"

Health care is the most recent industry to be transformed by the digital wave. It is the digital wave that has facilitated personalized health care, and that is catalyzing change. As such, we believe insights derived from the wave's previous impacts are relevant to health care, providing confidence in the power and future successes of PM.

In 2004, we previously concluded that "a sectoral shift in health care (drug development and treatment) was underway. The current practice of 'population-based' treatment is shifting towards 'personalized medicine'" (Brain Resource Company 2004) We also drew attention to the highly constructive, proactive role undertaken by the U. S. Food and Drug Administration's (FDA) initiatives in anchoring and catalyzing this shift (see Woodcock 2004).

PM is the culmination and convergence of the many advances in the life sciences developed over previous decades, which, through computing power, can now be harnessed into new tools. This is best exemplified in genetics, where the initial power of DNA and related discoveries are now being unleashed in products for the mass market. This is supported by computer processing power, which collects and processes the massive datasets required to extract meaningful information. The use of computers also underpins the growth of neuroimaging technologies.

PM is showing many of the hallmarks of the previous industrial transformations that arose from the digital revolution, and has its same share of enthusiasts and skeptics. In a presentation, one of the authors (Segal 1998) discussed the transformations then underway in the telecommunications sector (Figure 14-2). Substituting "PM" for "data wave," these same conclusions can be applied today.

There are many similarities between current and older attitudes surrounding PM. The following is from a Federal Communications Commission (FCC) report titled "Digital Tornado: The Internet and Telecommunications Policy":

> The growth potential of the Internet lends itself to both pessimistic and optimistic expectations. The pessimist, having struggled through descriptions of legal uncertainties, competitive concerns, and bandwidth bottlenecks, will be convinced that all these problems can only become worse as the Internet grows. The optimist, on the other hand, recognizes that technology and markets have proven their ability to solve problems even faster than they create them. (Federal Communications Commission 1997).

Data wave observations

- *It will be massive*
- *Data is a unifying technology with the sum greater than the parts*
- *Boundaries hard to define*
- *Changes relationship with customer*
- *History says forecast will be conservative*

Figure 14-2 Data wave observations relating to the transformations previously underway in the telecommunications sector.

Previous Transformations of the Digital Wave

The first impact of the digital wave was the automation of business administration, which transformed manual systems, including payroll, customer order processing, and inventory management. The catalyst for change was the invention of the mainframe computer in the 1960s, enabling computer-processing power to become more readily available to big businesses. IBM was the main player in this market.

This transformation led to the next phase, imparting both greater power and consequences than the initial impact. The computer industry was itself transformed first by the personal computer (PC), and then the networked PC. This saw many rapid shifts in value (Figure 14-3), with the dominant value credited to new competitors entering the market (Edholm 2000).

The next wave occurred in the 1990s, transforming communications and offering a wide range of retail commerce and entertainment services (including telecommunications, Internet, mobile telephony, entertainment, auctions, shopping, banking, etc.). The types of changes were broad, ranging from internal production efficiencies, to new product opportunities, to product replacement. The Internet provided a distribution network with unprecedented reach and access to many people. The critical driver was the ability to personalize the experience. No longer did one have to accept only what was offered at the local mall,

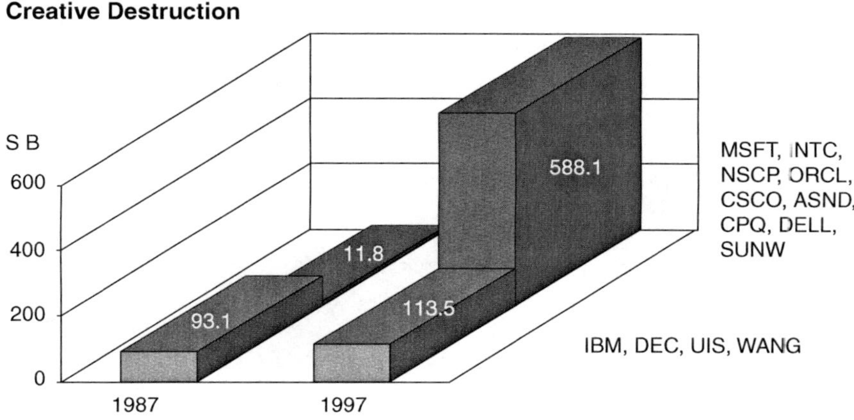

Figure 14-3 Changes in value resulting from transformations in the computer industry.

or available only through the local newspaper. This was well articulated by George Gilder: "The critics of the Internet are mostly skeptical about the value of choice. But choice validates freedom and substantiates individuality. Choice accords with the inexorable genetic diversity of humans". Source: *Forbes ASAP*- December 1, 1995.

The commercial attractions are best described by Chris Anderson's theory of the "long tail" (Anderson 2006), which posits that our culture and economy are increasingly shifting away from a focus on a relatively small number of "hits" (mainstream products and markets) at the head of the demand curve, and toward a huge number of niches in the tail, where many products are in low demand—the key being that the tail can collectively make up a market share that rivals or exceeds the relatively few current bestsellers and blockbusters. An example he cited is iTunes, where almost every one of their 6 million or so tracks in inventory sells at least once per quarter.

A Snapshot of Winners and Losers from the Above Transformations

Transformations are complex, with many factors influencing outcomes. Accordingly, the following are some of our generalized observations of factors that appeared influential (full disclosure: Dan Segal was an equities telecommunications analyst for Salomon Smith Barney/Citigroup through the major part of the 1990s). These may have relevance for strategic consideration as part of the PM transformation.

- Technology transformations follow the usual rules of economics, but the scale of impact forces management to examine and reexamine each business assumption, to avoid jumping at shadows, so to speak.
- The Internet boom did not change the need for a business to have content, distribution, and a payment-receival system.
- Companies needed both distribution and content, with at least one needing to be unique (for example, Google has unique content and common Internet distribution, telecom companies have unique distribution networks).
- Content needs to be monetizable (Google benefits from the unique synergy between search content and advertising revenue), and growth in customer numbers alone does not assure profits (e. g. see many telecommunication and Internet companies).
- The digital wave can radically alter old established rules. One example is the music industry which, while having content, experienced a diminishing ability to get paid, as copyright law became hard to enforce once the distribution floodgates opened.
- Traditional companies and industries faced a daunting challenge, as they struggled to foresee change while making the requisite radical decisions to adapt. Bill Gates was one who did just that in changing the entire focus of Microsoft in order to adapt to the Internet. Most, however, were slow to respond (for example, many telecommunications companies remained carriage providers, as opposed to benefiting from the value of what was carried).
- Companies with unique content and capable management, even where they moved slowly, saw through change (News Corporation), although some ended up with different business models to when they started (IBM).
- Half-hearted new initiatives generally resulted in write-offs. Mergers and acquisitions and cost-cutting generally provided, at best, only temporary or superficial solutions. These initiatives, in the main, eroded shareholder value (see Worldcom/MCI, ATT/TCI, Time Warner/AOL, Telstra/PCCW).

- The blocking power of vested interests, when faced by the might of the consumer, has only temporary effect.
- The best defence was innovation, with cost-cutting and mergers and acquisitions as its nemeses.

This was summed up by Warren Buffet, discussing Internet valuations just before the February 2000 collapse: "The key to investing is not assessing how much an industry is going to affect society, or how much it will grow, but rather determining the competitive advantage of any given company and above all the durability of that advantage. The products or services that have wide sustainable moats around them are the ones which deliver rewards to investors" (*Fortune* Magazine, Nov 22, 1999).

The Digital Wave Is Now Beginning to Transform Health Care

Table 14-2 summarizes the rolling nature of this wave, and also the progressive build-up of power as each sector is transformed. That is, communications relied on the changes that preceded it in business administration, and retail relied on changes from both of those previous transformations. Similarly, health care builds upon changes in all three previous transformations, with one significant difference: the U. S. government is incentivising change through payments for adopting electronic health records, and penalties for failing to adopt them.

Personalized Medicine Is Evolving

Over time, PM will mean many things and cover a wide scope. As with previous industry transformations, it is likely to overpromise, then under-deliver for a period, before dwarfing the initial promise. For example, some have started to expect that PM will have the capability to develop a specific drug treatment to cater to an individual's unique characteristics, as opposed to identifying what drug treatment from those commonly available is best suited to an individual's characteristics. While the former is appealing, it is currently beyond our reach.

Strong Economic Contributions

In light of the above-mentioned need for increased health care efficiencies, it is worth noting the economic benefits of the previous technology wave (see Figure 14-4), and Edholm 1997-98). While we are not claiming that health care would enjoy the same degree of gains, there are still significant gains to be made from the efficiencies PM introduces, including direct cost savings and productivity gains from improved treatment.

US Government Regulation to Promote Personalized Medicine

Given the high degree of regulation of the health care industry, progress would be difficult without regulatory support. The FDA continues to play a leading role in PM, in many ways driving the thought process ahead of the drug manufacturers.

(i) Guidance on Pharmacogenomics Data - issued November 3, 2003
(ii) Innovation or Stagnation? Challenge and Opportunity on the Critical Path to New Medical Products - issued March 16, 2004
(iii) Critical Path Opportunities Report & Opportunities List–March 2006,

Table 14-2 Wave of Transformations Flowing through a Number of Sectors Impacted by Digitization

	Transformed Sector			
	Business Administration	*Communication*	*Retail/ Entertainment*	*Healthcare*
Catalyst	Semiconductors	Computers	Relational Databases	Relational databases & new life science technologies
Inflection point	PC Specification	Internet Protocol	WWW & Data mining software	Electronic Health Records, Clinical Decision Support, Comparative Effectiveness Research and Predictive Markers
Key paradigm	Moore's Law[1]	Metcalfe's Law[2]	Anderson's Long Tail	Personalized Medicine
Key player/monetization	Microsoft/ software	Cisco/ hardware	Google/ advertising	Health services/US Govt. Reforms Incentives
Key strategy concept	Embrace & Extend	Acquisition & Development	Contextual adverts / Anything, anytime	Interoperability, Data
Key obstacles	Technology limitations	Regulatory Framework	Payment systems, Digital rights	Powerful vested interests, fear factor about change, Traditional disempowerment of consumers
Key Benefit	Dramatic productivity improvement	Dramatic productivity improvement	Dramatic productivity improvement	Dramatic productivity improvement & Risk Management
Key player market value	$350 billion	$200 billion	$220 billion	$Trillions (Government, Managed Care, EHR, Diagnostics, Pharma)

[1] Moore's Law: the number of transistors that can be placed on an integrated circuit doubles every two years, this observation by Intel co-founder Gordon Moore in 1965 has held since.

[2] Metcalfe's Law: the value of a telecommunications network is proportional to the square of the number of users of the system, Robert Metcalfe.

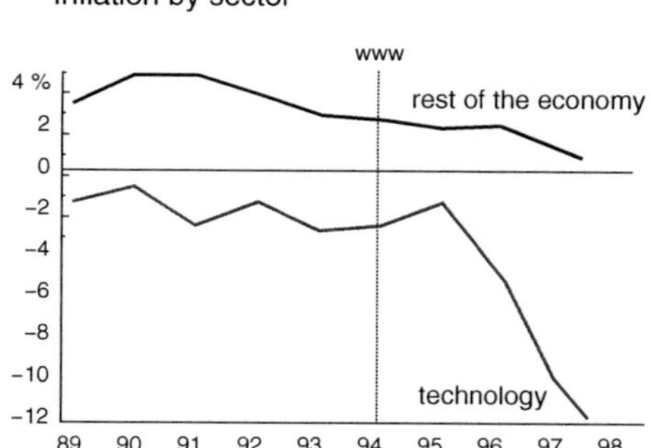

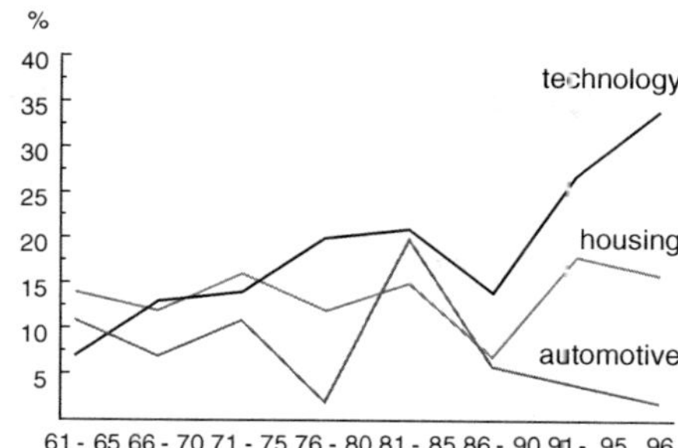

Figure 14-4 Impacts of the transformation of the technology sector in the 1990s on economic growth.

(iv) Critical Path Opportunities Initiated During 2006–Press release February 2007
(v) Pharmacogenomic Data Submissions—Companion Guidance, August 2007
(vi) FDA Clears Genetic Lab Test for Warfarin Sensitivity–September 2007 (but note that this did not ensure a commercial success, with Medicare refusing to provide coverage).

In 2008, Dr Thomas Insel, director of the National Institute of Mental Health (NIMH) issued the agency's strategic plan, which established the direction for this transformation:

New maps and new mapping tools for the human genome, for instance, have transformed our understanding of how individuals genetically vary from each other and how these variations can put some people at increased risk for certain illnesses. Neuroimaging tools to visualize the brain have given us an unprecedented view of brain activity, providing a new understanding of its development and a picture of how specific networks of cells change with experience. One goal of this Strategic Plan is to translate these and other advances to what the National Institutes of Health (NIH) calls the "4 P's" of research: increasing the capacity to Predict who is at risk for developing disease; developing interventions that Pre-empt (or interrupt) the disease process; using knowledge about individual biological, environmental, and social factors for Personalized interventions; and ensuring that clinical research involves Participation from the diversity of people and settings involved in health care. National Institute of Mental Health, 2008.

What is most timely, however, is the impetus being provided by reforms of the American health care system. Acting as the change driver, the American Recovery and Reinvestment Act of 2009 will devote $20 billion to establishing electronic health records, and $1.1 billion to research that compares the effectiveness of different medical options. Of particular note, to qualify for these incentives, the user needs to demonstrate that electronic health records (EHR) are being used "meaningfully." This includes the mandated use of Clinical Decision Support rules that are relevant to the user's specialty. This ensures a focus on improving health outcomes, and not simply on technology change for the sake of change.

These components have a direct impact on developing PM. The Affordable Health Care for America Act and the Mental Health Parity and Addiction Equity Act of 2008 have dramatically increased coverage by more than 30 million individuals and will demand new solutions for providing this expanded coverage.

Progress in comparative effectiveness research (CER) is a key part of any system that facilitates a move from payment for activity to one that rewards value. The relevance of

CER to mental health is paramount, given the paucity of objective benchmarks in current clinical and drug R & D use.

The most important and fundamental part of these reforms, however, is the electronic health record (EHR) incentive, mentioned earlier and discussed more in the following section.

Standardized Databases Integrating New Technologies

We believe standardized databases are among the most important factors in expanding access to personalized medicine, in that they anchor and facilitate a wide range of both existing and new technologies. Databases allow converging evidence to be identified and new insights to be extracted, providing the evidence base for any emergent PM solutions. Health care databases are evolving through many phases that sometimes overlap:

1. Phase 1: Disease and generic treatment information. This phase is now quite mature in terms of the plethora of Web sites providing health information.
2. Phase 2: Personal data repositories (or EHRs). This phase is currently being implemented through the Meaningful Use rule in the HITECH Act (Health Information Technology for Economic and Clinical Health Act which forms part of the American Recovery and Reinvestment Act of 2009), along with many other government initiatives. Also of note is Microsoft's and Google's entrance into the arena with their health record initiatives.
3. Phase 3: Active personal data repositories, including more complex medical data, such as that acquired using newer technologies, and clinical decision support systems. Both have the power to provide new, personalized markers and also to evaluate personal response factors, including to drugs, thereby leading to improved treatment.

Clinical Decision Support

The increasing number of decision support systems that are emerging best exemplifies the power of computing and databases. These systems integrate patient data with a wide variety of other generalized data, from claims data to clinical guidelines, to current research findings. Not only can they provide input into the case, they can also contextualize against options that are covered by the patient's insurance policy. These systems also allow clinicians to access information useful in developing a cost-benefit analysis (e.g., software can provide a differential diagnosis and determine what tests should be done or provide clarification on the degree of certainty of a disease).

Personalized Medicine Is Not Only About Genetics

To date, in many ways, the brain-based PM emphasis on genetics alone has generated few outcomes. Genetics has also been the focus of intensive R & D by drug companies. A recently completed large-scale genetics study found that "for any given trait there will be few (if any) large effects, a handful of modest effects, and a substantial number of genes generating small or very small increases in disease risk" (Wellcome Trust Case Control Consortium 2007). This led to the following observation: "If that is the case, the existing paradigm of drug discovery and development requires radical rethinking, as does the concept of personalized medicine based solely on genetics. Exit the silver bullet and enter the carronade (the carronade was a short-barrel, large-diameter cannon that could be loaded with a canister of 500 musket balls)"CT Dollery: Beyond Genomics. Nature 82 4 (2007).

Further, as has been pointed out in many studies to date, as well as by other contributors to this book, assessing genetics in isolation can be unproductive, given the large numbers of candidate genes that can be identified, and the daunting challenge of determining which genes are responsible for what developments. Genomics considered in the context of brain mechanisms and along with cognition markers, are likely to be more productive. This was best summed up by the Mayflower Action Group Call to Action, promoting the need for:

1. Establishing an integrated and standardized approach, and harnessing large numbers of researchers to study the most impacting gene and brain markers;
2. Finding ways to predict who will respond to treatment; and
3. Integrating Web-based screening measurements and Clinical Decision Support Systems into health records (Koslow, Williams and Gordon 2010).

Business Case Considerations

PM covers a broad spectrum of tools. PM and subsequent improvements in information technology have more value in some situations and less in others, with different impacts on different users. What is clear is that better information is both available and profitable when selectively used.

At some level, all groups benefit from PM. The following scenarios would be addressed and improved upon:

1. Patients currently suffer from delays in finding the right treatment, which also leads to poor compliance;
2. Clinicians experience time pressures and inefficiencies in the treatment process;
3. Government payers are directly responsible for the cost of funding and the massive cost to society;
4. Private payers/managed care organizations have to manage rising costs, while maximizing efficiencies; and
5. Pharmaceutical companies may face higher failure rates from clinical trials, growing costs, and diminished sales impact due to poor effectiveness and compliance.

We believe that the success or failure of PM rests with all constituent groups, but an increasing role for the consumer is likely to be particularly important. The technology continues to show value. However, like any industry, without widespread distribution, great technologies can stagnate. Health care is also highly controlled and regulated, so without support, reaching the tipping point will be that much harder.

Most health care systems cannot meet consumer demand. Thus, increasing efficiencies, achieved through PM and thus alleviating pressures, should not have a negative impact on return to the service provider. To illustrate this point, a recent report about medical tourism suggests that if every U. S. resident who could go abroad for treatment actually went, the savings on total medical costs would be about 5 percent, or less than half the existing growth rate (Williams and Seus 2009). Those who will suffer the most negative impact from PM are those selling inefficient solutions (such as those selling a drug that does not work for the majority of users). Losers will also include those risking development of new solutions (including marker discovery) that fail.

Personalized Medicine, Cost-Benefit Considerations, and Patient Benefit

PM's introduction is compelling where there is an increased patient benefit at no additional cost, or a cost reduction with no deterioration in patient care. That includes whether the cost of a PM-based test is priced at a level below the current process that it replaces—for example, below the cost of a saved clinical visit or additional drug prescription. There is also the additional practical issue of whether the test is something easy to administer and roll out.

The decision is far more complex when benefit to the patient and cost both increase because of PM. This, then, requires the less objective decision be made as to whether the improved patient benefit is cost-justified within overall budgetary constraints. There will, of course, be many situations in which one solution has widespread benefit and others in which PM does not add sufficient benefit to patient care, nor contribute to any cost savings. For example, it would be very difficult to justify any PM testing on the effects of aspirin.

Warfarin: A Billion-Dollar, Preventive Medicine Cost Savings

In the United States, approximately 2 million patients begin taking warfarin each year. But warfarin is the second most common drug, after insulin, resulting in emergency room (ER) visits for adverse drug events. The challenge is to find the optimal dose, which varies greatly from person to person. If the dose is too high, users are subject to increased risk of serious bleeding, and if too low, subject to increased risk of stroke.

The FDA recently updated warfarin's drug labeling to explain that people's genetic makeup may influence how they respond to the medication. This affords the opportunity for health care providers to use genetic tests to improve their initial estimate of what is a reasonable warfarin dose for individual patients.

The savings from using genetic testing for the two genes in warfarin therapy (*CYP2C9* and *VKORC1*) have been estimated at $1.1 billion annually: $1.15 billion in reduced bleeding costs plus $675 million in reduced treatment costs, and $700 million in diagnostic costs) (McWilliam, Lutter, and Nardinelli 2006) . These assumptions did not, however, factor in any indirect savings arising from the value of the health improvements among warfarin users. They also did not factor in volume discounts in the cost of each genetics test (assumptions included: genetic testing allowing 85,000 users to avoid bleeding events and 17,000 strokes annually; direct costs of $39,500 per stroke; direct costs of bleeding of $13,500; costs of genetic testing of about $700 million [2 million tests x $350 per test]; net savings made up of $1.15bn in reduced bleeding costs, $675m in reduced stoke costs, and$700m testing costs). The Centers for Medicare and Medicaid Services have stated that there is not enough evidence showing improvements in a patient's health by using these tests; further studies are warranted (Centers for Medicare and Medicaid Services 2009). Among complaints cited are that the results are not available in time for the physician to begin treatment. In our view, a major part of this decision is to call into question what levels of evidence are required for what recommendation. This raises the broader issue of coverage and the rules that apply. This is a topic that is beyond the scope of this article, especially given the current deliberations over mental health parity.

Pharmaceutical Companies

We will consider the impact on pharmaceutical research & development (R&D) and sales divisions separately.

Research and Development

As discussed above, there is a paucity of new drugs to replenish those losing their patent protection. It takes about ten to fifteen years in total to develop one new medicine and make it available to patients on the market. The average cost to research and develop each successful drug is commonly estimated to be between $800 million and $1 billion. This number includes the cost of the many drug failures: for every 5,000 to 10,000 compounds that enter the R & D pipeline, ultimately only one receives FDA approval.

The identification of genetic markers will offer R & D efficiencies by allowing the pharmaceutical companies to better select drug trial participants. This will deliver more quantifiable end points against which progress of a trial can be assessed, and help determine relative performance between candidate drugs in the inventory. Figure 14-5 shows the power of having a database marker baseline against which relative performance can be measured. There is also the very substantial issue of understanding placebo responders.

Having reliable trial screening endpoints (markers) can reduce cost. Better quality information lets you see which paths are working more effectively, as well as those that are not. Given the high costs, eliminating a non-performing drug candidate from trials more quickly can deliver significant savings. There is also a need to ensure that progress is aligned with the above-mentioned FDA guidelines, or guidelines in any country in which trials are being carried out or in which the drug needs approval.

Table 14-3 presents a simplified example of a drug data matrix for brain-related drugs. For example, this matrix has hundreds of variables across the four key processes (thinking, self-regulation, emotions, and feelings) in the Brain Resource International Database.

Herceptin: The $2.5 Billion, Preventive Medicine Revenue Accelerator

The "poster child" for PM is the antibody drug Herceptin (trastuzumab), which effectively treats breast cancer in women who have uniquely presented with overexpression of a cell surface protein, known as human epidermal growth factor receptor 2 (HER2). Because potential participants had to present with this unique overexpression, a smaller, faster trial was undertaken. Most significantly, as the life time for patent protection is finite, the faster

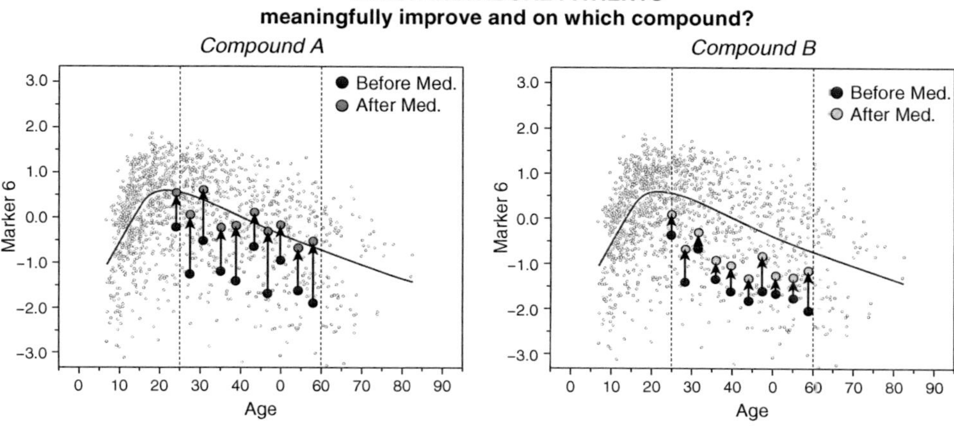

Figure 14-5 Compound A: Better responders to a compound. Compound B: Statistically significant but inferior response.

Table 14-3 Drug Data Matrix for Brain-related Drugs, Showing How Individual Drugs Can Affect Patients

Measure	*Normative group*	*Disease 1*	*Disease 2*	*Drug 1*	*Drug 2*
Thinking	√	√	x	-	√
Self Regulation	√	x	x	√	-
Emotions	√	√	√	-	-
Feeling	√	x	x	√	-

The check marks represent a normal response and x a problem. This simple example shows that only by having complete data matrices can deficits be identified, whether in disease treatment or in drug performance. Drug 1 is ideally matched to benefiting patient with Disease 1 (normalizing Self Regulation and Feelings) and a combination of Drug 1 and 2 are needed to normalize disease 2.

path to market meant a longer patent coverage, protecting a sales cycle worth billions of dollars (Frueh 2005).

Of course, there are also benefits of having the HER2 test in the marketplace. The relative cost (around $400 per test) is small compared to the benefits: a faster treatment response, more targeted treatment (that is, precious time is not wasted giving the wrong treatment to known nonresponders), and cost-efficiency. The cost of the test is relatively low compared to treatment costs, which can run into tens of thousands of dollars per patient per year.

Sales and the Pro-Business Case

Skeptics of PM believe that pharmaceutical companies have much to lose from its implementation in terms of lost revenues due to more targeted use of medications. We believe this risk is overstated and is relatively small compared to other sales risks from regulatory changes or litigation costs. To the contrary, our view is that PM could actually enhance sales.

It is worth pointing out that many drugs are not exposed to this risk. For example, Herceptin increased returns—from sales of a drug that, without PM, may not have ever been approved—and warfarin reduces indirect costs (but not necessarily drug sales). However, there will be circumstances where losses in sales do occur. Then, the issue is whether these lost revenues, perhaps due to patients who no longer take the drug because there is no predicted benefit for them, can be recouped elsewhere.

The underlying assumption is that the that are made available drugs do work—albeit not for all patients all the time. There are several methods by which a drug (call it Drug A) can recoup losses:

1. by capturing spillage from other drugs with no predicted patient benefit where Drug A would benefit;
2. through new patients entering the market because of an increased probability of finding a solution. The patient's greatest concern is, "Will this drug hurt me?" Increasing confidence on the part of patients will translate into new customers considering treatment, where before PM, they would not have considered treatment; and
3. a positive impact from increased compliance by those benefiting. The faster a patient can find a solution, the more wedded they will be to using this solution for a longer time.

A simple calculation highlights the possibility that this potential risk of sales erosion could actually be smaller than many people think, and could actually deliver better returns, on average. The loss in sales from PM is tied to the percentage of responders to a drug. Where this is between 20 and 60 percent, we estimate that the amount of sales that could be lost (or recouped) to be somewhere between 5 and 20 percent. The key assumption is that those who benefit from a drug take it for twelve months, while those showing no benefit discontinue it after one month. Clearly, if a nonresponder takes a drug for the same period of time as a responder, the loss in sales will be greater. Hence, there are situations where PM losses could be substantial, with a marginal ability to recoup them. This calculation also assumes no significant customer acquisition costs, which would offset the loss in sales. This scenario is not likely, given the massive advertising and marketing budgets of pharmaceutical companies.

Conclusion

Personalized medicine is an unstoppable wave, gaining strength from many fortuitous confluences among technology, regulation, and consumer need. PM is using technology to do things better. This includes personalizing health care to reflect individual differences. Personalized medicine offers solutions that fill consumer need; have solid, reproducible scientific results; are easy to use; are cost-effective, and will gain traction among payers. We believe that the tide has already turned with the commercialization of new technologies that already add value.

Successful identification of markers will have widespread benefit. Markers can better enable clinicians to target a patient's disease and help match the right drug to that person, thereby delivering significant cost and treatment efficiencies. The efficiencies of clinical drug trials could also be enhanced, by making it possible for sponsors to enroll only those patients with the target condition, thereby better assessing progress. Further, the data may also point to new drug candidates for future drug development.

New business cases are emerging to drive the ongoing discovery of new PM markers, EHR efficiencies, and continuum of care solutions. Government incentives, together with the power of data from EHR, further strengthen the transformation. Although we cannot predict exactly what this transformation will look like, we know it will be monumental.

References

1995. *Forbes ASAP*.

Anderson, C. 2006. *The long tail*. New York: Hyperion.

Brain Resource Company. 2004. Enabling personalized medicine. http://www.brainresource.com/uploads/BRC_PersonalMedDetail_Apr04.pdf (accessed June 16, 2010).

Burrill, G. S. 2009. Biotech 2009: Life sciences navigating the sea change.

Centers for Medicare and Medicaid Services. U. S. Department of Health and Human Services. 2009. CMS Proposed Decision Memo for Pharmacogenomic Testing for Warfarin Response (CAG-00400N). http://www.cms.hhs.gov/mcd/viewdraftdecisionmemo.asp?from2=viewdraftdecisionmemo.asp&id=224& (accessed June 16, 2010).

Centers for Medicare and Medicaid Services. U. S. Department of Health and Human Services. National health expenditures 2008 highlights. http://www.cms.hhs.gov/NationalHealthExpendData/downloads/highlights.pdf (accessed June 17, 2010).

Dollery, C. T. 2007. Beyond genomics. *Nature* 82: 4.

Edholm, M. 1997–98. Presentation based on data from Bureau of Economic Analysis.

Edholm, M. 2000. www.predicom.com

Federal Communications Commission. 1997. Digital tornado: The Internet and telecommunications policy.

Frueh, W. 2005. FDA: Personalized medicine: What is it? How will it affect health care?

Kessler, R. C., W. T. Chiu, O. Demler, and E. W. Walters. 2005. Prevalence, severity, and comorbidity of 12-month DSM-IV disorders in the national comorbidity survey replication. *Archives of General Psychiatry* 62: 617–27.

Koslow, S. H., L. M. Williams, and E. Gordon. 2010. Personalized medicine for the brain. *Molecular Psychiatry*. 2010: 1–2.

Loftus, P. Meter set to run on J&J's cash-back deal. *The Wall Street Journal,* Oct. 10.

Marlowe, J. F. 2002. Depression's surprising toll on worker productivity. *Employee Benefits Journal*, March 2002 .

A. McWilliam, X. X., R. Lutter and C. Nardinelli. 2006. Health care savings from personalizing medicine using genetic testing: The case of Warfarin, AEI-Brookings Joint Center for Regulatory Studies Working Paper.

Moore, T. J., M. R. Cohen, and C. D. Furberg. 2007. Serious adverse drug events reported to the Food and Drug Administration, 1998–2005. *Arch Intern Med* 167: 1752–59.

O'Connor, E. A., E. P. Whitlock, B. Gaynes, and T. L. Beil. 2009. Screening for depression in adults and older adults in primary care: An updated systematic review. Evidence Synthesis No. 75. AHRQ Publication No. 10-05143-EF-1. Rockville, Maryland: Agency for Healthcare Research and Quality.

Segal, Dan. 1998. Presentation to Telstra Corporation senior management.

Stewart, W. F. et al: 2003. Cost of lost productive work time among U.S. Workers with depression. *JAMA*, June 18, 2003.

Tai-Seale, M., T. Mcguire, C. Colenda, D. Rosen, and M. Cook. 2007. "Two-Minute Mental Health Care for Elderly Patients: Inside Primary Care Visits,". *Journal of American Geriatric Society* 55: 19903–1911, 2007.

Watson and Wyatt. 1998. Staying at work survey.

Wellcome Trust Case Control Consortium. 2007. Genome-wide association study of 14,000 cases of seven common diseases and 3,000 shared controls. *Nature* 447: 1038.

D. Williams, J. Seus. Medical Tourism: Implications for participants in the US Health Care system.

Woodcock, J. et al. 2004. U. S. Food and Drug Administration. Innovation or stagnation? challenge and opportunity on the critical path to new medical products.

Section 5

Conclusion and Recommendations

15 Accelerating the Future of Personalized Medicine

Evian Gordon, PhD, MBBCH and Stephen H. Koslow, PhD

Having reached the final chapter, you now have an informed opinion of the core focus of this book, which has been to detail the range of genomic markers and brain markers across a spectrum of major brain-based disorders. We have learned where the research field is in terms of relevant brain markers, and what the key factors are to accelerate this critical field. The chapters have presented the most current, cutting-edge knowledge on personalized medicine.

We reiterate here some of the key points that personalized medicine emphasizes, who will be most affected by the implementation of personalized medicine, and key factors likely to accelerate its adoption into mainstream clinical practice.

Personalized medicine promises:

- Targeted predictive treatments;
- Objective measures of disease onset, diagnosis, successful treatment, side effects, prognosis, outcomes testing, and measures of cure; and
- Research insights unraveling disease etiology and prevention.

Personalized Medicine is critical for:

- Patients/Consumers;
- Clinicians;
- Diagnostic development;
- The drug regulatory system;
- Health care delivery systems; and
- Pharmaceutical discovery stakeholders.

Personalized medicine has gained tremendous popularity of late, however, it is not new. For decades, the field of pharmacogenetics has been investigating the genomic uniqueness of individual diseases, and variable actions of pharmacological agents due to unique gene isomers in individuals. As mentioned earlier, major impetus for the currency of personalized medicine came to the fore with the 2003 completion of the Human Genome Project, which opened the door to personalized medicine. There is still much to learn about genes and their regulation, expression, and relation to diseases.

You have read that much has gone into delineating prominent genes in brain disorders. Rarely, however, has a gene yet to be identified that has led to the useful understanding and treatment of a brain disorder. There are many plausible explanations for the modest success in the current approaches to brain-based personalized medicine. One key issue is that many diseases are multifactorial and involve many genes and epigenetic factors. Additionally, whole genome studies do not measure the expression of the genes in specific organs within the body.

In essence, the complexity of the brain is likely to require a shift from a single or multiple gene-marker focus (and proteins and metabolites), to an integrated approach where *direct* brain-related, information-yielding brainmarkers and cognition markers are also taken into account. One should include, in addition to genes, objective and quantitative measures of brain structure MRI and DTI and function and functional brain outcomes, including fMRI, PET, EEG, ERPs, cognitive assessment, emotional assessment, and autonomic function (such as heart rate variability).

We conclude by focusing on ten factors that seem likely to expedite the assimilation of personalized medicine into mainstream clinical treatment of brain disorders.

Accelerate "Brainmarker" Discovery

Accelerating the discovery of gene and functional brain markers and their translation into clinical practice is dependent on enhancing all processes, from molecular to functional brain studies. This acceleration will be enhanced by defining "niche busters" that are effective in groups of people, rather than everyone with the disorder. This includes the discovery of drugs that are targeted to a small, well-defined group, by using genes and functional brain measures in addition to identifying clinical groups according to their traditional "signs and symptoms," which should nevertheless be done as well.

Codevelopment of BrainMarkers with New Medications

There is a growing realization that there is value in exploring new markers in conjunction with new drugs. This codevelopment approach, coupled with patient genetic and functional brainmarker groupings, will significantly enrich clinical trials. It should also accelerate conditional/interim approvals, and reduce post-market label warnings or recalls by defining unique responsive subpopulations of diseases. These factors will enhance the likelihood of successful, well-matched drugs to individual patients.

Replication and Validation of New Biomarkers in Real-World Patient Groups

Given the high rate of "failure to replicate" genomic studies, a growing number of high-impact international journals are insisting upon replication of genomic studies prior to publication. This augers well for accelerating the most effective study designs for delineating gene-brain-behavior linkages that is most likely to have predictive treatment validity. We need to add to this the caution that clinical studies need to include a representative real-world sample of patients, including those with comorbidities.

Focus on the Extent of Clinical Utility

The strengths of the evidence of clinical utility defined as how likely the results of a Personalized Medicine outcome will change a clinician's treatment recommendation. This is predicted by health care analysts (for example, see Price Waterhouse Cooper's 2010 Personalized Medicine Report) to be *the* key driver for payers' decisions about coverage and reimbursement. In simple terms this means that it is not enough to just find new markers, but it is critical to find markers that really make a big clinical difference. The extent of diagnostic sensitivity and specificity should be reported in all personalized medicine studies, not just group results. These measures should examine issues of diagnosis, treatment prediction, treatment response, side effects, disease onset, and disease remission.

Facilitate FDA Registration and reregulation

The FDA continues to explore how to optimize the regulations involved in evidence generation in diverse ways, including through drug labeling and diagnostic test regulation.

Facilitate Clinical Workflow Implementation of Biomarker Usage

Money will always be a factor in health care, but reimbursement is not the primary concern of clinicians and physicians adopting personalized medicine. Clinicians value a seamless connection to their daily workflow of patient care (and into their electronic health records, for those who have them). Clinicians will more readily adopt a new technique in their time-pressured practice, if it is simple to do, easy to interpret, saves them time, makes their job easier, is reimbursable, and of clear benefit for the patient.

Integrate Electronic Health Records (EHRs) and Clinical Decision Support Systems (CDSS) into Clinical Practice

Integrated data sets need be established to allow interoperable data connectivity with electronic health records (EHRs). The American Recovery and Reinvestment Act of 2009 provides $36 billion to create an infrastructure that will include a national, interoperable EHR system. A crucial criterion for this funding is meaningful use of the EHR data. This places an emphasis on having clinical decision support systems (CDSS) solutions embedded in the EHR. A number of chapters highlighted the use of CDSSs as a first pragmatic step in Personalized Medicine. Clinical decision support systems (including those in EHR systems) can be extended in a step-by-step manner to incorporate new brain-based personalized medicine marker discoveries.

The focus on meaningful use will further accelerate the adoption of biomarker and clinical decision support systems and solutions. This focus on clinical solutions (rather than merely IT efficiency), is completely aligned with the core goal of personalized medicine.

Fix Reimbursement Models to include "Pay for Performance"

The change in reimbursement models from volume-based to value-based rewards clinicians for obtaining good outcomes and improving wellness. This is an incentive driver toward a more personalized medical system.

Encourage the Development and Use of Standardized Methods, Protocols, and Data Aggregation into Databases

Standardized methods and protocols are key drivers for more readily making data available to inform us on what really works, and the "personalized extent of clinical effectiveness, comparative effectiveness and cost effectiveness." Databases are powerful platforms of aggregated information. Databases go beyond merely collecting diverse information. The true value of standardization lies in their statistical power of numbers and in exploring interrelationships and integration between and among biological systems. Developing common data standards further increases the likelihood of finding the most effective outcomes. This does not preclude the innovation of new methods or diverse databases, but rather serves the goals of finding the most significant personalized brain markers.

Embrace the "Consumer Revolution"

The shift in focus from illness to wellness is part of a broader trend toward consumer-focused health care. This consumer focus has been enabled by easy Web access to health information that was once available only to medical professionals. According to the search engine Ask.com, 70 percent of adults use the Internet as a primary resource for medical and health information. This information access is creating educated health care consumers, and shifting the traditional balance between patients and providers. Direct-to-consumer marketing is already beginning to occur in the absence of reimbursement. Personalized medicine is being driven as much by consumers themselves, seeking the best solutions to their medical problems, as by any other stakeholder.

Personalizing medicine in brain research must be a long-term commitment with some ongoing short-term outcomes. The brain is too important not to have the collective attention of those who understand aspects of this enormously complex system and care about the great human toll of its disorders. A first focused effort at this goal was the meeting on "Personalized Medicine For The Brain: A Call For Action" held in Washington D. C. in October of 2009. The brief report of key recommendations from this meeting is in appendix I. The path to more effective treatment is the rigorous collection of information about genes and brain function at every available level of knowledge.

No one has a monopoly when it comes to finding the right diagnosis and treatment for the right person at the right time. This is a unique time in medicine. This is a time of unprecedented confluence of factors, including:

- Health care reform;
- Digitalization of health care, notably through electronic health records;
- Mental health parity;
- Standardized methods and protocols;
- Standardized databases; and
- Consumer empowerment by the Internet.

All stakeholders should have a sense of cooperative humility in listening to and learning from each other, as we seek to understand how to translate and implement this new knowledge with the paradigm shift into the practice of personalized medicine.

References

Koslow, S. H, Williams, L. M. and Gordon, E. 2010. Personalized medicine for the brain: A call for action. *Molecular Psychiatry* 15: 229–30.

Appendix

Personalized Medicine for the Brain
A Call for Action

Stephen H. Koslow, Leanne M. Williams, and Evian Gordon

Disorders of the human brain, such as depression, schizophrenia, and addiction, are the cause of immeasurable human suffering. Because they are largely chronic and strike in youth, brain disorders lead to greater disability and loss of productivity than any other category of illness. On October 24-25, 2009, leaders from the worlds of research, medicine, industry, government, and philanthropy convened at the Mayflower Hotel in Washington D.C. to launch an initiative fostering personalized medicine for the brain. The Mayflower Action Group Initiative was instigated by BRAINnet, a new, nonprofit foundation that provides a database on the human brain using standardized methods.

The Mayflower Action Group advocates the following actions to make personalized medicine for the brain a reality:

1. Integrate

Study the brain as a system. The brain is highly connected and enormously complex, so it must be studied as a system. Studying genes alone is not enough. Multiple levels of information—from genes to brain structure, brain function, cognitive performance and symptoms—must be brought together. Ultimately, the brain's actions must be captured in real time.

Look at variables in combination. Study composite effects rather than factors in isolation. For example, studying variations in multiple genes will better explain the long-term molecular effects of early childhood trauma and how innate biology interacts with the environment to heighten individuals' predisposition to depression.

Dismantle silos. In addition to bringing together data, it is urgent to bring together people: geneticists and neuroscientists, clinicians and the pharmaceutical industry, funders

Molecular Psychiatry advance online publication, 12 January 2010; doi:10.1038/mp.2009.147

of healthcare and funders of research, regulators and policy makers, and crucially, patients. Much more can be achieved by aligning all stakeholders.

2. Standardize

Collect and bring together standardized data. Standardized measurement methods allow data to be pooled and compared. Making aggregated information accessible can reveal how to make real differences now. Consistent, standardized measurements will shed light on the defining characteristics of disorders and for the first time allow researchers to compare the basis of seemingly disparate disorders.

Reconsider diagnostic classification systems. Disease will increasingly be defined at the level of genes and brain biology and may lead to whole-scale changes in the categorization of disorders.

3. Represent real populations

Clinical studies should represent patients as a whole, not just "pure" cases, so that results in fact apply to real people. In brain disorders, co-morbidity is the norm. Yet studies on depression, for example, routinely leave out those who abuse substances, have post-traumatic stress or anxiety disorders, a known risk of suicide, or physical ailments, like diabetes or heart disease.

4. Meet real-world needs

Distill information for clinical practice. Massive amounts of heterogeneous information must be translated into specific guidelines and measurement tools for health care and done so rapidly. Patients should benefit from the efficacy and safety that the full range of existing knowledge can support. For example, genetics can identify who will benefit from medications for addiction, and heart rate variability training can be used to reduce stress and chronic pain, but such knowledge is rarely put into practice and when it is, the transition is slow.

Meet the consumer revolution. While doctors, politics, and health care agencies are slow to change, consumers have already harnessed the power of instantaneous and widespread access to knowledge through the internet and increasingly demand treatments that will work best for them with the least side effects. In the case of depression, for example, there is no way to predict which patient will respond to which therapy the first time. That needs to change.

5. Harness the power of numbers

Use databases. Databases should bring together complex information obtained in rigorously controlled and standardized ways. Large pools of layered data, made widely accessible, reveal connections in information. Data, gathered well and replicated, becomes the arbiter of what works and how well. For example, cognitive problems in schizophrenia predict disruptions in patients' lives, but treatments rarely cure these problems. Variations in illness course and treatment response may well reflect differences in gene expression, and large-scale analysis is needed to drill down to that level of understanding.

Solutions. We need to implement the current solutions that have demonstrated significant potential to screen for risk and predict treatment outcomes. Solutions providing benefits now need to be more clearly delineated from those requiring more research. A move to electronic health records can be paired with new tools and solutions. Web-based screening measurements and treatment algorithms should be integrated with health records.

Personalizing medicine in brain research must be a long-term commitment. The brain is too important not to have the collective attention of those who understand aspects of this enormously complex system and care about the great human toll of its disorders. The path to more effective treatment is the rigorous collection of information about the brain at every available level of knowledge.

The BRAINnet database is one effort toward this goal. It collects data using highly standardized methods agreed upon by a group of expert users. Researchers are invited to obtain free access to data about the human brain at BRAINnet.net. BRAINnet currently includes 200 collaborators and data from 10,000 people and is growing rapidly. Researchers can use BRAINnet data to answer their own questions and its standardized methods to acquire new data. BRAINnet is one approach to encouraging large, strategically-designed studies that will help speed the translation of basic science into medicine that makes a personalized difference in human lives.

Organized and Prepared by: Stephen H. Koslow, PhD and Leanne Williams, PhD, BRAINnet, Evian Gordon Brain Resource Ltd. with Karin Jegalian, Ph.D. Science Writer/ Editor

BRAINnet Foundation is grateful to Brain Resource Ltd. for sponsoring this meeting and delivering the consented data and database to BRAINnet Foundation for transparent governance and use by the scientific community.

Participants: Edward Abrahams, PhD (Personalized Medicine Coalition); Edward Allera (Buchanan Ingersoll, Rooney); Larry Alphs, MD, PhD(Ortho-McNeil Janssen Scientific Affairs); Eugene Baker, PhD (OptumHealth Behavioral Solutions), Charles M. Beasley, Jr., MD (Eli Lilly and Company); Linda S. Brady, PhD (NIH); H. Westley Clark, MD, JD, MPH, CAS, FASAM (HHS); Paula Clayton, MD (American Foundation for Suicide Prevention); John P. Docherty, MD (New York-Presbyterian/Weill Cornell); Katie Doll (OptumHealth Behavioral Solutions); Richard Givertz, PhD (Alliant University); Evian Gordon, MD, PhD (Brain Resource Ltd); Steven Grant, PhD (NIH); John F Greden, MD (University of Michigan); Henry Harbin, MD, John Hollister, MD (NARSAD); Shitij Kapur, PhD (Institute of Psychiatry, London); Helen Karuso (Australian Trade Commission, Washington DC); Stephen H Koslow, PhD (BRAINnet); Julio Licinio, MD (Australian National University); Jeffrey Alan Lieberman, MD (Columbia University), Norman Moore, MD (East Tennessee State University); Katherine T. Moortgat, PhD (MDV-Mohr Davidow Ventures);Charles B. Nemeroff, MD, PhD (University of Miami); Charles P. O'Brien, MD, PhD (University of Pennsylvania); Herbert Pardes, MD (New York - Presbyterian Hospital); Joseph Parks, MD (Missouri Department of Mental Health); Robert Paul, PhD (University of Missouri – St Louis); Joseph Perpich, MD, JD (JG Perpich, LLC); Frida E. Polli, PhD (Massachusetts Institute of Technology); John V.W. Reynders, PhD (Johnson & Johnson); Rhonda Robinson-Beale, MD (OptumHealth Behavioral Solutions); Deborah Runkle, PhD (American Association for the Advancement of Science); Sunil Sachdev, MD (OptumHealth Behavioral Solutions); Steven Secunda, MD (Steven Secunda Associates,

Pennsylvania); Sue Siegel (MDV-Mohr Davidow Ventures); Steven Silverstein, PhD (University of Medicine & Dentistry New Jersey); Madhukar H. Trivedi, MD (University Of Texas Southwestern Medical School); George Viamontes, MD, PhD (OptumHealth Behavioral Solutions & University of Missouri -Columbia School of Medicine); Phillip Wang, MD (NIH); Stephen Whisnant (United States Institute of Peace); David Whitehouse, MD (OptumHealth Behavioral Solutions); Peter C. Whybrow, MD (UCLA Center for Health Sciences Neuropsychiatric Institute); Leanne Williams, PhD (BRAINnet & University of Sydney Medical School).

For additional information contact BRAINnet@brainnet.net

Conflict of interest

SHK employed as the Research Director for the American Foundation for Suicide Prevention, serves as a Director for BRAINnet, receives a consultation fee for work with Brain Resource as a science consultant, and is a small equity holder (stock options) in Brain Resource. LMW is a small equity holder in Brain Resource and Chair of BRAINnet, and has received fees from Brain Resource for projects and consultancy unrelated to the Washington meeting. EG is the CEO and Chairmen of Brain Resource Ltd and has significant equity and stock options in the company. Dr SH Koslow is at the BRAINnet Foundation, 250 W 93rd Street, Suite 12E, New York, NY 33412, USA. E-mail: stephen.koslow@brainnet.net Dr LM Williams is at the BRAINnet Foundation, Sydney Medical School, University of Sydney at Westmead Hospital, Sydney, Australia. E-mail: lea.williams@brainnet.net Dr E Gordon is at Brain Resource Ltd, Headquarters at Sydney, Australia and San Francisco, USA. E-mail: eviang@brainresource.com

Contributor Disclosure Statements

Edward Abrahams
E.A. has no conflicts of interest to disclose.

Nili Avidan
N.A. has not disclosed any conflicts of interest.

Eugene Baker
To E.M.B.'s knowledge, all of his possible conflicts of interest and those of his coauthors, financial or otherwise, including direct or indirect financial or personal relationships, interests, and affiliations, whether or not directly related to the subject of the chapter, are listed below. He works closely with Brain Resource and is employed by OptumHealth Behavioral Solutions to facilitate deployment of Brain Resource capabilities within OptumHealth products.

Jacob S. Ballon
J.S.B. has no conflicts of interest to disclose.

Elisabeth B. Binder
E.B.B. has received grant support from NARSAD, Doris Duke Foundation, and the NIMH.

Edward T. Bullmore
E.T.B. has not disclosed any conflicts of interest.

Opher Caspi
O.C. has not disclosed any conflicts of interest.

Simon D. Clarke
S.D.C. has not disclosed any conflicts of interest.

William R. Crum
W.R.C. has no conflicts of interest to disclose.

Ulrich Ettinger
U.E. has no conflicts of interest to disclose.

Alex Fornito
A.F. has not disclosed any conflicts of interest.

Richard Gevirtz
R.G. has no conflicts of interest to disclose.

Charles F. Gillespie
C.F.G. has no conflicts of interest to disclose. He is funded by NIMH & NARSAD only. He has received grant support from the National Alliance for Research on Schizophrenia and Depression (NARSAD) and the NIH (K23 MH082256).

Ragy R. Girgis
R.R.G. has received research support from Janssen and Lilly through APIRE and a travel stipend from Lilly, Forest, and Elsevier Science through the Society of Biological Psychiatry.

Evian Gordon
E.G. is the founder of and receives income as the Chief Executive Officer and Chairman for Brain Resource Ltd. He has stock options in Brain Resource Ltd.

Ronald Grunstein
Funds were received by R.G.'s department in the past 5 years for sponsored clinical trials by Sanofi-Aventis, Merck, Actelion, GlaxoSmithKline, and Cypress Bioscience. Equipment support was received from ResMed, Somnomed, Phillips Respironics, and DiagnoseIT. R.G. is a scientific consultant and stockholder in DiagnoseIT, and minority shareholder (< 20K stock) in Brain Resource Company.

Paul E. Holtzheimer
P.E.H. has received grant support from the Dana Foundation, Greenwall Foundation, and the NIH (K23 MH077869), and a NARSAD Young Investigator Award.

Matthew J. Kempton
J.S.B. has no conflicts of interest to disclose.

Michael R. Kohn
M.R.K. has not disclosed any conflicts of interest.

Stephen H. Koslow
S.H.K. is Research Director at the American Foundation for Suicide Prevention; scientific consultant and holder of stock options in Brain Resource Ltd.; Director of the BRAINnet Foundation; Acting CEO and holder of an equity position in Enhanced Pharmaceuticals; and CEO of Biomedical Synergy Consulting.

Izabella Lejbkowicz
I.L. has not disclosed any conflicts of interest.

Jeffrey A. Lieberman
J.A.L. serves on the Advisory Board of Bioline, GlaxoSmithKline, Intracellular Therapies, Eli Lilly, Pierre Fabre, Psychogenics, and Wyeth. He does not receive financial compensation or salary support for his participation as an advisor. He receives grant support from Allon, Forest Labs, Merck and Pfizer; and he holds a patent from Repligen.

Ellen M. Migo
E.M.M. has no conflicts of interest to disclose.

Ariel Miller
A.M. has not disclosed any conflicts of interest.

Charles B. Nemeroff
C.B.M. is a member of the Board of Directors at AFSP, Novadel Pharma, and Mt. Cook Pharma. He is a member of the Scientific Advisory Board at Cenerx, Pharmaneuroboost, AstraZeneca, NARSAD, AFSP, and Skyland Trail Equity. He holds stock options or stock in Cenerx, Pharmaneuroboost, Reevax, and Novadel Pharma.

Tamar Paperna
T.P. has not disclosed any conflicts of interest.

Herbert Pardes
H.P. has not disclosed any conflicts of interest.

Alan F. Schatzberg
A.F.S. is a named inventor on pharmacogenetic and glucocorticoid antagonist use patents. He is a Co-founder of Corcept Therapeutics and has equity in Pfizer.

Dan Segal
D.S. is the COO of Brain Resource Ltd (a provider of scalable solutions for clinical care and optimal wellbeing) and holds equity and stock options in the company. He is a co-founder of Medpharma Partners LLC (a provider of biopharmaceutical and healthcare consulting services).

Michael Silver
M.S. has no conflicts of interest to disclose.

Elsebeth Staun-Ram
E.S.-R. has not disclosed any conflicts of interest.

David Whitehouse
D.W. has not disclosed any conflicts of interest.

David E. Williams
D.E.W. has no conflicts of interest to disclose.

Leanne M. Williams
L.W. has received consulting fees and stock options in Brain Resource Ltd and is a stockholder in Brain Resource Ltd. She has received Advisory Board fees from Pfizer.

Steve C.R. Williams
S.C.R.W. has no conflicts of interest to disclose.

Index

Note: Page references followed by "*f*" and "*t*" denote figures and tables, respectively.